Lecture Notes in Economics and Mathematical Systems

473

Springer
Berlin
Heidelberg
New York
Barcelona
Hong Kong
London
Milan
Paris
Singapore
Tokyo

Stefan Helber

Performance Analysis of Flow Lines with Non-Linear Flow of Material

 Springer

Author

Prof. Stefan Helber
Technical University of Clausthal
Institute of Business Administration and Economics
Julius-Albert-Str. 2
D-38678 Clausthal-Zellerfeld

Library of Congress Cataloging-in-Publication Data

Helber, Stefan, 1963-
    Performance analysis of flow lines with non-linear flow of
material / Stefan Helber.
        p.   cm. -- (Lecture notes in economics and mathematical
systems ; 473)
    Includes bibliographical references (p.).
    ISBN-13: 978-3-540-65954-9   e-ISBN-13: 978-3-642-95863-2
    DOI: 10.1007/978-3-642-95863-2
    1. Production management--Mathematical models. 2. Production
engineering--Mathematical models. 3. Stochastic analysis.
I. Title. II. Series.
TS155.H374 1999
658.5'001'5118--dc21   99-23638
                                                        CIP

ISSN 0075-8442
ISBN-13: 978-3-540-65954-9

Typesetting: Camera ready by author

"Gedruckt mit Unterstützung der Deutschen Forschungsgemeinschaft"

As "Habilitationsschrift" printed by recommendation of Fakultät für Wirtschafts-wissenschaften of the University of Munich with the support of Deutsche Forschungsgemeinschaft.

SPIN: 10699877   42/3143-543210 - Printed on acid-free paper

Preface

This book describes methods for the analysis of flow lines with limited buffer capacity. The flow of material is occasionally disrupted due to machine failures or quality problems. The focus is on flow lines with non-linear flow of material, for example because of assembly operations or quality inspections. We describe mathematical approximation techniques that can be used to estimate production rates and inventory levels for a given configuration of machines and buffers.

These technical performance measures are linked to cash flows in order to assess the investment in a flow line from an economic point of view. This economic perspective shows that buffer allocation is an extremely important aspect of flow line design.

This book has been accepted as a habilitation thesis by the Department of Business Administration at the *Ludwig-Maximilians-Universität* in Munich, Germany. Support for this thesis came from many people, and it came in many ways. I am most grateful to Hans-Ulrich Küpper, my advisor in Munich. He has been a source of constant moral support, even when my work departed from his primary research interests. Hans-Ulrich always forced me to ask for the economic implications of my technical results. With the benefit of hindsight, his (at that time) annoying persistence proved to be excellent guidance. Hermann Meyer zu Selhausen (also at the University of Munich) served as a referee and made helpful suggestions. The advice to start studying stochastic manufacturing systems came—like a lot of valuable advice before—from Horst Tempelmeier (University of Cologne). While I believe that Horst knew how tedious this would be, I am grateful that he did not tell me at that time. Otherwise I might have done something else during the last years. It is not possible to study flow lines without discovering that one of the major experts in this field is Stan Gershwin. He invited me several times to work with him at M.I.T. where his questions and suggestions shaped this thesis to an extent that can hardly be overrated. In his hospitality, Stan provided not only innumerable good ideas, but also food, housing, and lots of interesting discussions. During my time in Munich, I also experienced the friendship of many other people with different research interests. I wish to thank Daniela Triebel, Yvette and Christian Hofmann, Birgitta Wolff, and Tanja Ripperger. I am also indebted to the unknown Bavarian taxpayer for supporting me financially for three years under the *Bayerischer Habilitationsförderpreis 1994* and to the *Deutsche Forschungsgemeinschaft* for financially supporting the publication of this habilitation thesis.

Finally, I wish to thank my wife Maura and my son Alexander for helping me to keep this whole project in perspective and for giving me reasons to spend my evenings at home.

Stefan Helber

Table of Contents

1. Introduction

Manufacturing systems that produce few product types in large numbers often consist of a set of serially arranged work stations. Each product unit flows through the system along the same pre-specified path and each work station performs the same operation on the arriving product units over and over again. Since the design of the manufacturing system is determined by the flow of the produced objects, the systems are called *flow lines*. This design principle is based on a high degree of division of labor and leads to economics of scale. However, due to the division of labor, problems in one part of the system can easily have negative effects on other parts of the system. These problems can lead to disruptions in the flow of material as the amount of material between any two adjacent work stations is usually limited. For this reason, the system has to be seen as a whole, i.e. a *systems perspective* is required when deciding about design and control of a flow line. The major decisions that affect the profitability of a flow line are made during the design phase, i.e. before the system is actually installed. These effects have to be taken into account early in the design process of the manufacturing system.

This thesis deals with some of those problems that lead to *random disruptions* in the flow of material. These disruptions can be due to random processing times, machine failures, and/or quality problems in the production process, i.e. the production of bad parts. To some extent, they can be reduced by installing buffers between adjacent work stations. However, these buffers can be expensive and it is usually not obvious how to distribute them over the whole system such that the investment is as profitable as possible. If one is interested in an optimal system design, the selection of machines and the allocation of buffers should be determined simultaneously. The analysis is not limited to systems with a purely linear flow of material. Assembly and disassembly systems with a converging or diverging flow of material are treated as well as systems with rework loops for bad parts.

In practice, simulation is often used to predict the performance of a planned system—if one is at all aware of the importance of random disruptions in flow lines. Simulation can be time consuming, especially when a large number of very similar configurations has to be analyzed and compared. This problem arises when the optimal buffer allocation is sought and may prevent an optimal system design based on simulation models. For this reason, analytical models of stochastic manufacturing systems have been developed for several decades. They allow for an extremely quick approximate performance analysis of a planned flow line. Using analytical techniques, it is often possible to identify possibly attractive configurations and to exclude unattractive configurations from further consideration. Simulation can then be used for the fine-tuning of the system for the remaining attractive configurations. In this thesis, analytical techniques for flow lines with non-linear flow of material are developed and compared to simulation models in order to demonstrate their strengths as well as their limitations.

When designing a flow line, it is not only important to have powerful numerical techniques. A system planner also needs some intuition about what is going on in a stochastic manufacturing system, intuition about what is driving the performance of the system. The more complex a system is, the more difficult it is to develop this intuition. For this reason, some artificial systems are studied to focus on those effects that are due to non-linearities in the flow of material. The technical performance measures such as production rates and inventory levels have to be related to cash flows in order to achieve an economically sound system design.

Section 2 treats issues and methods that occur in the context of flow line analysis. The emphasis is on the economic impact of flow line design. Section 3 treats systems with assembly operations. Rework and scrapping processes due to the production of bad parts are modeled in Section 4 and 5. Conclusions and suggestions for further research are given in Section 6. Several of the analytical techniques developed in Sections 3 to 5 require relatively long formal derivations that are of interest only to those who want to extend these models and techniques. For these readers, the complete derivations are given in a set of appendices.

2. Issues, Goals, and Methods of Flow Line Analysis

2.1 Variability and the Performance of Flow Lines

Figure 2.1 illustrates a flow line with a linear flow of material. The squares indicate work stations or machines M_i and the circles represent buffers $B_{i,q}$ of limited capacity $C_{i,q}$ between adjacent machines M_i and M_q. The flow of the discrete material follows the directed arcs. Throughout the thesis we assume that an *input machine* without preceding machines like Machine M_1 has always (raw) material available, i.e. it is never *starved*. In a similar way we assume that an *output machine* without succeeding machines like Machine M_6 can always dispose the processed parts, i.e. it is never *blocked*. Due to these modeling assumptions, the flow line can be considered as isolated from the surrounding parts of the manufacturing or logistics system.

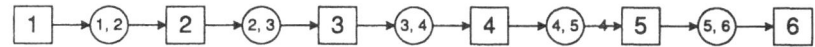

Fig. 2.1. Flow Line with Linear Flow of Material

Whenever a machine has processed a part, it tries to dispose the processed part into its downstream buffer (unless it is an output machine where processed parts eventually leave the system). After disposing the processed part, the now empty machine tries to take its next part out of its input buffer (unless it is an input machine where new parts enter the system). The major technical performance measures in flow line analysis are the *production rate*, i.e. the number of parts produced per time unit, and the *inventory level* at any production stage.

If the workload or processing time were identical and deterministic for all the machines, the movement of parts would be perfectly synchronized.[1] In such a system, all machines would always be operating, i.e. there would be no idleness. Furthermore, no part would ever have to wait between any two adjacent machines, i.e. the flow time in the system would equal the sum of the processing times at the machines. There would be no need for buffers as

---

[1] The workload per station is determined in planning step called *line balancing*. See Scholl (1995) for a review of the literature on line balancing.

they would always be empty. From a technical point of view, this would be a perfect flow line with 100% capacity utilization and zero waiting time for the processed parts. The production rate would be the inverse of the processing time and the inventory level between any two machines could be zero.

In reality, however, it is not always possible to have a perfectly synchronized flow of material. There may be various causes of variability that lead to disruptions in the flow of material. Whenever human operators perform tasks repeatedly, their processing time will usually vary. There may be systematic variations: processing times may decrease when the human operator learns on the job or they may increase if he or she gets tired. In addition to these systematic variations that can to some extent be predicted, there are often unpredictable variations around some average processing time.

If a task is performed by a machine, the machine may occasionally experience a failure that may require a repair. The repair time may be subject to random variations, especially if the repair is performed by human operators. This constitutes another class of unpredictable variations.

Assume, for example, that due to a failure of Machine M_3 in Figure 2.1 the flow of material into Buffer $B_{3,4}$ is interrupted. In this case, Machine M_4 will sooner or later be starved and Machine M_2 will sooner or later be blocked as Buffer $B_{3,4}$ will become empty and Buffer $B_{2,3}$ will become full. If the repair of Machine M_3 lasts long enough, the complete system comes to a halt.

Another class of disruptions is due to quality problems in the production process. If a bad part is produced, it may require rework or it may have to be scrapped. Rework or scrapping can be performed at dedicated machines that process bad parts only, leading to non-linearities in the flow of material that make flow line analysis even more difficult. The analysis of these systems is one of the major topics of this thesis.

To some extent, the disruptions in the flow of material can be reduced by inserting buffers between adjacent machines. Consider the two-machine system depicted in Figure 2.2.

Fig. 2.2. Two-Machine System

Assume that processing times are deterministic and equal at the two machines and can be taken as the time unit. Assume furthermore that a machine can only fail if it operates and both the mean time to failure and the mean time to repair are 10 time units. This holds for both machines, i.e. the system is *homogeneous*. Given these parameters, each machine is - in the long run - up 50% of the time and down the other 50% of the time and the production rate is 0.5 *if the machine operates in isolation.* However, if they operate together in a two-machine system as depicted in Figure 2.2, Machine M_1 is occasionally blocked and M_2 is occasionally starved which reduces the production rate.

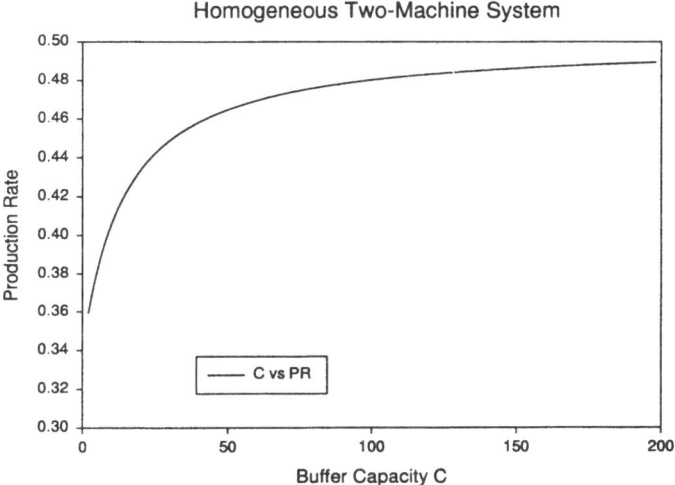

Fig. 2.3. Production Rate as a Function of Buffer Space

Figure 2.3 shows how the production rate of the two-machine system depends on the number of buffer spaces between Machines M_1 and M_2.[2] If the buffer is very small, a significant fraction of the production rate is lost due to the disruptions in the flow of material. However, the marginal increase of the production rate decreases as the buffer capacity increases and the production rate approaches an upper limit as the buffer capacity approaches infinity. The upper limit is given by the isolated production rate of the slowest machine in the system. In this example, both machines have the same isolated production rate of 0.5 parts per time unit. If buffers require scarce resources such as floor space or some type of equipment, the question arises of how much buffer space to allocate.

A common conjecture in the literature on flow line analysis is that this shape holds for systems with limited buffer capacity in general, i.e. that adding buffer space can never lead to a decrease in the production rate. If a system has more than one buffer, one may ask how to allocate a given total buffer capacity among the different buffers.

---

[2] See Gershwin and Schick (1980) and Gershwin (1994, p. 76-93) for a formal description of the model and the numerical technique used to generate the graph.

If the distributions of the completion times[3] (including failures and repairs) at the machines are identical at all machines, the buffer space should be allocated according to the shape of a bowl turned upside down. Due to this *bowl phenomenon*,[4] the first and last buffers should be relatively small and the buffers in the middle should be larger. The reason for this shape is that the machines in the middle of the line can be both blocked and starved. The last machines, however, will rarely be blocked and the first machines will rarely be starved. For this reason, less buffer space is required at the first and the last production stages.

This rule of thumb applies only if

- the processing times at all the machines have the same distribution and
- buffer space at all the production stages is equally costly.

Both conditions may not always be met. Furthermore, only the *shape* of the allocation of a given total number or buffer spaces is addressed. How large their *total number* should be remains an open question. These problems can only be solved using specialized approaches.[5] They should take the economic consequences with respect to *cash flows* into account if one is interested in an economically sound system design.

2.2 Non-Linearities in the Flow of Material

The focus of this thesis is on flow lines with non-linear flow of material. The models and techniques for performance evaluation and system optimization can all be used for the special case of linear flow lines as the example in Figure 2.1.

2.2.1 Assembly and Disassembly Operations

One class of non-linearities in the flow of material can arise when products are assembled. An example is the production of automobiles where different components of a car may be sub-assembled in dedicated parts of the assembly plant. These subassemblies are then combined to create the final product. This leads to a converging flow of material as in the artificial system[6] depicted in Figure 2.4.

In this network, Machines M_3, M_6, and M_7 perform assembly operations. For each operation, Machine M_3 has to take one part out of Buffer $B_{1,3}$ and one part out of Buffer $B_{2,3}$. These parts are combined during the operation at Machine M_3 to produce a subassembly. The process at Machine M_6 is

---

[3] Gaver (1962)
[4] Conway et al. (1988)
[5] Schor (1995), Kuhn (1998)
[6] Gershwin (1994)

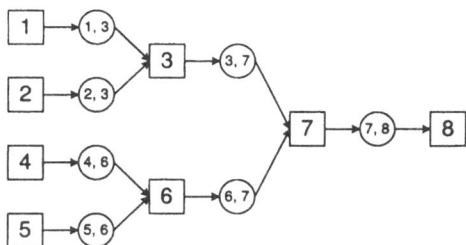

Fig. 2.4. Eight-Machine Assembly Network

similar. The subassemblies coming from Machines M_3 and M_6 are assembled at Machine M_7. Each assembly operation can only be performed if all the required parts are available. For this reason, the production rate is identical for all the machines in the assembly network and a long-lasting failure of a single machine can force the whole system down.

Very similar systems can be used to take previously assembled products apart. This may, for example, happen to old automobiles or computers where some of the different components may be recycled. In this case, the flow of material is diverging as depicted in Figure 2.5.

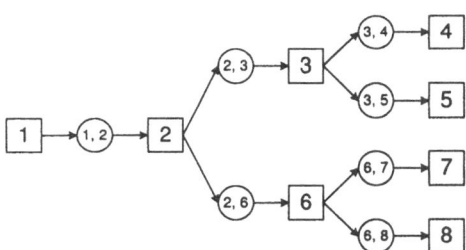

Fig. 2.5. Eight-Machine Disassembly Network

In this case, a disassembly machine like M_2 in Figure 2.5 sends one dis-assembled component to Machine M_3 and one to M_6 for each operation it performs. It cannot perform a disassembly operation if it is blocked because one of the buffers receiving the disassembled parts is full.

Assembly and disassembly (A/D) operations may also occur together, i.e. in the same network. The networks in Figures 2.4 and 2.5 are tree-structured, i.e. they do not contain loops. The analysis of A/D-networks in this thesis is limited to tree-structured systems. Systems with loops due to rework opera-tions are studied in the next section, but these systems do not perform A/D operations. However, once methods are available for both A/D operations and rework loops, it appears to be relatively straightforward to combine these two methods in order to analyze systems with A/D operations *and* loops. [7]

2.2.2 Split and Merge Operations and Rework Loops

A second class of non-linearities in the flow of material can arise because of quality problems in the production process, i.e. if some of the parts at some stage of the production process do not always meet pre-specified quality requirements. These bad parts may need some special treatment in a rework process, or they may have to be scrapped. If scrapping or rework is performed at dedicated machines or work stations, a machine that performs an inspection may have multiple alternative successors. This situation is depicted in Figure 2.6.

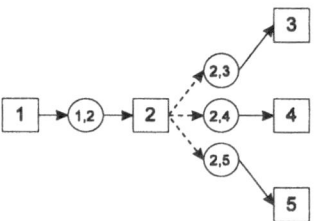

Fig. 2.6. System with a Split Operation

In this system, there is one machine, M_2, that has *multiple alternative* immediate successors. After processing a part, Machine M_2 sends this part to *one* of its immediate successors. Thus, if at time t a part is sent from Machine M_2 to M_3, then nothing is sent to M_4 and M_5. We use this concept of split operations to model a probabilistic routing of parts due to random quality properties such as being "good" or "bad."

We assume that the routing decision is made *after* the part has been processed, i.e. at the end of the inspection process. Thus, if Machine M_2 does not process a part because it is down, starved, or blocked, no routing decision is made. We model the routing decision as the flip of a multi-sided coin. That is, the choice of the succeeding machine for any part is random and independent of the state of the surrounding machines and buffers and the choices for previous parts. (This may be a simplification since in reality, machine failures and the production of bad parts may be correlated.) Formally, if $D(i)$ is the set of buffers immediately downstream of Machine M_i, a part processed by Machine M_i is sent to M_q with probability $d_{i,q}$ where

---

[7]   See Dallery and Frein (1989) and Frein et al. (1996) for the analysis of systems with loops, but without A/D operations. In these systems, the loop is not due to a rework process, but to a *fixed* number of pallets circulating in a manufacturing system.

$\sum_{(i,i)\in D(i)} d_{i,q} = 1$. The directed arcs in Figure 2.6 between Machine M_2 and its immediately succeeding buffers are broken to indicate that they represent *alternative* routings. We will use these alternative routings to model phenomena such as scrapping or rework of bad parts.

The split in the flow of material after Machine M_2 in Figure 2.6 should not be confused with a disassembly operation. In a disassembly operation as depicted in Figure 2.5, there is no choice between different downstream buffers as *each* downstream buffer receives one part whenever an operation has been performed by the disassembly machine. The split operation in this section, however, is due to a *random choice* between alternative routings and only one of the alternative downstream buffers receives the part. Note that the routing depends solely on the random quality properties of the respective part: If we have produced a good part and the downstream buffer holding good parts is full while the buffer holding bad parts is not full, the good part remains at the workspace of its current machine. This machine is now blocked until a buffer space for good parts becomes available.

The second phenomenon that we introduce into the analysis of production lines is the *merge operation* shown in Figure 2.7. Here, Machine M_3 has two immediate predecessors, M_1 and M_2. Whenever Machine M_3 performs an operation, it takes a part out of either Buffer $B_{1,3}$ or $B_{2,3}$. Machine M_3 is starved if both buffers are empty.

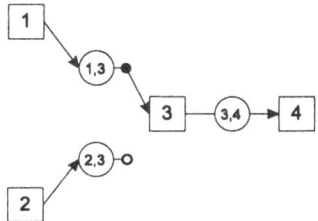

Fig. 2.7. System with a Merge Operation

We assume that from the point of view of Machine M_3, parts coming from M_1 and M_2 are identical. We have to specify how a merging machine chooses between its two input buffers. For the sake of simplicity, we assume that a merging machine uses a *priority ranking*. In Figure 2.7, the priority one buffer $B_{1,3}$ is always chosen unless it is empty. If it is empty, a part is taken from the priority two buffer $B_{2,3}$ (if $B_{2,3}$ is non-empty). In the graphical representation in Figure 2.7, the priority one buffer $B_{1,3}$ is depicted with a closed connection to its downstream machine, whereas the connection between the priority two buffer $B_{2,3}$ and the downstream machine is open.

Note that Machine M_2 does not perform an assembly operation, as there is a *choice* between two input buffers that hold the same part type (from the perspective of Machine M_2). In an assembly operation, a part is taken from each input buffer as *different* part types are *matched* in the assembly process.

If we allow for machines where the flow of material splits or merges, we can model systems like the one shown in Figure 2.8. In this system, there are two alternative input buffers immediately upstream of Machine M_2. Buffer $B_{8,2}$ containing the reworked parts has priority over Buffer $B_{1,2}$. Giving the higher priority to reworked parts leads to a lower total inventory and decreases the probability of deadlock situations.

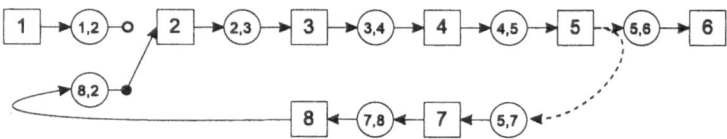

Fig. 2.8. System with a Loop in the Flow of Material

In the system in Figure 2.8, bad parts may be detected at Machine M_5. At Machines M_7 and M_8, all bad parts receive some treatment, such that previous processing steps can be repeated. These parts are fed back into the line at Machine M_2.

Assembly/disassembly and split/merge operations can occur together. In this thesis, however, they are treated in isolation. If one is able to analyze both assembly/disassembly and split/merge operations for a given type of system, it appears to be relatively straightforward to combine these building blocks of network analysis.

2.3 Economic Design Problems in Flow Line Analysis

The design of a flow line is to a large extent determined by the technology used in the production process for processing, storing, and transporting the product units. There may, however, be alternative technically feasible systems that differ with respect to their economic consequences.

Two kinds of economic choices appear to be important,

- the choice between alternative combinations of machines and
- the allocation of buffer space.

Several machines may be available to perform a given operation or set of operations. These machines may differ, both with respect to their price and/or operating cost and with respect to speed, reliability, and the quality of their output. It may furthermore be possible to install buffers at some or all production stages, in order to reduce the negative effects of disruptions in the flow of material due to machine failures, quality problems etc.

Both problems should be treated together. The reason is that, for example, a less expensive and less reliable machine *together* with some buffer spaces may be economically superior to a more expensive and more reliable machine without buffers. These problems can only be analyzed if the system is treated as a whole.

When designing a new line, knowledge of several quantities is needed. These quantities include

- the required production rate,
- the cost and scrap value of machines and buffers,
- the speed, reliability, and quality properties of the machines,
- the cost of raw material entering the line,
- the revenue due to finished products leaving the line,
- the time span over which the system is expected to be used, and
- the interest rate.

Many of these quantities can only be estimated. This induces a large amount of uncertainty. Estimating the mean time to failure ($MTTF$) and the mean time to repair ($MTTR$) of a machine can be relatively easy, if a machine of the same type is already operating under comparable conditions. It can be extremely difficult for a newly developed machine. Cost and revenue of raw material and finished products may change after the system has been installed, and the time span over which the system can be operated may depend on several external factors.

For this reason, one should have a tool to quickly explore the range of conditions under which a possible system will be profitable, and to find among the technologically feasible alternatives those that will be most profitable.

Having such powerful tools is especially important if one wants to analyze several different scenarios. This is the first necessary step to deal with the inherent uncertainty of long-term investment decisions.[8]

The first type of problem to be studied below addresses the selection of both machines and buffers. In some cases, however, a system may already exist or machines may have already been selected. Some buffers may have already been installed and one may ask for an optimal reconfiguration of some or all buffer spaces. This constitutes a second type of problem. The next two subsections treat these two closely related problems.

2.3.1 Cash-Flow Oriented Machine Selection and Buffer Allocation

If a new line is being planned, there may be several technologically feasible combinations of machines. For each of these combinations of machines, a planner may ask for

---

[8] This hint is due to Horst Tempelmeier (private conversation).

- the production rate of the system as a function of the buffer allocation,
- the optimal allocation of buffers for each possible production rate,
- the profitability of each such configuration, and
- the most profitable among the possible configurations.

To answer these questions, the following quantities have to be estimated for each of the possible designs defined by a given combination of machines:

- The initial investment A [$] in machines, buffers, and work-in-process inventory at time 0,
- the (scrap) value L [$] of machines and buffers at time T,
- the (continuous) cash flow c [$/year] between time 0 and time T that depends on the
- expected production rate PR [parts/production cycle] as a function of the buffer allocation $\underline{C}$,
- the expected work in process $\bar{n}_{i,q}$ [parts] for each buffer $B_{i,q}$, as a function of the buffer allocation $\underline{C}$,
- the number of production cycles[9] Y per year,
- the per unit cost of raw material uc [$/part] and the per unit revenue ur [$/part] for the finished products,
- the interest rate i [%/year], and
- the length T [years] of the time span over which the system is expected to operate.

The analytical methods used in this thesis to evaluate a transfer line design provide only estimates of average values of production rates and inventory levels for the *steady state* of the system between time 0 and T.[10] If the system starts to operate at some moment in time before time 0 (in order to be *in* steady state at time 0), all machines are up and idle, all buffers are empty, and the system is *not* in steady state. It takes some time for the system to reach steady state, but analytical techniques that describe this *transient* behavior of the system are not available.

During this initial transient phase of unknown length, two things happen: First, in-process inventories build up between adjacent machines, due to the (random) dynamics of the system. At the end of the initial transient phase, there is an expected inventory $\bar{n}_{i,q}(\underline{C})$ in Buffer $B_{i,q}$ that depends on the chosen vector $\underline{C}$ of buffer capacities. Second, some parts are completed that can be sold, yielding some positive net cash flow during this initial transient phase. If the transient phase is long, there will be a lot of such positive net cash flow, and if it is short there will be little.

---

[9] A production cycle is a unit of time chosen such that, for example, one product unit can be processed and leave the system during one production cycle.

[10] In steady state, the *probabilities* of the different system states do not change over time. However, the system state (machines being up or down, buffer levels) changes over time according to the dynamics of the system.

We consider long-term investment decisions in systems that produce a large number of product units per year over several years. Therefore, it appears to be reasonable to neglect this initial transient phase. If we assume as an approximation that the system reaches steady state immediately, i.e. the lengths of the transient phase is zero, there is a negative cash flow $uc\,\bar{n}_{i,q}(\underline{C})$ to build up the expected inventory $\bar{n}_{i,q}(\underline{C})$ in Buffer $B_{i,q}$ in steady state, but no positive cash flow as no parts are completed and sold during this infinitesimally short period of time. Given that the length of this initial transient phase is zero, the initial negative cash flow occurs at time 0^-.

Figure 2.9 depicts the structure of the expected cash flow over time. At time 0, there is an initial negative cash flow $-A$.

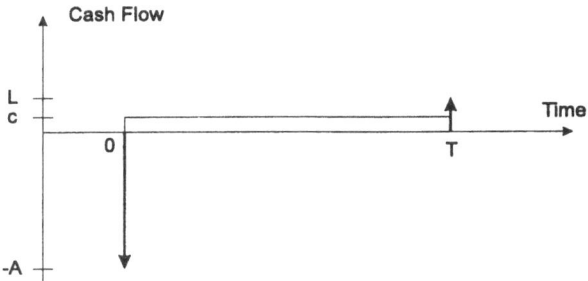

Fig. 2.9. Cash Flow Related to a Flow Line Investment

The initial investment A is the sum of the investment A_{M_m} in machines M_m, in buffer spaces, and in the initial work in process (WIP).

$$A = \sum_{m=1}^{M} A_{M_m} + \sum_{\forall(i,q)} \left(A_{B_{i,q}} C_{i,q} + uc\,\bar{n}_{i,q}(\underline{C}) \right) \tag{2.1}$$

The investment in buffer spaces for each buffer $B_{i,q}$ is assumed to be the product of the cost $A_{B_{i,q}}$ [\$] per unit buffer space times the chosen size $C_{i,q}$ of Buffer $B_{i,q}$. We assume that the system starts to operate at time 0 with the average work in process $\bar{n}_{i,q}(\underline{C})$ in each Buffer $B_{i,q}$ as it reaches steady state immediately. The average WIP of Buffer $B_{i,q}$ depends on the chosen vector of buffer sizes $\underline{C}$. If each unit of raw material costs uc, the initial investment in WIP for Buffer $B_{i,q}$ at time 0 is $uc\,\bar{n}_{i,q}(\underline{C})$.

At time T the system is in steady state, i.e. there is some expected inventory $\bar{n}_{i,q}(\underline{C})$ in Buffer $B_{i,q}$. Now a second transient phase follows during which no additional raw material enters the system. The remaining in-process inventory is processed until all the material leaves the system. As an approximation, we assume again that this final transient phase is infinitesimally short, compared to the years over which the system operates in steady state.

If we assume that all these final parts can be sold, this leads to an expected revenue $ur\,\overline{n}_{i,q}(\underline{C})$ at time T^+ for the material in each buffer.

At time T, there may be some positive cash flow L reflecting the scrap value of machines and buffers and some revenue due to selling the remaining WIP. It depends on the scrap value L_{M_m} for each machine M_m, the scrap value per unit buffer space $L_{B_{i,q}}$ times the chosen buffer capacity $C_{i,q}$, and the revenue $ur\,\overline{n}_{i,q}(\underline{C})$ for selling the WIP:

$$L = \sum_{m=1}^{M} L_{M_m} + \sum_{\forall(i,q)} \left(L_{B_{i,q}} C_{i,q} + ur\,\overline{n}_{i,q}(\underline{C})\right) \qquad (2.2)$$

Between time 0 and T, there should be some positive continuous cash flow that is due to the value added to the processed products. This positive cash flow c [\$/year] depends on the expected value $PR(\underline{C})$ of the production rate, which is in turn a function of the buffer allocation $\underline{C}$. The continuous cash flow has some fixed component c_{fix}, for example for wages, and a variable component that depends on the value added per product unit $(ur - uc)$, the production rate $PR(\underline{C})$ per cycle, and the number of production cycles Y per year:

$$c = -c_{fix} + (ur - uc)\,PR(\underline{C})\,Y \qquad (2.3)$$

Given the initial investment A at time 0, the final revenue L at time T, the expected value of the continuous cash flow c between time 0 and T, and the interest rate i, the expected NPV of the investment in the flow line is

$$E\{NPV\} \quad = \quad -A + Le^{-iT} + \int_0^T ce^{-it}dt. \qquad (2.4)$$

Since the expected value of the continuous cash flow c does not depend on t, we find

$$E\{NPV\} \quad = \quad -A + Le^{-iT} + c\left[-\frac{1}{i}e^{-it}\right]_0^T$$

$$= \quad -A + Le^{-iT} + \frac{c}{i}\left[1 - e^{-iT}\right]. \qquad (2.5)$$

The expected NPV can be easily calculated for each possible design using Equation (2.5), if one has a procedure to determine the expected production rate $PR(\underline{C})$ and the expected work-in-process $\overline{n}_{i,q}(\underline{C})$ for a given combination of machines and allocation of buffers $\underline{C}$. This can be done either by simulation or by analytical approaches like those developed in this thesis.

The following example shows how important the optimization of the buffer allocation can be. Consider the hypothetical six-machine flow line depicted in Figure 2.10. Assume that each machine costs $160,000 and has a scrap value of $16,000. Each buffer space costs $2,000 and has a scrap value of $200. A unit of raw material costs $200 and a unit of the finished product leads to a revenue of $230. Processing times are deterministic (5 min.) and identical for all machines. They constitute the production cycle. The system operates for 24,000 cycles per year.

Machines are identical with respect to their failure and repair time distributions. Times to failure are assumed to be geometrically distributed with a $MTTF$ of 100 cycles, times to repair are also geometrically distributed with a $MTTR$ of 10 cycles. Given these distributions, each machine is operational $\frac{MTTF}{MTTF+MTTR} \approx 91\%$ of the time if it operates in isolation. The system operates for 4 years and the annual fixed cost is $220,000. The interest rate is 10% per year.

Fig. 2.10. Flow Line with Linear Flow of Material

If only two buffer spaces are installed between any two adjacent machines, the average production rate of the system is approximately 0.68 parts per cycle.[11] The maximum average production rate of this system is approximately 0.91 parts per cycle. This is the upper limit of the production rate for an infinitely large number of buffer spaces between any two adjacent machines.

If the system operates with only two buffer spaces between any two adjacent machines at a production rate of 0.68 parts per cycle, the net present value (NPV) of the investment in the system is approximately $-24,000, i.e. this investment destroys value.

Increasing one buffer by one space leads to an additional expense for this buffer space, which has a negative effect on the NPV. It does, however, also result in an increased production rate and an increased net revenue. This is due to the additionally produced parts, which may more than compensate for the investment in the additional buffer space.

Figure 2.11 depicts the expected value of the NPV as a function of the production rate. To determine this curve, buffer sizes were increased one by one using a greedy approach such that they led to the largest increase (or smallest decrease) of the NPV. The figure shows that by adding buffer spaces, the NPV of the investment can be brought up to $151,000. At this point, the average production rate is approximately 0.81 parts per cycle. It is possible to increase the production rate further by adding more buffer space. However,

---

[11] To calculate this and the following production rate estimates, the model and decomposition equations in Gershwin (1987) was used. The equations were solved as proposed in Dallery et al. (1988) and Burman (1995, p. 84-87).

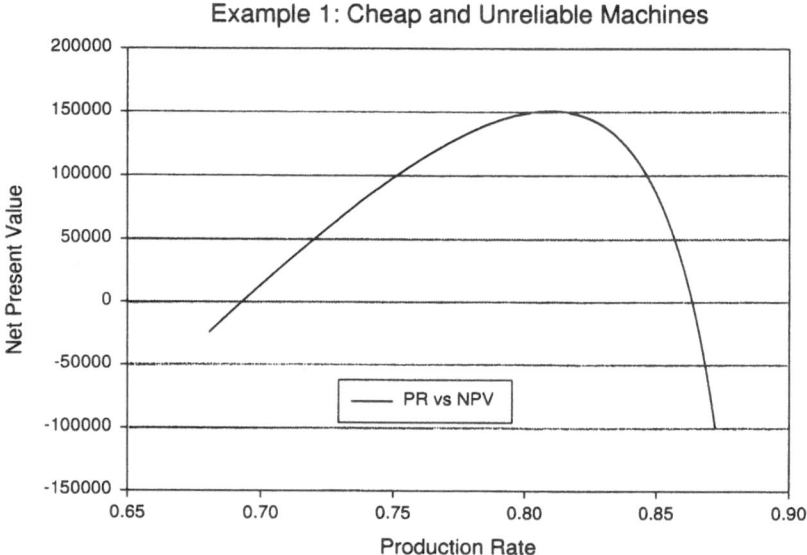

Fig. 2.11. Net Present Value for the First System

this leads to a *decrease* of the NPV, which approaches minus infinity as the production rate approaches the limit of 0.91 parts per cycle.

Each point on the PR - NPV curve in Figure 2.11 corresponds to a unique buffer allocation. All these buffer allocations are *efficient* in the sense that it is not possible to achieve a higher NPV by redistributing the buffer space. The maximum of the NPV is reached for a total of 80 buffer spaces between the machines. They are distributed according to the pattern (11, 18, 21, 19, 11), i.e. the first buffer can hold 11 parts, the second 18 and so on. This shape of the buffer space is an example of the bowl phenomenon mentioned on Page 6.

Figure 2.11 shows that the NPV depends heavily on the buffer allocation and the resulting production rate. Allocating buffer space can therefore be extremely important from an economic point of view. Using analytical techniques like those developed in this thesis, it is relatively straightforward to determine these figures for different technically feasible combinations of machines.

Assume that an alternative set of machines is available for the system in Figure 2.10. Assume that each of these machines costs $200,000 instead of $160,000 and has an estimated scrap value of $20,000. The alternative machines are not only more expensive, but also more reliable, i.e. failures occur less frequently. The mean time to failure for all machines is twice as long as for the first example, i.e. we have $MTTF = 200$. The isolated production rate of these machines is approximately $\frac{200}{200+10} = 95.2\%$, compared to 90.9% for the first example.

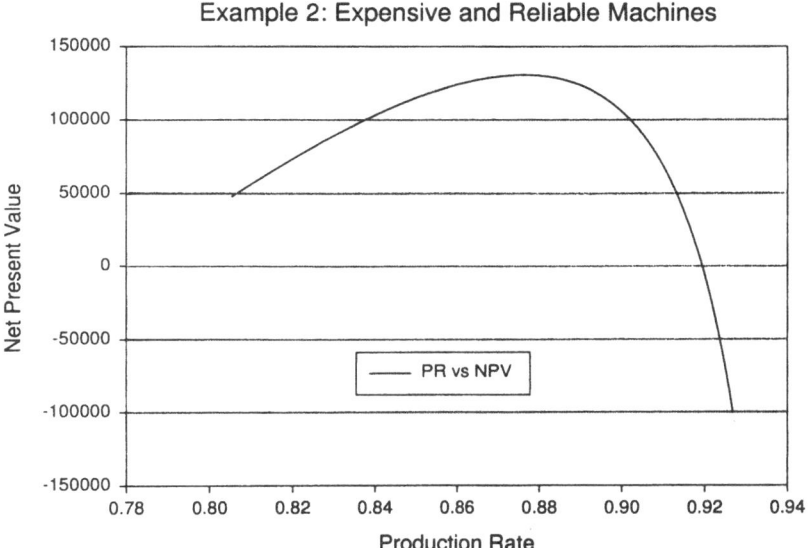

Fig. 2.12. Net Present Value for the Second System

Figure 2.12 depicts again the NPV as a function of the production rate, which results by adding buffer space in order to achieve the maximum increase of the NPV. The shape of the curves in Figure 2.11 and 2.12 is similar, but the most profitable production rate for the second system is higher (0.88 parts per cycle, as opposed to 0.81 for the first example). The corresponding buffer space allocation is (6, 14, 16, 14, 6), i.e. the optimal configuration for the second system uses less total buffer space than the one for the first system. This example exhibits a trade-off between the investment in machines and buffer: more reliable and more expensive machines call for fewer buffers and *vice versa*.

When designing a flow line, there will usually be a target production rate that may be subject to some uncertainty. Figure 2.13 shows the NPV functions for both technically feasible sets of machines.

This figure can be valuable in many ways. It can be used to identify the most profitable combination of machines for each target production rate. It

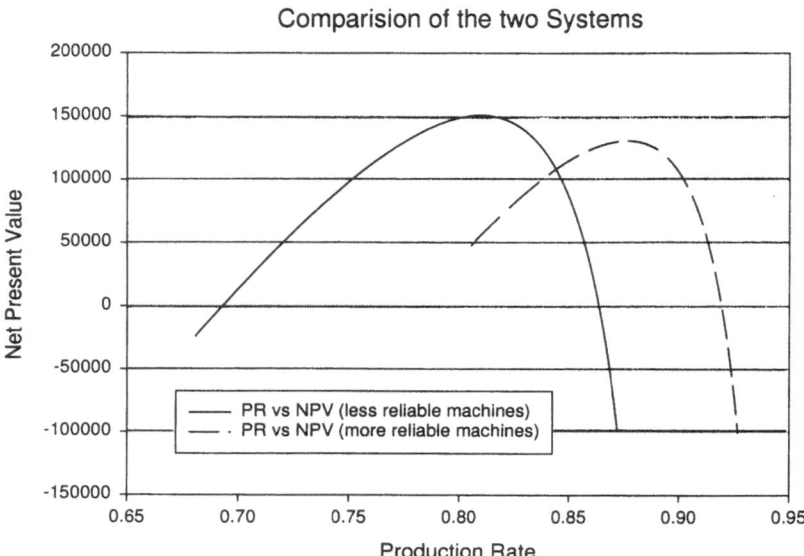

Fig. 2.13. NPV Functions for the Two Sets of Machines

can also be used to find the production rate, the corresponding combination of machines and buffer allocation that leads to the highest possible NPV. Knowledge about the shape of these curves helps to understand how sensitive the economic impact of a system design is with respect to changes of the target production rate or other parameters. These curves can therefore be a helpful instrument when designing a flow line.

The problem of finding the *optimal* buffer allocation for a *given* combination of machines can be expressed as follows:

$$E\{NPV\} \quad = \quad -A + Le^{-iT} + \int_0^T ce^{-it}dt \rightarrow \text{ max!} \qquad (2.6)$$

subject to Equations (2.1), (2.2), and (2.3). Due to the shape of the curves in Figure 2.3 and 2.11, it is possible to use powerful gradient techniques like those proposed in Schor (1995) and Gershwin and Schor (1997) to determine the optimal buffer allocation without computing the complete function of the NPV over the production rate.

Some remarks about this model appear to be worthwhile. All the economic parameters, such as the price of machines A_{M_m} and buffer spaces $A_{B_{i,q}}$ are directly related to cash flows. Cash flows are observable quantities[12] as opposed, for example, to manufacturing *costs*, which have to be *defined* and are meaningful only with respect to the corresponding definition. Since the model is based directly on cash flows, there is no need to calculate quantities such manufacturing costs per production stage based on whatever depreciation of machines and equipment. It is not even *possible* to calculate these cost quantities precisely without solving the buffer allocation problem first. The reason is that the manufacturing cost as it is traditionally defined[13] includes some component that is due to the depreciation of the machines and buffers. This component depends on the production rate that in turn depends on the buffer allocation. It is therefore not possible to calculate a per part manufacturing cost first and to solve the buffer allocation problem later. However, there is not even a need to use any of these derived quantities as all the required quantities can be directly observed or at least estimated.

A second remark is closely related. Flow lines are usually designed to produce a large number of identical or similar product units over some time span, often several years. The average WIP at any moment in time is usually only a small fraction of the total system output over the time span the system operates. Equation (2.1) for the initial investment A

$$A = \sum_{m=1}^{M} A_{M_m} + \sum_{\forall(i,q)} \left(A_{B_{i,q}} C_{i,q} + uc\,\overline{n}_{i,q}(\underline{C}) \right)$$

contains a component $uc\,\overline{n}_{i,q}(\underline{C})$ that corresponds to the initial WIP. It reflects the assumption that the system starts to operate with the average WIP, which leads to some negative cash flow for the required raw material at time 0^-.[14] At time T, there is some positive cash flow L (2.2)

$$L = \sum_{m=1}^{M} L_{M_m} + \sum_{\forall(i,q)} \left(L_{B_{i,q}} C_{i,q} + ur\,\overline{n}_{i,q}(\underline{C}) \right)$$

with a component $ur\,\overline{n}_{i,q}(\underline{C})$ which is due to selling the expected remaining WIP after closing down the flow line. Since the average WIP is extremely small compared to the total system output, these cash flows are small compared to the expenses for machines and buffers and the revenues for the total system output.

Omitting these terms related to the WIP from Equations (2.1) and (2.2) has almost no impact on

- the optimal NPV of the project and
- the optimal allocation of buffer spaces.

---

[12] Küpper (1991, p. 53)
[13] Schweitzer and Küpper (1998, p. 176)
[14] See Page 13.

In other words, for the given assumption about cash-flows, the capital tied up in the average WIP can often be *ignored* when deciding about the flow line design. If it can be ignored, the accuracy of analytical techniques with respect to the estimated average WIP is not important for the given problem.

Saying that the capital tied up in the WIP can be ignored does not mean that WIP is not important. On the contrary, there can be a positive relation between WIP and the production rate if the WIP serves to compensate for stochastic disruptions in the flow of material. The production rate, in turn, is extremely important with respect to the profitability and optimal design of a flow line. This is because the total revenue depends on the production rate. Analytical techniques for the performance evaluation should therefore be accurate with respect to the estimated production rates, but they 'may' be inaccurate with respect to the estimated WIP for this particular type of investment problem.[15]

2.3.2 Cash-Flow Oriented Buffer Allocation for a Given Set of Machines

The previous section showed that buffer allocation and machine selection are two closely related problems that should be treated together when planning a new flow line. However, buffer (re-)allocation may also be an issue for an already existing system. An economically rational decision has to take the impact on cash flows into account.

2.3.2.1 Adding Buffer Spaces to an Already Existing System. Assume in a first example that the first set of machines (less expensive and reliable) as introduced in the previous subsection was selected. However, a thorough analysis of the required buffer space was omitted and each buffer was given two buffer spaces for whatever reason. The analysis in Figure 2.11 indicates that this design destroys value as it has a negative NPV of approximately $-24,000 at time 0. After operating the flow line for six months it becomes apparent that both the production rate and the cash flow due to processed parts are too small to make the investment profitable.

The *initially* optimal buffer configuration[16] with the pattern (11, 18, 21, 19, 11) is now - with a remaining lifetime of 3.5 years - no longer the optimal solution. The reason is that there is now less time left to 'pay' for the buffers. The maximum possible *increase* of the expected NPV at time 0.5 that can still be reached by adding buffer spaces is approximately $144,531.[17] It is

---

[15] There are other problems where it can be very important to have a precise estimate of the in-process inventory, see Footnote 7 on Page 8.

[16] See Page 16.

[17] This can be determined by deriving a figure comparable to Figure 2.11 for $T = 3.5$ years and asking for the maximum possible increase of the NPV due to the optimal buffer allocation.

related to the buffer pattern (9, 17, 19, 17, 9). This is the optimal pattern for the remaining lifetime of the system of 3.5 years. It uses less buffer space, as there is now less time left for the system to 'pay' for the buffers. The NPV increase of \$144,531 at time 0.5, i.e. after six months, corresponds to an NPV increase of $\$144,531e^{-0.1\cdot0.5} = \$137,482$ at time 0. The initial value of the investment at time 0 with only two spaces per buffer is \$-24,186. Improving the system by adding buffer spaces after six months results in a discounted value at time 0 of $\$-24,186 + \$144,531e^{-0.1\cdot0.5} = \$113,296$.

Thus, the investment is still profitable. The maximum NPV for the buffer pattern (11, 18, 21, 19, 11), installed at time 0, is \$150,904. The difference $\$150,904 - \$113,296 = \$37,608$ is the economic value that was lost in this system by installing the buffer spaces six months too late.

2.3.2.2 Reallocation of the Existing Buffer Space. In some cases it may be possible to reallocate the given buffer space in order to make the flow line more profitable. Assume that the system with the less expensive and reliable machines was designed according to the buffer pattern (11, 18, 21, 19, 11) that is at least close to optimal[18] for the original assumptions about mean times to failure and repair. Now assume that after 6 months, the first two machines in the line turn out to be much less failure-prone than expected, i.e. they have a $MTTF = 1000$ instead of 100 production cycles. All the other parameters remain unchanged.

In this case, the isolated production rates of the first two machines are approximately 0.99 instead of 0.91, assuming again a minimum of two physical buffer spaces. This calls for a different allocation of the given buffer space according to the pattern (2, 2, 25, 30, 21), i.e. the buffer spaces move towards the less reliable machines. Reallocating the buffer space leads to an increase in the estimated expected production rate from 0.8284 to 0.8446 parts per production cycle. The increase in the NPV at time 0.5 for the remaining 3.5 years is approximately \$35,000. Given the initial expected NPV of about \$151,000 at time zero, which led to the decision to install the system, this is a major possible increase that can be reached if one is aware of the importance of the buffer (re-)allocation problem.

In a real-world context, there may be some additional constraints. In some cases, it may not be possible to change the size of all the buffers, or there may be maximum or minimum buffer sizes. Given the smooth shape of the production rate as a function of the buffer size as depicted in Figure 2.3 on Page 5, it does not appear to be too difficult to meet such constraints.

---

[18] We cannot guarantee optimality because we use an approximate technique for the performance evaluation and we only conjecture that adding buffer spaces can never lead to a decrease of the production rate.

2.4 Methods of Performance Analysis

Two different methods can be used for the performance analysis of stochastic manufacturing systems:

- simulation methods and
- analytical methods.

These two methods appear to be confused frequently.[19] The difference between simulation and analytical methods is basically the difference between trying and thinking. Very often, simulation is the only approach that leads to any results at all. This happens if a system has to be studied in great detail and analytical techniques for the particular system are not available. However, the question remains to which extent time-consuming simulation experiments can be replaced by quick and (more or less) accurate analytical techniques. This question becomes important, whenever building and running simulation models requires a lot of time and resources.

The difference between the two approaches can be explained using the following example related to machine reliability.[20] Assume that a single machine of the type used in the previous examples operates in isolation, i.e. without preceding or succeeding machines. Since it operates in isolation, it can never be starved or blocked and the fraction of time the machine is up depends only on its own stochastic failure and repair processes.

Assume that the machine can fail at the beginning of a production cycle with probability $p = \frac{1}{6}$, if it was up at the end of the previous production cycle. The machine is repaired with probability $r = \frac{4}{6}$ at the beginning of a production cycle, if it was down at the end of the previous cycle. This machine is always either up, operating, and waiting for the next failure, or it is down, not operating, and waiting for the next repair.

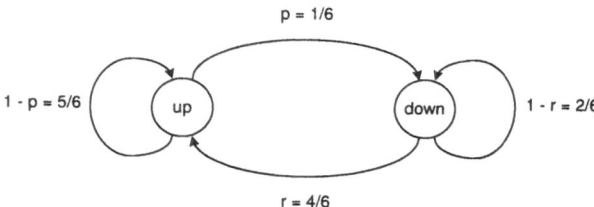

Fig. 2.14. State Space and Possible Transitions of the Single Machine Reliability Problem

The two possible states of the system and the possible transitions with their corresponding probabilities are depicted in Figure 2.14. In a perfor-

[19] Gershwin (1994, p. xviii)
[20] Gershwin (1994, p. 26-30)

mance analysis one might be interested in the fraction of time the machine is up.

To simulate the behavior of the system over time, a die and a sheet of paper are in principle sufficient. We assume that the system is initially up and simulate the state of the machine over 20 production cycles as depicted in Figure 2.15. If the machine is up and the die shows a 1 (with probability $\frac{1}{6}$), this indicates a random failure. If it is down and we throw a number smaller than 5 (with probability $\frac{4}{6}$), it indicates a random repair.

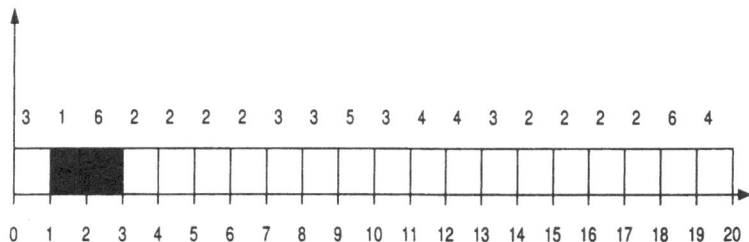

Fig. 2.15. Manual Simulation of Single Machine over 20 Production Cycles

In Figure 2.15, a white bar indicates that the machine is up during a production cycle and a black bar represents a failure. In this short simulation run the machine is up 90% of the time. A new simulation run may lead to different results. If the simulation runs are much longer, the results will be more similar and give us more confidence in their accuracy. In practice, one would use a computer program instead of a die and a sheet of paper to simulate a system like this. As a matter of principle, the result of a simulation experiment is a *set of numbers* that needs to be interpreted carefully. If a similar system with slightly different parameters is to be analyzed, the whole process has to be repeated.

It straightforward to evaluate[21] this system analytically. The machine can be in two states: up or down. The state of the machine during production cycle t depends only on the state in the previous cycle $t-1$ and the failure and repair probabilities p and r. The probability that the machine is up at time t is the probability that it was up in the previous cycle times the probability $(1 - p)$ that it did not fail plus the probability that it was down times the probability r that it was repaired, so:

$$\text{prob[up at time } t] = (1 - p)\,\text{prob[up at time } t - 1] \qquad (2.7)$$
$$+ r\,\text{prob[down at time } t - 1]$$

Since we ask for the steady-state behaviour of the system where the probabilities do not change over time, we omit the time index to find

---

[21] Gershwin (1994, p. 26-30)

$$\text{prob[up]} = (1 - p) \text{ prob[up]} + r \text{ prob[down]} \tag{2.8}$$

and a similar equation holds for the probability of the machine being down:

$$\text{prob[down]} = p \text{ prob[up]} + (1 - r) \text{ prob[down]} \tag{2.9}$$

Note that these equations are not independent. However, since probabilities must add up to one

$$\text{prob[up]} + \text{prob[down]} = 1, \tag{2.10}$$

Equation (2.10) and either (2.8) or (2.9) constitute a linear set of equations with the following solution:

$$\text{prob[up]} = \frac{r}{r + p} \tag{2.11}$$

$$\text{prob[down]} = \frac{p}{r + p} \tag{2.12}$$

For the given values of failure and repair probabilities p and r, the probability that the machine is up is $\frac{4/6}{4/6+1/6} = 80\%$. A sufficiently long simulation run comes close to these values.

An analytical method leads to an equation or a set of equations as those derived above and a procedure to evaluate these equations. Deriving the equations can be more time-consuming than building a simulation model, but solving them for a given set of parameters is usually less time-consuming than a simulation run. If a lot of similar systems have to be analyzed, analytical techniques can be faster.

In this example, it was relatively easy to analyze the system, due to its Markov property.[22] However, as systems get more complex, it can become extremely challenging to derive formulas to predict the system behavior.[23] In this situation simulation is often the last resort. Expert knowledge is required to understand what can be done using analytical methods,[24] whereas it can be straightforward to develop *some* simulation model of a system using a commercial software package.[25] However, simulation models of a real world

---

[22] The Markov property says that given the knowledge about the current state of the system, knowing the (previous) history does not help to predict the future behavior, i.e. the system is memoryless.

[23] Burman et al. (1998, p. 22)

[24] Kuhn and Tempelmeier (1997)

[25] See Banks (1996) and Swain (1997) for a brief review of currently available commercial simulation software.

system can differ with respect to their level of detail and expert knowledge
is again required to understand the relationship between the level of detail
on the one hand and the usefulness and model-building effort on the other.
It is not trivial to build valid and credible simulation models and to interpret simulation output.[26] Like the development of analytical techniques this
requires solid knowledge of statistics and probability theory.

2.5 Two-Machine Decomposition of Flow Lines

All the models developed and analyzed in this thesis are Markov process
models as these are analytically more tractable than other models. However,
even for a Markov process model of a flow line like the one introduced in
Section 2.3.1, the state space of the Markov process model is so large that
we cannot compute its steady-state probabilities as easily as for the simple
example in Figure 2.14. The reason is that there can be an extremely large
number of *combinations* of different machine states and buffer levels, if there
are more than two machines and one buffer.

The state space for a two-machine, one-buffer system like the one depicted
in Figure 2.16, however, is relatively small. If a two-machine system has C
buffer spaces between the two machines, and each of the two machines can
hold a part, the total or *extended* buffer space related to the two-machine
system N is

$$N = C + 2. \tag{2.13}$$

A part is assumed to be stored at the upstream machine if this machine
is blocked.

Number of possible states: 2   (N+1)   2

Fig. 2.16. Number of Possible States in a Two-Machine System

Each of the two machines in a two-machine system can be either up or
down, i.e. there are $2 \cdot 2 = 4$ different combinations of machine states. The
(extended) buffer can be empty or hold up to N parts, that is there are $N+1$
buffer levels and the total size of the state space is therefore $4(N+1)$. Some of
these states may be transient, depending on the assumptions about failures
etc.

---

[26] Law and Kelton (1991, p. 298-324 and p. 522-581)

The computational effort to solve a two-machine model of the type given above is negligible. Solving the model in the context of performance evaluation means to determine all steady state probabilities and to compute performance measures such as production rate and inventory levels.[27]

To determine performance measures for flow lines with non-linear flow of material, we decompose them into a set of two-machine transfer lines of the same type, i.e. with unreliable machines and limited buffer capacity. The reason for this approach is that the virtual two-machine systems arising in the decomposition can be solved easily.

For each buffer $B_{j,i}$ between two machines M_j and M_i, we introduce a virtual upstream machine $M_u(j, i)$ that represents to 'an observer in the buffer of the original line'[28] the flow of material into this buffer. A virtual downstream machine $M_d(j, i)$ represents the flow out of this buffer in the original line. The flow line depicted in the upper part of Figure 2.17, for example, is decomposed in three two-machine systems corresponding to the buffers in the original system as shown in the lower part of Figure 2.17.

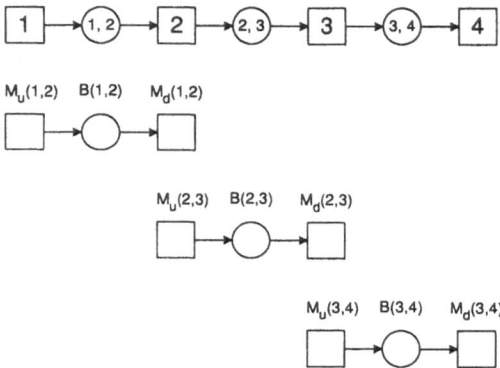

Fig. 2.17. Example of a Decomposition

We have to determine parameters describing processing times, times to failure and to repair of these virtual machines for each line $L(j, i)$ in the decomposition. Given these parameters, we can efficiently compute the steady-state probabilities of all the virtual two-machine lines using a suitable two-machine model. Performance measures of the original system can then be estimated from the steady-state probabilities of these two-machine models.

The goal of the derivations in the remainder of this thesis is therefore to determine the parameters of the virtual machines in the decomposition for various types of non-linearities in the flow of material. This leads to systems of equations that have to be solved simultaneously as the perspectives of

---

[27] See Gershwin (1994, p. 59-132) for the analysis of various two-machine flow lines.
[28] Dallery and Gershwin (1992)

different observers in different buffers of the original system are interrelated. To solve these equations, a version of the iterative DDX-algorithm is used.[29]

Several of the decomposition equations are approximations. For this reason, the performance measures determined by the decomposition are also approximations and their accuracy has to be checked against a simulation model.

2.6 Review of the Literature

Several researchers have studied unreliable transfer lines and/or assembly/disassembly (A/D) systems with limited buffer capacity. A comprehensive survey has been given by Dallery and Gershwin (1992). Their review includes the literature on reliable two-machine transfer lines, on transfer lines without buffers, as well as longer lines with more than two machines and A/D systems. Earlier reviews are Koenigsberg (1959), Buxey et al. (1973) and Buzacott and Hanifin (1978). Current textbooks covering these topics in detail are Altiok (1996), Buzacott and Shanthikumar (1993), Papadopoulus et al. (1993), as well as Gershwin (1994) which gives a thorough introduction into how to derive these models.

The large number of publications on the performance analysis of multistage transfer lines with limited buffer capacity can be classified according to the following criteria:

- the decomposition principle
- the modeling of machine failures
- the modeling of processing times
- the layout of the manufacturing network

We discuss some of the work that is most closely related to this thesis along these criteria.

Decomposition principle. Given that many analytical models of stochastic manufacturing systems with limited buffer capacity cannot be analyzed exactly, due to the size and structure of the state space, a large number of approximate techniques has been developed. They can be classified according to the underlying decomposition principle as[30]

- one-machine decompositions,
- two-machine decompositions, and
- two-machine aggregations.

---

[29] Dallery et al. (1988)

[30] Bürger (1997, p. 44)

In a *one-machine decomposition*, a single-machine queuing model is introduced for each machine in the original system and the parameters, especially the processing times, of this single-machine model are modified in order to reflect the impact of the other components of the system.[31]

In a *two-machine decomposition*, the building block is a two-machine model with limited buffer capacity, for which it is often possible to determine all steady state probabilities very quickly. This iterative approach has been discussed in Section 2.5. It has led to a large number of quick and precise procedures for the analysis of longer lines. Compared to a one-machine decomposition, there are often more parameters to capture the impact of the surrounding machines. Very often, these models are Markov process models where quantities like processing times or times to failure and repair are exponentially or geometrically distributed, or they have a phase-type distribution. Under these conditions it is often possible to compute all steady-state probabilities for the two-machine models and to determine performance measures that approximate those of the real system.

When analyzing the related work with respect to two-machine models and decomposition approaches, we can distinguish[32]

- Markov processes with discrete state and discrete time,
- Markov processes with discrete state and continuous time, and
- Markov processes with mixed state and continuous time.

In the first two cases, the state is discrete, since discrete parts are produced and machines can be either operational (up) or under repair (down). Time is divided into discrete periods in the first case or treated as continuous in the second. In the third group of Markov processes, one assumes that continuous material is produced in continuous time (which leads to a continuous buffer level), but machine states are discrete.

Table 2.1 gives an overview of two-machine models and decomposition approaches for the case of unreliable machines and limited buffer capacity.

The approach used in de Koster (1989) is based on a successive *aggregation* of subsystems until a single two-machine model results that allows to estimate performance measures. This approach can also be used to model non-linear flow of material. However, there is no obvious aggregation sequence: For a three-machine transfer line $M_1 \rightarrow B_{1,2} \rightarrow M_2 \rightarrow B_{2,3} \rightarrow M_3$, one might aggregate the first two machines to a new virtual machine, yielding a new, two-machine transfer line $M_{1,2}^{agg} \rightarrow B_{2,3} \rightarrow M_3$ that can be analyzed exactly. This leads to a production rate estimate for the whole line and to a buffer level estimate for the second buffer. To determine a buffer level estimate for the first buffer, one has to aggregate the second and third machine yielding $M_1 \rightarrow B_{1,2} \rightarrow M_{2,3}^{agg}$, and this may lead to a second production rate estimate that *differs* from the one for the first aggregation. This problem might be

---

[31] An example is the approach by Hillier and Boling (1967).
[32] Dallery and Gershwin (1992)

Type of Process	Analysis of Two-Machine Models	Approximate Decomposition Approaches
Discrete State/ Discrete Time	Artamonov (1977) Buzacott (1967) Buzacott and Hanifin (1978) Gershwin and Schick (1983) Helber (1995) Okamura and Yamashina (1977) Yeralan and Muth (1987)	Dallery et al. (1988) Gershwin (1987) Gershwin (1991) Helber (1998)
Discrete State/ Continuous Time	Buzacott (1972) Gershwin and Berman (1981) Sastry and Awate (1988)	Choong and Gershwin (1987) Gershwin (1989) Helber (1998) Jeong and Kim (1998)
Mixed State/ Continuous Time	Bohuslav and Stoyan (1967) Gershwin (1994) Gershwin and Schick (1980) Sevast'yanov (1962) Wijngaard (1979) Zimmern (1956)	Burman (1995) Dallery et al. (1989) Di Mascolo et al. (1991) Petigk (1987)

Table 2.1. Two-Machine Models and Approximation Approaches

the reason why decomposition approaches appears to be more popular so far than aggregation techniques.

Modeling of machine failures. Random failures of machines or equipment are often considered to be either

- time dependent or
- operation dependent.

Time dependent failures (TDFs) occur irrespective of the time the system has been operating since the last repair, for example due to power failures. Operation dependent failures (ODFs) can only occur if a machine operates, for example because a tool breaks. Data by Hanifin reported in Buzacott and Shanthikumar (1993, p. 232) suggests that failures at single stations tend to be operation dependent, whereas failures of a whole line tend to be time dependent. Failures of single stations occurred more frequently in the case studied by Hanifin. Buffers can only protect against isolated failures of single machines, but not against failures of the whole system. Since the distribution of buffers is a major economic decision variable in the design of transfer lines, we assume ODFs.[33]

If one assumes that machine failures are a major source of disruptions in the flow of material, there are in principle two ways to model these random

---

[33] See also Footnote 1 in Gershwin (1994, p. 61).

failures and repairs. They can either be included in the probability distribution of the processing time in a model where machines are reliable, or they can be explicitly modeled as random changes of the machine states.

In the first case, the mean (and often also the standard deviation) of the real processing time distribution is inflated to include some fraction of average repair times in the *completion time*[34] of each part. It depends on the distribution of processing times, times to failure and to repair how well matching only the first two moments works. Bürger (1997, p. 107) uses this approach to apply a procedure for the analysis and optimization of reliable transfer lines to unreliable transfer lines.

The second approach is more direct as the state of each machine is explicitly modeled. This introduces additional degrees of freedom that can lead to more precise performance estimates. It is common to assume exponentially or geometrically distributed times to failure and to repair. If there are different types of failures with mean times to repair of different orders of magnitude, it is possible to use a *generalized exponential distributions*[35] to capture these different types of failures.

Modeling of processing times. The approach to model processing times is closely related to those for times to failure and to repair. In a real system, processing times may be either deterministic or random.

If a single product type is produced and machines perform the same operation over and over again, processing times may be perfectly deterministic. However, if human operators perform a task, there will usually be some variability in their processing time.

Furthermore, the product units may be similar, though not identical, in a system that has some degree of flexibility and produces several variants of a product simultaneously. In this case, the workload placed on a machine by two successive parts may differ. From the perspective of the machine, this can be modeled as a random processing time.

Even if processing times are deterministic, including random failures and repairs as described above leads to random completion times.

In this thesis, random failures and repairs are always modeled explicitly. In Section 3 we treat the case of random processing times for an assembly/disassembly system with unreliable machines. Deterministic processing times are studied in two models of transfer lines with split and merge operations in Sections 4 and 5.

In models with deterministic processing time and unreliable machines, there are two fundamentally different approaches:

- the discrete material-discrete time approach and
- the continuous material-continuous time approach.

---

[34] Gaver (1962)
[35] Dallery (1994)

If processing times at all machines are identical and deterministic, they can be taken as the time unit in a *discrete material-discrete time* model. The common assumption here is that the state of the system changes only at the beginning or at the end of a period, so that only these distinct moments in time have to be observed. In these models, only two parameters per machine have to be taken into account that describe failures and repairs. If a transfer line is well-balanced, i.e. the work content per work station is identical, this may be a reasonable assumption. It is also the starting point for our analysis of systems with split and merge operations in Section 4.

However, it is not always possible or desirable to have a perfectly balanced system, i.e. processing times will often *differ* between machines, even though they are deterministic.

In this important case, there is no common processing time that can be taken as a time unit. In a discrete material model with deterministic and machine-specific processing times, the process does not possess the Markov property: Assume that on two adjacent machines two operations start simultaneously. If processing times are deterministic and one machine completes its operation while the other one is still working, we *know* how long the remaining processing time of the second machine is. Thus, the knowledge about the history of the process, i.e. about how long a discrete part has already been processed, helps to predict when the operation on the second machine will be completed. In this situation the process is not memoryless, i.e. it does not possess the Markov property and cannot be analyzed as 'easily' as a Markov process model.

In order to deal with this problem, the model of discrete material is replaced by a model of a liquid and the machines are thought to operate like pumps or valves through which a fluid flows continuously in a *continuous time-continuous material* model. This type of model again possesses the Markov process property and it provides a way to deal with machine-specific processing times. In Section 5, a continuous material model for split and merge operation is developed that generalizes work by Burman (1995).

In some applications, it is actually relatively natural to use continuous material models. Bohuslav and Stoyan (1967) develop a two-machine model where the 'machines' actually represent unreliable conveyor belts used to transport material like coal. These belts are buffered using a bunker between the belts. Petigk (1987) extends these methods to consider the case of multiple stages and gives qualitative design rules.

Layout of the manufacturing network. Most of the literature deals with serial arrangements of machines. The literature on flow lines with non-linear flow of material is extremely limited. Harrison (1973) showes that assembly systems based on GI/G/1 queues with unlimited buffer capacity are unstable in the sense that infinite inventory builds up over time. Ferschl (1991) studies a model with limited buffer capacity and Poisson arrivals at an 'assemblage' process that requires zero time. Since the assemblage station is never blocked, one of the input buffers is always empty. The method to determine the steady-state probabilities appears to be limited to small buffer sizes and up to three input buffers.

Gershwin (1991) extends the decomposition for a flow line in Gershwin (1987) to assembly and disassembly systems. These models assume deterministic and identical processing times at all machines. The processing times are taken as the time unit in models of the discrete state-discrete time type. Di Mascolo et al. (1991) allow for machine-specific deterministic processing time. This requires a model of the mixed state-continuous time type. In this model, the material is modeled as a continuous fluid that is processed at machine-specific rates. Section 3 of this thesis deals with stochastic processing times in unreliable assembly/disassembly systems. A month before the submission of this thesis, Jeong and Kim (1998) presented a decomposition method for assembly/disassembly systems with exponentially distributed processing times, times to failure and to repair. Since the same model is analyzed in Section 3, we compare the results briefly.

The few papers that address scrapping of bad parts in the context of performance analysis of production lines differ with respect to the assumptions about scrapping of bad parts. Okamura and Yamashina (1977) assume in a two-machine model that whenever a stage breaks down, its current part is scrapped. Shanthikumar and Tien (1983) develop a two-machine model where parts are scrapped with some probability when their current machine fails. Jafari and Shanthikumar (1987) extend this two-machine model to longer transfer lines and present an approximation technique to determine production rates and buffer levels. The scrapped parts leave the line immediately and can never be reworked. Some papers consider scrapping and/or rework for two- and multi-stage[36] systems with unlimited buffer capacity.

Bürger (1997) develops a decomposition approach for reliable and unreliable linear transfer lines that allow for scrapping of parts at each machine. Scrapped parts leave the system and cannot be reworked. He does not consider dedicated machines for scrapping or rework operations, i.e. it is not possible that the main line is blocked due to a failure of a rework or scrapping machine. His approach is based on a decomposition into a set of GI/GI/Z-*stopped arrival* queueing system. Machine failures and repairs are modeled using the completion time approach by Gaver (1962).

Gopalan and Kannan (1994) present a two-machine zero-buffer model in which bad parts can be reworked or scrapped. Rework takes place at the machine where the bad part is produced and it starts immediately.

Pourbabai (1990) describes a model with more than two machines and non-zero buffers, but assumes that if blocking occurs, the blocked workpieces are permanently lost.

---

[36] Yu and Bricker (1993)

Many models of unreliable transfer lines with limited buffer capacity assume that the repair time of any machine is independent of the state of the other machines in the network. In reality, however, there is usually a limited number of operators available to repair a machine that is down. Assuming independent repair times is equivalent to assuming that a machine never has to wait for an operator or maintenance personnel. In a recent habilitation thesis, Kuhn (1998) develops several models to analyze the impact of the maintenance system on the performance of transfer lines. One building block is a two-machine transfer line including an operator to repair machines. At any moment in time, only one machine can be repaired. Kuhn studies the first come-first served protocol as well as priorities for repairs and patrolled systems. He develops integrated models that are again evaluated using a decomposition approach and analyzes the trade-off between the number of buffers and the size of the repair crew required for a given production rate.

In another habilitation thesis, Schmidbauer (1995) introduces a decomposition of a transfer line with stochastic demand into a set of two-buffer, one-machine subsystems with a stochastic demand indicator behind the second machine.[37] This demand indicator allows to determine service levels. He also considers a random routing due to a split operation, but he only models the case of two instead of n output buffers and he does not analyze merge operations. He treats the case of identical deterministic processing times at all machines in a model of the discrete time-discrete material type.

Yeralan and Tan (1997c) describe a one-machine single-buffer continuous-material system with input and output valves, which is the building block of a decomposition approach developed in Yeralan and Tan (1997b). In this approach, the parameters of the valves are modified during the decomposition, whereas in most other approaches, the parameters of the machines are modified. They also develop a flexible methodology for the analysis of two-machine lines in Yeralan and Tan (1997a) that is to be extended to multi-station systems.

We are not aware of papers that explicitly consider a random routing of parts and loops in the flow of material due to rejects, rework, or scrapping of bad parts at dedicated work stations in the presence of limited buffers and unreliable machines.

The work most closely related to the research reported below is the transfer line model in Gershwin (1987) and a Ph.D. thesis on flow line analysis (Burman (1995)). Burman shows how to formulate decomposition equations in a way that leads to a dramatically improved convergence behavior.

Optimization. The major part of the literature focuses on the performance evaluation in a rather technical sense, i.e. seeking production rates, inven-

---

[37] See also Schmidbauer and Rösch (1994).

tory levels etc. There is relatively little work published on the design and optimization of transfer lines.[38]

Some authors use simulation for the performance analysis of any given design, but this tends to limit the applicability to small systems as simulations are more time consuming than analytical approaches. Others concentrate on the algorithms to determine optimal or suboptimal designs. We use a gradient algorithm similar to those developed in Schor (1995) to determine (sub-)optimal buffer allocations.

Optimization problems are occasionally formulated in terms of profit functions that are to be maximized,[39] or one ignores monetary quantities at all and seeks the minimal number of buffer spaces to meet some target production rate.[40] In profit-oriented models, it is common to use *holding cost* parameters reflecting the interest on the capital tied up in in-process inventories.

Instead of using dubious cost parameters, our NPV maximization model[41] uses the assumptions on discounted cash flows directly, without falling back on (derived) cost parameters. From a mathematical point for view, this may lead to models that have the same structure, which may be why the merely mathematically interested do not appear to care very much for this difference. However, from a management perspective, it is most important to tie the observable technical quantities like production rates to the observable economic quantities like cash flows instead of assuming the existence of some derived 'cost'-parameters which may or may not include the relevant discounted cash flows.[42]

---

[38] See Schor (1995, p. 15-20) and Gershwin and Schor (1997) for a recent review of the literature on optimization.

[39] See, for example, Schor (1995, p. 86) or Hillier et al. (1993).

[40] See, for example, Schor (1995, p. 27).

[41] See Section 2.3 on Page 10.

[42] See the remarks on Page 19.

3. Assembly/Disassembly Systems with Random Processing Times

3.1 Discrete and Continuous Time Models

Assembly and disassembly operations. Consider a manufacturing system like the hypothetical one depicted in Figure 3.1. The network is tree-structured since there is exactly one sequence of machines and buffers that connects any two machines in the network, irrespective of the direction of the connecting arcs.

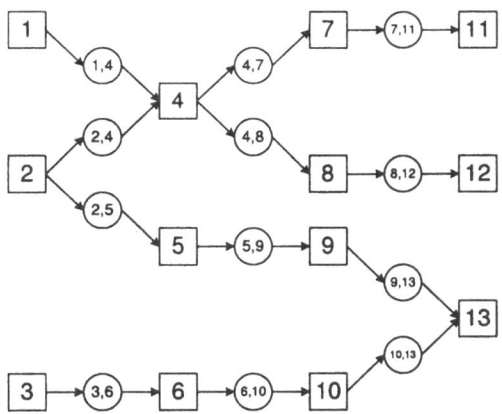

Fig. 3.1. Hypothetical Assembly/Disassembly System with 13 Machines

Machines without preceding (succeeding) machines are input (output) machines. In Figure 3.1, Machines M_1, M_2, and M_3 are input machines, and M_{11}, M_{12}, and M_{13} are output machines. The system produces discrete parts. Supply of input machines with raw material is assumed to be unlimited. The storage area downstream of the output machines is also assumed to be unlimited. Between two adjacent machines M_i and M_q is a buffer $B_{i,q}$ with limited storage capacity $C_{i,q}$. Some of the machines (in the example M_{13}) perform *assembly operations*, i.e. they require exactly one part from each immediate upstream machine that are matched during the assembly operation. A machine may also (like M_2) disassemble parts with exactly one component of the disassembled part being sent simultaneously to each immediate downstream machine. It is even possible that a machine (like M_4) performs both assembly and disassembly operations. Such a machine takes one part from each of its input buffers (in the example, Buffers $B_{1,4}$ and $B_{2,4}$), performs the operation, and puts one part into each downstream buffer ($B_{4,7}$ and $B_{4,8}$). A network of machines that performs assembly and/or disassembly operations is called an assembly/disassembly system (A/D system).

Blocking and starving. A machine can perform an operation if none of its input buffers is empty and none of its output buffers is full. Otherwise, the machine is starved or blocked, respectively. We assume *blocking after service*: a machine is blocked if a part has been processed, but at least one output buffer is full and the part has to remain at the workspace of the machine.

Blocking or starving due to randomness in processing times, as well as machine failures and repairs, can propagate through the complete network. Suppose, for example, that Machine M_{10} fails. If Machine M_{10} remains down long enough, its upstream machines will be blocked and its downstream machine M_{13} will be starved. Starving of M_{13} due to M_{10}, however, leads to blocking of M_9, M_5, and M_2. As a result, all the other machines downstream of M_2 will be starved. That is, the failure of one machine can stop the complete system. Since the capacity of the buffers is limited, all machines process the same number of parts in the long run. We assume that a machine can fail only while it is processing a part, i.e. failures are operation dependent.

Discrete vs. continuous time model. In this section on the analysis of A/D systems, we study two closely related models. The models differ with respect to the representation of time. In the first model, time is discrete, i.e. the state of the system can only change at certain moments in time. In the second model, time is continuous. The two models are so closely related that they lead to the same results if the length of the time unit approaches zero. Since the modeling assumptions, the notation, and the derivation is very similar for both models, we present the discrete-time model in detail and give results for the continuous time model only where they differ from the first one. Table 3.1 compares the assumptions regarding processing times, times to failure and to repair for both models.

In both cases, processing times are random and machine specific. The random processing time case may be relevant when a flexible assembly system assembles a variety of part types simultaneously with each part type placing a different workload on the machines or when human operators perform a task.[1]

---

[1] Human operators often have processing times with a coefficient of variation (CV) smaller than one, the CV of exponentially distributed random variables. In this case, our models overestimate processing time variability and hence underesti-

	Discrete time model	Continuous time model
Processing times	Geometrically distributed with mean $1/u_i$	Exponentially distributed with mean $1/\mu_i$
Times to failure	Geometrically distributed with mean $1/p_i$	Exponentially distributed with mean $1/p_i$
Times to repair	Geometrically distributed with mean $1/r_i$	Exponentially distributed with mean $1/r_i$

Table 3.1. Discrete vs. Continuous Time Model

Random events in the discrete time model. Time is considered to be discrete. We assume that the smallest processing time of any part at any machine is fixed and known in advance. We use this smallest possible processing time as the time unit. Thus, processing a part takes at least one period on our time scale. Due to the mechanism of production and transportation, parts can only move from one machine to the next at the end of a period. Transportation time is zero.

Each machine M_i can be in two states that can only change at the beginning of periods. Let $\alpha_i(t)$ indicate the state of Machine M_i at time t. If $\alpha_i(t) = 1$, Machine M_i is operational or up; if $\alpha_i(t) = 0$, it is under repair or down.

When machine M_i is up and neither blocked nor starved at the beginning of a period, it completes an operation by the end of this period with probability u_i, $0 < u_i < 1$. In this case, it takes one part from each immediate upstream buffer and places one part into each immediate downstream buffer. Thus, processing times follow a geometric distribution with a mean processing time of $1/u_i$ periods.

Machine failures are *operation dependent*: If Machine M_i is operational and neither blocked nor starved at the end of period t, it fails at the beginning of period $t + 1$ with probability p_i, i.e. times to failure are geometrically distributed with $MTTF_i = 1/p_i$.

When a machine fails, the current part stays at this machine until the machine is repaired and resumes the operation. If a machine is down at the end of period t, is is repaired at the beginning of period $t + 1$ with probability r_i. Thus, times to repair are also geometrically distributed with with $MTTR_i = 1/r_i$. Since a machine cannot fail when it is starved or blocked and the workpiece remains at the machine while the machine is repaired, the machine can never find itself blocked or starved after the repair.

mate the production rate, i.e. they provide a conservative estimate of the production rate.

Random events in the continuous time model. In the continuous time version of the model, processing times are assumed to be exponentially distributed, i.e. the probability that a part is processed at Machine M_i during a small time interval of length δt is $\mu_i \delta t$ and the average rate at which parts are processed is $1/\mu_i$.

Failures are again operation dependent. Times to failure are exponentially distributed with rate p_i, i.e. the probability that an operating machine fails between time t and time $t + \delta t$ is $p_i \delta t$. Repairs are also exponentially distributed with rate r_i.

Relationships to previous models. In the previous models of A/D systems by Gershwin (1991) and Di Mascolo et al. (1991), processing times were assumed to be deterministic. The discrete-time model developed in this thesis[2] is a generalization of Gershwin (1991) since it allows for a machine specific probability for an operational machine M_i to complete an operation during a period. If the average processing time $1/u_i$ in the discrete-time model approaches one, this model leads to both analytical and numerical results that are very similar to those in Gershwin (1991). The continuous-time model of the A/D system is a generalization of the n-machine transfer line model in Choong and Gershwin (1987).

In order to use the two-machine decomposition technique to determine performance measures, an exact solution procedure for the corresponding two-machine system is necessary. For the continuous time model, we use the model and solution procedure in Gershwin and Berman (1981). For the discrete time model, a similar solution procedure is developed in this thesis.

The main contribution of this work is to allow for non-deterministic processing times in A/D networks and to develop fast and accurate algorithms with a highly reliable convergence behavior. Based on these algorithms, decision support systems can be developed that are directed at the economic design of A/D systems.

Jeong and Kim (1998) published a decomposition for the continuous time model presented here and in Helber (1998) immediately before the submission of this thesis. Since they study exactly the same model, we compare the two approaches and their results briefly after deriving the decomposition equations.

3.2 Exact Solution of a Two-Machine Subsystem

The decomposition approach used in this thesis to analyze flow lines requires solution procedures for the corresponding two-machine transfer line. For the discrete time model with geometric processing times, times to failure and to

---

[2] See also Helber (1998).

repair, the required procedure is presented in this section.[3] The two-machine line and the corresponding machine and buffer parameters for the discrete time case are depicted in Figure 3.2.

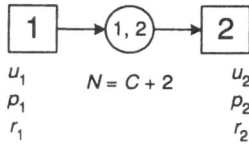

Fig. 3.2. Two-Machine Line with Machine and Buffer Parameters

Given the processing, failure and repair probabilities u_i, p_i, and r_i of the two machines M_1 and M_2, as well the number of buffer spaces C between the machines, we have to determine the steady-state probabilities of all possible system states.

Let α_i indicate the state of Machine $M_i, i = 1, 2$. If $\alpha_i = 1$, Machine M_i is operational or up; if $\alpha_i = 0$, it is under repair or down. The physical storage or buffer between M_1 and M_2 has a capacity of C parts.

We distinguish this physical buffer *between* the machines from the storage of the whole system. The storage of the transfer line includes the work areas of M_1 and M_2. It may contain an integer number n of parts with $0 \leq n \leq N$, where $N = C + 2$ is the *extended* storage size including the workspaces of the two machines.

A part is assumed to be stored at the work area of Machine M_1 if it has been processed by Machine M_1, but finds the storage already containing C parts and one part currently occupies the work area of Machine M_2. In this case, the new storage level n is $N = C + 2$ and M_1 is blocked. However, a part in the work area of the first machine that is still being processed or waiting for the first machine to be repaired is not assumed to be stored in the extended storage related to the buffer as it is not yet available for Machine M_2. Thus, the storage level can drop to 0 when M_2 is starved. Machine M_1 is never starved due to the unlimited supply of raw material. Machine M_2 is never blocked since its downstream storage area is unlimited. Thus, the state of the system is given by the vector $s = (n, \alpha_1, \alpha_2); n = 0, ..., N; \alpha_1 = 0, 1; \alpha_2 = 0, 1$. Let $\mathbf{p}[n, \alpha_1, \alpha_2]$ denote the steady-state probability of the system being in state s. We assume $C \geq 2$ (or $N \geq 4$) because transfer lines with this property have a common mathematical structure that we can exploit. Excluding systems with less than two physical storage places C is not a severe restriction. Very often, more than two physical storage places are necessary to achieve an acceptable system performance.

---

[3] See also Helber (1995).

3.2.1 Performance Measures

The steady-state expected production rate PR_i of machine M_i is the probability that this machine completes processing a part at the end of an *arbitrary* period.

$$PR_1 = u_1 \text{ prob}[\{\alpha_1(t+1) = 1\} \text{ and } \{n(t) < N\}] \qquad (3.1)$$
$$PR_2 = u_2 \text{ prob}[\{\alpha_2(t+1) = 1\} \text{ and } \{n(t) > 0\}] \qquad (3.2)$$

It must be up and neither blocked nor starved at the beginning of the respective period. Since the flow of parts is linear and parts are not destroyed or created at either machine, the production rates of both machines are equal:

$$PR_1 = PR_2 \qquad (3.3)$$

This conservation of flow equation states that the long-run output of both machines is equal.

Another quantity of interest is the isolated production rate IPR_i of each machine:

$$IPR_1 = u_1 \frac{r_1}{p_1 + r_1} \qquad (3.4)$$
$$IPR_2 = u_2 \frac{r_2}{p_2 + r_2} \qquad (3.5)$$

The isolated production rate is the probability that the machine produces a part during a period if this machine operates in isolation so that it can never be starved or blocked. The minimum of IPR_1 and IPR_2 is an upper bound of the production rate of the system.

The expected in-process inventory $\bar{n}$ can be written as

$$\bar{n} = \sum_{n=0}^{N} \sum_{\alpha_1=0}^{1} \sum_{\alpha_2=0}^{1} n \, \mathbf{p}[n, \alpha_1, \alpha_2]. \qquad (3.6)$$

The probabilities p_b of blocking of Machine M_1 and p_s of starving of Machine M_2 are

$$p_b = \mathbf{p}[N, 1, 0] + \mathbf{p}[N, 1, 1] \qquad (3.7)$$
$$p_s = \mathbf{p}[0, 0, 1] + \mathbf{p}[0, 1, 1] \qquad (3.8)$$

Note that a machine cannot be blocked or starved when it is down, since it cannot fail if it is either blocked or starved.

3.2.2 Derivation of the Transition Equations

We assume that the extended storage N is greater than or equal to four. In this case, we can divide the system states with respect to the storage level n into non-empty sets of so-called lower boundary states, internal states and upper boundary states. After stating the transition equations, we first derive general expressions for the probabilities of internal states. Once these probabilities of internal states are known, we can compute those of the boundary states.

Internal states. The transition equations for probabilities $p[n, \alpha_1, \alpha_2]$ of states (n, α_1, α_2) with $1 < n < N - 1$ (referred to as internal states) have a common structure. In order to derive the transition equations, the transition diagrams in Figures 3.3 to 3.6 are used.

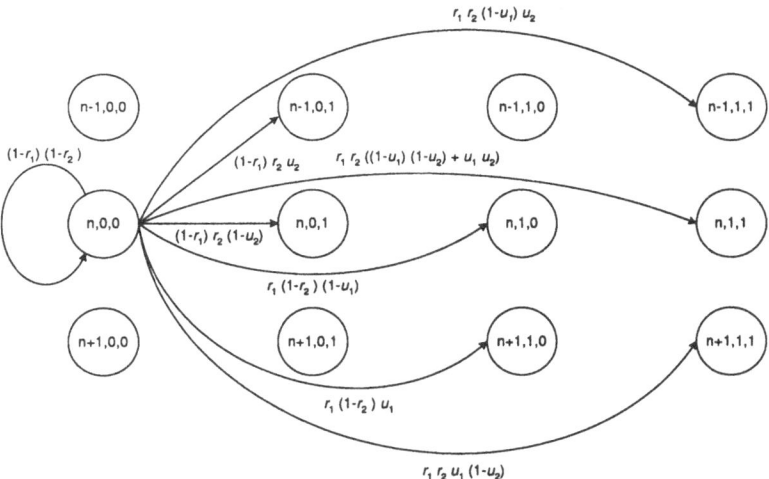

Fig. 3.3. Possible Transitions from State (n,0,0)

These diagrams show the possible transitions of the Markov process model during one time step. Consider in Figure 3.3 the transition from state $(n, 0, 0)$ at time t to state $(n - 1, 0, 1)$ at time $t + 1$. At time t, both machines are down. The first machine must not be repaired, with probability $(1 - r_1)$. The second machine must be repaired, with probability r_2, and it must complete a part at the end of period $t + 1$, with probability u_2. The probability of the transition from state $(n, 0, 0)$ to state $(n - 1, 0, 1)$ is therefore $(1 - r_1)r_2u_2$.

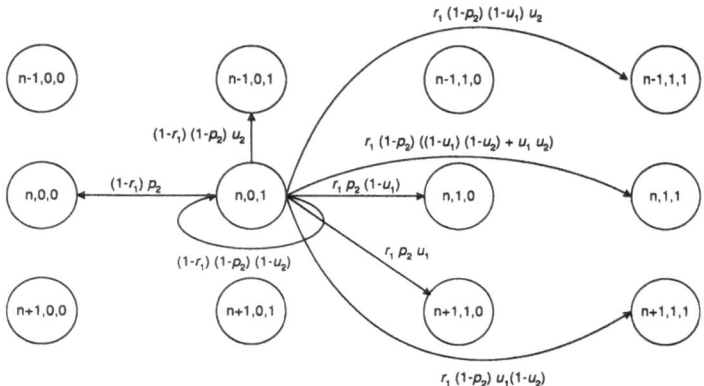

Fig. 3.4. Possible Transitions from State (n,0,1)

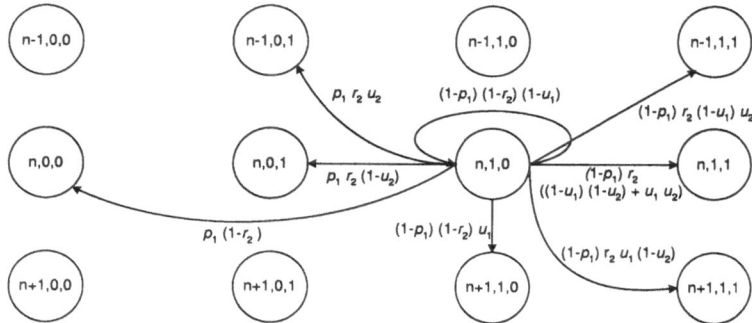

Fig. 3.5. Possible Transitions from State (n,1,0)

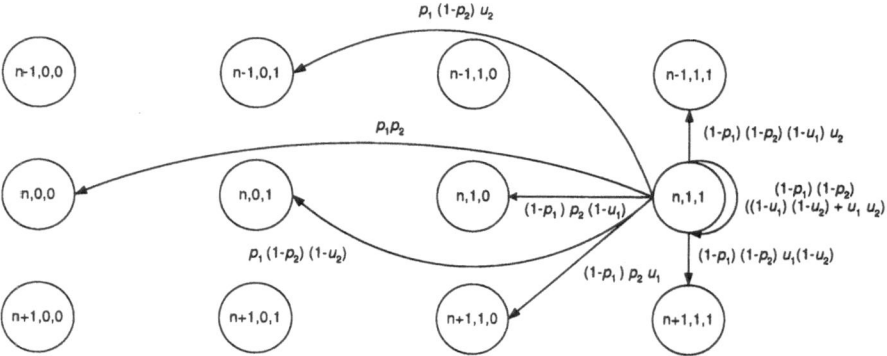

Fig. 3.6. Possible Transitions from State (n,1,1)

In order to derive the transition equations, we use the transitions diagrams to identify for each state the possible previous states and the respective transition probabilities.

Consider, for example, the first term on the right hand side of (3.9) related to state $(n, 0, 0)$:

$$
\begin{aligned}
\mathbf{p}[n, 0, 0] \quad = \quad & \mathbf{p}[n, 0, 0] \ (1 - r_1)(1 - r_2) + \mathbf{p}[n, 0, 1](1 - r_1)p_2 + \\
& \mathbf{p}[n, 1, 0] \ p_1(1 - r_2) + \mathbf{p}[n, 1, 1]p_1 p_2
\end{aligned}
\tag{3.9}
$$

If the system is in state $(n, 0, 0)$ at time t, neither the first nor the second machine may be repaired, with probability $(1 - r_1)(1 - r_2)$, in order to remain in this state. This can be seen from Figure 3.3.

The remaining internal transition equations are:

$$
\begin{aligned}
\mathbf{p}[n, 0, 1] \quad = \quad & \mathbf{p}[n, 0, 0] \ (1 - r_1)r_2(1 - u_2) + \\
& \mathbf{p}[n, 0, 1] \ (1 - r_1)(1 - p_2)(1 - u_2) + \\
& \mathbf{p}[n, 1, 0] \ p_1 r_2(1 - u_2) + \\
& \mathbf{p}[n, 1, 1] \ p_1(1 - p_2)(1 - u_2) \\
& \mathbf{p}[n + 1, 0, 0] \ (1 - r_1)r_2 u_2 + \\
& \mathbf{p}[n + 1, 0, 1] \ (1 - r_1)(1 - p_2)u_2 + \\
& \mathbf{p}[n + 1, 1, 0] \ p_1 r_2 u_2 + \\
& \mathbf{p}[n + 1, 1, 1] \ p_1(1 - p_2)u_2
\end{aligned}
\tag{3.10}
$$

$$
\begin{aligned}
\mathbf{p}[n, 1, 0] \quad = \quad & \mathbf{p}[n - 1, 0, 0] \ r_1(1 - r_2)u_1 + \\
& \mathbf{p}[n - 1, 0, 1] \ r_1 p_2 u_1 + \\
& \mathbf{p}[n - 1, 1, 0] \ (1 - p_1)(1 - r_2)u_1 + \\
& \mathbf{p}[n - 1, 1, 1] \ (1 - p_1)p_2 u_1 \\
& \mathbf{p}[n, 0, 0] \ r_1(1 - r_2)(1 - u_1) + \\
& \mathbf{p}[n, 0, 1] \ r_1 p_2(1 - u_1) + \\
& \mathbf{p}[n, 1, 0] \ (1 - p_1)(1 - r_2)(1 - u_1) + \\
& \mathbf{p}[n, 1, 1] \ (1 - p_1)p_2(1 - u_1)
\end{aligned}
\tag{3.11}
$$

$$
\begin{aligned}
\mathbf{p}[n,1,1] \;=\; & \mathbf{p}[n-1,0,0]\, r_1 r_2 u_1 (1-u_2) + \\
& \mathbf{p}[n-1,0,1]\, r_1(1-p_2)u_1(1-u_2) + \\
& \mathbf{p}[n-1,1,0]\, (1-p_1)r_2 u_1(1-u_2) + \\
& \mathbf{p}[n-1,1,1]\, (1-p_1)(1-p_2)u_1(1-u_2) \\
& \mathbf{p}[n,0,0]\, r_1 r_2[(1-u_1)(1-u_2)+u_1 u_2] + \\
& \mathbf{p}[n,0,1]\, r_1(1-p_2)[(1-u_1)(1-u_2)+u_1 u_2] + \\
& \mathbf{p}[n,1,0]\, (1-p_1)r_2[(1-u_1)(1-u_2)+u_1 u_2] + \\
& \mathbf{p}[n,1,1]\, (1-p_1)(1-p_2)[(1-u_1)(1-u_2)+u_1 u_2] + \\
& \mathbf{p}[n+1,0,0]\, r_1 r_2(1-u_1)u_2 + \\
& \mathbf{p}[n+1,0,1]\, r_1(1-p_2)(1-u_1)u_2 + \\
& \mathbf{p}[n+1,1,0]\, (1-p_1)r_2(1-u_1)u_2 + \\
& \mathbf{p}[n+1,1,1]\, (1-p_1)(1-p_2)(1-u_1)u_2 \qquad (3.12)
\end{aligned}
$$

Note that in equation (3.12) there are two possible transitions from states $(n,0,0)$, $(n,0,1)$, $(n,1,0)$, and $(n,1,1)$ to state $(n,1,1)$. Consider, for example, the transition from $(n,0,0)$. In either case, both machines have to be repaired (with probability $r_1 r_2$). The storage level n remains unchanged if either, with probability $(1-u_1)(1-u_2)$, none or, with probability $u_1 u_2$, both of the two machines complete an operation.

Lower boundary states. States with empty ($n = 0$) or almost empty ($n = 1$) storage are called lower boundary states. The transition equations related to states $(0,0,1)$ and $(0,1,1)$ are given in Figures 3.7 and 3.8.

States $(0,0,0)$ and $(0,1,0)$ are transient:

$$
\mathbf{p}[0,0,0] = \mathbf{p}[0,1,0] = 0 \qquad (3.13)
$$

State $(0,0,0)$ can only be reached from itself by means of a transition with no repair of any machine and from state $(0,1,0)$ through a failure of Machine M_1 and no repair of M_2. This can be seen from Figures 3.7 and 3.8 for transitions from states with buffer level $n = 0$ and from Figures 3.3 to 3.8 from states with $n = 1$.

The other lower boundary state $(0,1,0)$ can also be reached only from itself and from $(0,0,0)$ by means of a repair of M_1 and no repair of M_2. Both $(0,0,0)$ and $(0,1,0)$ can neither be reached from states $(0,0,1)$ and $(0,1,1)$ since a starved machine M_2 cannot fail, nor from a state with $n = 1$ since with M_2 being down, the buffer cannot become empty. Thus both $(0,0,0)$ and $(0,1,0)$ are transient.

Two of the lower boundary states, $(1,0,0)$ and $(1,0,1)$, have transition equations that are of internal form (3.9) and (3.10).

The transition equations of the remaining four lower boundary states are:

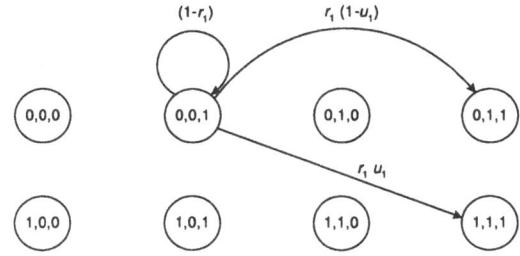

Fig. 3.7. Possible Transitions from State (0,0,1)

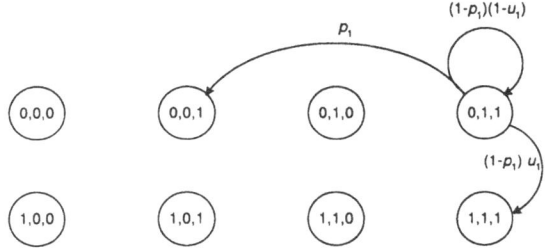

Fig. 3.8. Possible Transitions from State (0,1,1)

$$
\begin{aligned}
\mathbf{p}[0,0,1] \;=\; & \mathbf{p}[0,0,1]\,(1-r_1) + \mathbf{p}[0,1,1]\,p_1 \\
& \mathbf{p}[1,0,0]\,(1-r_1)r_2 u_2 + \mathbf{p}[1,0,1]\,(1-r_1)(1-p_2)u_2 + \\
& \mathbf{p}[1,1,0]\,p_1 r_2 u_2 + \mathbf{p}[1,1,1]\,p_1(1-p_2)u_2 \qquad (3.14)
\end{aligned}
$$

$$
\begin{aligned}
\mathbf{p}[0,1,1] \;=\; & \mathbf{p}[0,0,1]\,r_1(1-u_1) + \mathbf{p}[0,1,1]\,(1-p_1)(1-u_1) + \\
& \mathbf{p}[1,0,0]\,r_1 r_2(1-u_1)u_2 + \mathbf{p}[1,0,1]\,r_1(1-p_2)(1-u_1)u_2 + \\
& \mathbf{p}[1,1,0]\,(1-p_1)r_2(1-u_1)u_2 + \\
& \mathbf{p}[1,1,1]\,(1-p_1)(1-p_2)(1-u_1)u_2 \qquad (3.15)
\end{aligned}
$$

$$
\begin{aligned}
\mathbf{p}[1,1,0] \;=\; & \mathbf{p}[1,0,0]\,r_1(1-r_2)(1-u_1) + \mathbf{p}[1,0,1]\,r_1 p_2(1-u_1) + \\
& \mathbf{p}[1,1,0]\,(1-p_1)(1-r_2)(1-u_1) + \\
& \mathbf{p}[1,1,1]\,(1-p_1)p_2(1-u_1) \qquad (3.16)
\end{aligned}
$$

$$
\begin{aligned}
\mathbf{p}[1,1,1] \quad = \quad & \mathbf{p}[0,0,1]\, r_1 u_1 + \mathbf{p}[0,1,1]\,(1-p_1)u_1 \\
& \mathbf{p}[1,0,0]\, r_1 r_2[(1-u_1)(1-u_2)+u_1 u_2] + \\
& \mathbf{p}[1,0,1]\, r_1(1-p_2)[(1-u_1)(1-u_2)+u_1 u_2] + \\
& \mathbf{p}[1,1,0]\,(1-p_1)r_2[(1-u_1)(1-u_2)+u_1 u_2] + \\
& \mathbf{p}[1,1,1]\,(1-p_1)(1-p_2)[(1-u_1)(1-u_2)+u_1 u_2] + \\
& \mathbf{p}[2,0,0]\, r_1 r_2(1-u_1)u_2 + \\
& \mathbf{p}[2,0,1]\, r_1(1-p_2)(1-u_1)u_2 + \\
& \mathbf{p}[2,1,0]\,(1-p_1)r_2(1-u_1)u_2 + \\
& \mathbf{p}[2,1,1]\,(1-p_1)(1-p_2)(1-u_1)u_2
\end{aligned}
\tag{3.17}
$$

There are again two possible transitions, for example, from state $(1,0,1)$ to $(1,1,1)$.

Upper boundary states. States with an almost full ($n = N-1$ or full ($n = N$) storage are called upper boundary states. There is a symmetry between lower and upper boundary states: transition equations for upper boundary states can be derived from lower boundary states if we replace $\mathbf{p}[n,\alpha_1,\alpha_2]$, u_i, p_i, and r_i in lower boundary transition equations by $\mathbf{p}[N-n,\alpha_2,\alpha_1]$, u_{3-i}, p_{3-i}, and r_{3-i}.

3.2.3 Identities

Conservation of flow. The probabilities of a part passing through Machines M_1 and M_2 during an arbitrary period are equal:

$$
PR_1 = u_1 \sum_{\alpha_2=0}^{1} \sum_{n=0}^{N-1} \mathbf{p}[n,1,\alpha_2] = u_2 \sum_{\alpha_1=0}^{1} \sum_{n=1}^{N} \mathbf{p}[n,\alpha_1,1] = PR_2
\tag{3.18}
$$

The reason is that the flow of material is linear and parts are neither created nor destroyed at either machine.

Repair frequency equals failure frequency. In the long run, there must be a repair for each failure of a machine:

$$
r_1 \sum_{n=0}^{N-1} \sum_{\alpha_2=0}^{1} \mathbf{p}[n,0,\alpha_2] = p_1 \sum_{n=0}^{N-1} \sum_{\alpha_2=0}^{1} \mathbf{p}[n,1,\alpha_2]
\tag{3.19}
$$

$$
r_2 \sum_{n=1}^{N} \sum_{\alpha_1=0}^{1} \mathbf{p}[n,\alpha_1,0] = p_2 \sum_{n=1}^{N} \sum_{\alpha_1=0}^{1} \mathbf{p}[n,\alpha_1,1]
\tag{3.20}
$$

Observe that a machine must be in the appropriate state to experience a failure or repair. A formal proof of (3.18), (3.19), and (3.20) is not yet available. However, in the numerical solution procedure described below we have to solve a linear system of equations that contains (3.18), (3.19), and (3.20) to compute steady state probabilities. This system can only be solved if (3.18), (3.19), and (3.20) hold for the given numerical set of parameters and N. If we can solve this set of equations, we know they hold at least for this current set of parameters and may therefore compute the desired performance measures. We used this two-machine model in decomposition approaches to analyze longer lines and A/D systems where we solved some hundred thousand different two-machine problems without ever encountering a case where (3.18), (3.19), and (3.20) did not apply. While from the theoretical point of view the formal proof of these equations is clearly desirable, we may well use these equations in our algorithm, as long as they apply to the current numerical set of parameters under study.

3.2.4 Derivation of the Solution

In this section, we derive the transition equations for the Markov process model and a specialized solution procedure similar to the one given in Gershwin and Berman (1981). The reason to develop a specialized solution procedure instead of using a standard approach to solve a linear system of equations is that the specialized approach can be modified in order to determine performance measure for *non-integer* buffer sizes. A non-integer buffer size does not have a meaningful physical interpretation if we produce discrete parts. However, in numerical optimization algorithms for buffer allocation that use fast gradient techniques, it turns out to be useful to determine performance measures for non-integer buffer sizes. This cannot be done if we use a standard approach to solve the set of transition equations explicitly.

3.2.4.1 Analysis of Internal States.
Following Gershwin and Berman, we assume that the internal equations (3.9)-(3.12) have a solution of the form

$$\mathbf{p}[n,\alpha_1,\alpha_2] = \sum_{j=1}^{J} c_j \xi_j(n,\alpha_1,\alpha_2) = \sum_{j=1}^{J} c_j X_j^n Y_{1j}^{\alpha_1} Y_{2j}^{\alpha_2} \qquad (3.21)$$

where c_j, X_j, Y_{1j}, and Y_{2j} are parameters to be determined. The analysis below is very similar to the one in Gershwin and Berman (1981). Replacing $\mathbf{p}[n,\alpha_1,\alpha_2]$ by $X_j^n Y_{1j}^{\alpha_1} Y_{2j}^{\alpha_2}$ in Equations (3.9), (3.10), and (3.11), we derive the following non-linear set of equations:

$$1 = (1-r_1)(1-r_2) + (1-r_1)p_2Y_2 + p_1(1-r_2)Y_1 + p_1p_2Y_1Y_2 \quad (3.22)$$

$$\begin{aligned}
Y_2 = \ & (1-r_1)r_2(1-u_2) + (1-r_1)(1-p_2)(1-u_2)Y_2 + \\
& p_1r_2(1-u_2)Y_1 + p_1(1-p_2)(1-u_2)Y_1Y_2 + \\
& (1-r_1)r_2u_2X + (1-r_1)(1-p_2)u_2XY_2 + \\
& p_1r_2u_2XY_1 + p_1(1-p_2)u_2XY_1Y_2
\end{aligned} \quad (3.23)$$

$$\begin{aligned}
Y_1 = \ & r_1(1-r_2)u_1X^{-1} + r_1p_2u_1X^{-1}Y_2 + \\
& (1-p_1)(1-r_2)u_1X^{-1}Y_1 + (1-p_1)p_2u_1X^{-1}Y_1Y_2 + \\
& r_1(1-r_2)(1-u_1) + r_1p_2(1-u_1)Y_2 + \\
& (1-p_1)(1-r_2)(1-u_1)Y_1 + (1-p_1)p_2(1-u_1)Y_1Y_2
\end{aligned} \quad (3.24)$$

Inserting (3.21) in (3.12) does not provide additional information, since the resulting equation is the product of (3.23) and (3.24). Equations (3.22), (3.23), and (3.24) can be simplified if we factor the terms on the right hand side and divide (3.22) and (3.23) by (3.24):

$$\begin{aligned}
1 &= (1-r_1+p_1Y_1)(1-r_2+p_2Y_2) & (3.25) \\
Y_2 &= (1-u_2+u_2X)(r_2+Y_2-p_2Y_2)/(1-r_2+p_2Y_2) & (3.26) \\
XY_1 &= (u_1+X-u_1X)(r_1+Y_1-p_1Y_1)/(1-r_1+p_1Y_1) & (3.27)
\end{aligned}$$

Equations (3.25) and (3.26) are used to eliminate Y_2. From the resulting equation and (3.27) we can next eliminate X. A considerable algebraic effort leads to the following fourth degree equation in Y_1

$$p_1^2(Y_1^3 + sY_1^2 + tY_1 + v)(1-p_2-r_2)(1-u_2)(p_1Y_1 - r_1) = 0 \quad (3.28)$$

with auxiliary variables $s, t, v,$ and w defined as follows:

$$w = p_1(1-p_2-r_2)(1-u_2) \quad (3.29)$$

$$\begin{aligned}
s = \frac{1}{w} \ & (p_1 - p_2 - p_1p_2 - 2r_1 + 2p_2r_1 - r_2 - p_1r_2 + 2r_1r_2 + \\
& u_1 - p_1u_1 - p_2u_1 + p_1p_2u_1 - r_2u_1 + p_1r_2u_1 - u_2 - p_1u_2 + \\
& p_2u_2 + p_1p_2u_2 + 2r_1u_2 - 2p_2r_1u_2 + 2r_2u_2 + p_1r_2u_2 - 2r_1r_2u_2)
\end{aligned} \quad (3.30)$$

$$\begin{aligned}
t = \frac{1}{p_1w} \ & (p_2r_1 - p_1p_2 - 2p_1r_1 + 2p_1p_2r_1 + r_1^2 - p_2r_1^2 - p_1r_2 + \\
& r_1r_2 + 2p_1r_1r_2 - r_1^2r_2 - p_2u_1 + p_1p_2u_1 - r_1u_1 + \\
& 2p_1r_1u_1 + p_2r_1u_1 - 2p_1p_2r_1u_1 - r_2u_1 + p_1r_2u_1 + r_1r_2u_1 - \\
& 2p_1r_1r_2u_1 - p_1u_2 + p_1p_2u_2 + r_1u_2 + 2p_1r_1u_2 - p_2r_1u_2 - \\
& 2p_1p_2r_1u_2 - r_1^2u_2 + p_2r_1^2u_2 + r_2u_2 + 2p_1r_2u_2 - \\
& 2r_1r_2u_2 - 2p_1r_1r_2u_2 + r_1^2r_2u_2)
\end{aligned} \quad (3.31)$$

$$v = \frac{1}{p_1 w} (p_2 r_1 + r_1^2 - p_2 r_1^2 + r_1 r_2 - r_1^2 r_2 - p_2 r_1 u_1 - r_1^2 u_1 + \quad (3.32)$$
$$p_2 r_1^2 u_1 - r_1 r_2 u_1 + r_1^2 r_2 u_1 + r_1 u_2 - p_2 r_1 u_2 - r_1^2 u_2 +$$
$$p_2 r_1^2 u_2 + r_2 u_2 - 2 r_1 r_2 u_2 + r_1^2 r_2 u_2)$$

From the last term on the left side of Equation (3.28) we see that one solution to (3.28) is

$$Y_{11} = \frac{r_1}{p_1} \qquad (3.33)$$

Applying this result to Equation (3.25), we find

$$Y_{21} = \frac{r_2}{p_2} \qquad (3.34)$$

and from (3.33) and (3.34) in (3.26) or (3.27) we see that

$$X_1 = 1 \qquad (3.35)$$

The remaining three solutions to (3.28) are[4]

$$Y_{12} = 2\sqrt{-\frac{a}{3}} \cos\left(\frac{\phi}{3}\right) - \frac{s}{3} \qquad (3.36)$$

$$Y_{13} = 2\sqrt{-\frac{a}{3}} \cos\left(\frac{\phi}{3} + \frac{2\phi}{3}\right) - \frac{s}{3} \qquad (3.37)$$

$$Y_{14} = 2\sqrt{-\frac{a}{3}} \cos\left(\frac{\phi}{3} + \frac{4\phi}{3}\right) - \frac{s}{3} \qquad (3.38)$$

with auxiliary variables

$$a = \frac{1}{3}\left(3t - s^2\right) \qquad (3.39)$$

$$b = \frac{1}{27}\left(2s^3 - 9st + 27v\right) \qquad (3.40)$$

$$\phi = \arccos\left(-\frac{b}{2\sqrt{\frac{-a^3}{27}}}\right) \qquad (3.41)$$

The corresponding values of Y_{22}, Y_{22}, and Y_{24} are again determined via (3.25). The values of X_2, X_3, and X_4 are next computed from (3.26) or (3.27).

---

[4] See Bronstein and Semendjajew (1983, Sec. 2.4.2.3, p. 131)

Since we have found four solutions to equations (3.25), (3.26), and (3.27), the general expression for the steady-state probabilities of the internal states is as follows

$$\mathbf{p}[n, \alpha_1, \alpha_2] = \sum_{j=1}^{4} c_j \xi_j(n, \alpha_1, \alpha_2) = \sum_{j=1}^{4} c_j X_j^n Y_{1j}^{\alpha_1} Y_{2j}^{\alpha_2} \qquad (3.42)$$

where we still have to determine the parameters c_j.

3.2.4.2 Analysis of Boundary States. There is a total of 16 boundary states in the model. The transition equations of four of them $((1,0,0), (1,0,1),$ $(N-1,0,0),$ and $(N-1,1,0))$ are of internal form (3.9) - (3.12), i.e. their steady-state probabilities can be computed from equation (3.42) even though they are boundary states.

Since $\mathbf{p}[1, 0, 0]$ and $\mathbf{p}[1, 0, 1]$ are given from (3.42), the corresponding equations (3.9) and (3.10) related to states $(1, 0, 0)$ and $(1, 0, 1)$ constitute a linear system of two equations in two unknowns $\mathbf{p}[1, 1, 0]$ and $\mathbf{p}[1, 1, 1]$ with the following solution:

$$\mathbf{p}[1, 1, 0] = \frac{u_1 - 1}{p_1} \left(\mathbf{p}[1, 0, 0](r_1 r_2 - r_1 - r_2 + p_1 r_2) + \qquad (3.43) \right.$$
$$\left. \mathbf{p}[1, 0, 1](p_2 - p_1 p_2 - r_1 p_2) \right)$$

$$\mathbf{p}[1, 1, 1] = \frac{1}{p_1 p_2} \left(\mathbf{p}[1, 0, 0](r_1 + r_2 - r_1 r_2) + \qquad (3.44) \right.$$
$$\mathbf{p}[1, 0, 1](r_1 p_2 - p_2) +$$
$$\left. \mathbf{p}[1, 1, 0](p_1 r_2 - p_1) \right)$$

Given $\mathbf{p}[1, 0, 0]$, $\mathbf{p}[1, 0, 1]$, $\mathbf{p}[1, 1, 0]$, and $\mathbf{p}[1, 1, 1]$, Equations (3.14) and (3.15) constitute another linear system that is solved by

$$\mathbf{p}[0, 0, 1] = \frac{u_2}{r_1 u_1} \left(\mathbf{p}[1, 0, 0](p_1 r_2 + r_2 u_1 - p_1 r_2 u_1 - r_1 r_2 u_1) + \qquad (3.45) \right.$$
$$\mathbf{p}[1, 0, 1] \; (p_1 - p_1 p_2 + u_1 - p_1 u_1 - p_2 u_1 +$$
$$p_1 p_2 u_1 - r_1 u_1 + r_1 p_2 u_1) +$$
$$\left. \mathbf{p}[1, 1, 0] p_1 r_2 + \mathbf{p}[1, 1, 1](p_1 - p_1 p_2) \right)$$

$$\mathbf{p}[0, 1, 1] = \frac{1}{p_1} \left(\mathbf{p}[0, 0, 1] r_1 + \mathbf{p}[1, 0, 0](r_1 r_2 u_2 - r_2 u_2) + \qquad (3.46) \right.$$
$$\mathbf{p}[1, 0, 1](p_2 u_2 - u_2 + r_1 u_2 - r_1 p_2 u_2) -$$
$$\left. \mathbf{p}[1, 1, 0] p_1 r_2 u_2 + \mathbf{p}[1, 1, 1](p_1 p_2 u_2 - p_1 u_2) \right)$$

The upper boundary steady-state probabilities are determined in exactly the same way as now $(N-1, 0, 0)$ and $(N-1, 1, 0)$ are of internal form and we may compute $\mathbf{p}[N-1, 0, 0]$ and $\mathbf{p}[N-1, 1, 0]$ from (3.42) and finally solve another two linear systems in two unknowns each to find:

$$\mathbf{p}[N-1,0,1] = \frac{u_2-1}{p_2} \left(\mathbf{p}[N-1,0,0](r_1p_2-r_1-r_2+r_1r_2)+ \atop \mathbf{p}[N-1,1,0](p_1-p_1p_2-p_1r_2)\right) \tag{3.47}$$

$$\mathbf{p}[N-1,1,1] = \frac{1}{p_1p_2} \left(\mathbf{p}[N-1,0,0](r_1+r_2-r_1r_2)+ \atop \mathbf{p}[N-1,0,1](r_1p_2-p_2)+ \atop \mathbf{p}[N-1,1,0](p_1r_2-p_1)\right) \tag{3.48}$$

$$\mathbf{p}[N,1,0] = \frac{u_1}{r_2u_2} \big(\mathbf{p}[N-1,0,0](r_1p_2+r_1u_2-r_1p_2u_2-r_1r_2u_2)+$$
$$\mathbf{p}[N-1,0,1]r_1p_2+$$
$$\mathbf{p}[N-1,1,0]\ (p_2-p_1p_2+u_2-p_1u_2-p_2u_2+$$
$$p_1p_2u_2-r_2u_2+p_1r_2u_2)+$$
$$\mathbf{p}[N-1,1,1](p_2-p_1p_2)\big) \tag{3.49}$$

$$\mathbf{p}[N,1,1] = \frac{1}{p_2} \big(\mathbf{p}[N,1,0]r_2+ \tag{3.50}$$
$$\mathbf{p}[N-1,0,0](r_1r_2u_1-r_1u_1)-$$
$$\mathbf{p}[N-1,0,1]r_1p_2u_1+$$
$$\mathbf{p}[N-1,1,0](p_1u_1-u_1+r_2u_1-p_1r_2u_1)+$$
$$\mathbf{p}[N-1,1,1](p_1p_2u_1-p_2u_1)\big)$$

Consider again the symmetry of upper and lower boundary values.

Since boundary states are expressed in terms of internal states in Equations (3.43) to (3.50), and since internal states are of the form

$$\mathbf{p}[n,\alpha_1,\alpha_2] = \sum_{j=1}^{4} c_j \xi_j(n,\alpha_1,\alpha_2), \tag{3.51}$$

Equations (3.43) to (3.50) hold for each solution $\xi_j(n,\alpha_1,\alpha_2)$ of the equations for internal states. The equation (3.43) corresponding to state $(1,1,0)$, for example, leads to

$$\sum_{j=1}^{4} c_j \xi_j(n,\alpha_1,\alpha_2) \tag{3.52}$$

$$= \frac{u_1-1}{p_1} \left(\sum_{j=1}^{4} c_j\xi_j(1,1,0)(r_1r_2-r_1-r_2+p_1r_2)+ \atop \sum_{j=1}^{4} c_j\xi_j(1,0,1)(p_2-p_1p_2-r_1p_2)\right)$$

Similar equations can be found to determine the terms $\xi_j(n,\alpha_1,\alpha_2)$ for the seven other boundary state probabilities. The terms $\xi_j(n,\alpha_1,\alpha_2)$ corresponding to transient states are all zero.

Now all steady-state probabilities have been related to equation (3.42). What remains to be done is to find appropriate values of the coefficients c_j in (3.42).

3.2.4.3 Determination of Coefficients c_j. To determine four coefficients $c_j, j = 1, ..., 4$, a linear system of equations in the four unknowns c_j can be solved. The following four equations can be derived by inserting (3.42) into the conservation of flow equation (3.18), the two repair frequency equals failure frequency equations for the two machines (3.19) and (3.20), and the condition that all probabilities sum up to one:

Conservation of flow

$$\sum_{j=1}^{4} \left[u_1 \sum_{n=0}^{N-1} \sum_{\alpha_2=0}^{1} \xi_j(n, 1, \alpha_2) - u_2 \sum_{n=1}^{N} \sum_{\alpha_1=0}^{1} \xi_j(n, \alpha_1, 1) \right] c_j = 0 \qquad (3.53)$$

Repair frequency equals failure frequency at Machine M_1

$$\sum_{j=1}^{4} \left[\sum_{n=0}^{N-1} \left(r_1 \sum_{\alpha_2=0}^{1} \xi_j(n, 0, \alpha_2) - p_1 \sum_{\alpha_2=0}^{1} \xi_j(n, 1, \alpha_2) \right) \right] c_j = 0 \qquad (3.54)$$

Repair frequency equals failure frequency at Machine M_2

$$\sum_{j=1}^{4} \left[\sum_{n=1}^{N} \left(r_2 \sum_{\alpha_1=0}^{1} \xi_j(n, \alpha_1, 0) - p_2 \sum_{\alpha_2=0}^{1} \xi_j(n, \alpha_1, 1) \right) \right] c_j = 0 \qquad (3.55)$$

Probabilities sum up to one

$$\sum_{j=1}^{4} \left[\sum_{n=0}^{N} \left(\xi_j(n, 0, 0) + \xi_j(n, 0, 1) + \xi_j(n, 1, 0) + \xi_j(n, 1, 1) \right) \right] c_j = 1 \qquad (3.56)$$

Note that the right hand side of the three of the four equations is zero. For this reason, it is relatively painless to solve this linear system of equations in c_j numerically.

3.2.5 The Algorithm to Determine Steady-State Probabilities and Performance Measures

The algorithm to compute the required steady-state probabilities $p[n, \alpha_1, \alpha_2]$ and performance measures PR and $\bar{n}$ consists of the following steps:

1. Compute auxiliary variables w, s, t, v, a, b, and ϕ from (3.29)-(3.32) and (3.39)-(3.40). Compute Y_{11} from (3.33) and $Y_{12}...Y_{14}$ from (3.36)-(3.38). Compute $Y_{21}...Y_{24}$ from (3.25) and $X_1...X_4$ from (3.26) or (3.27).
2. Determine the coefficients $c_j, j = 1, ..., 4$ in Equation (3.42) by solving the linear system of equations given by (3.53)-(3.56).
3. Use the c_j from Step 2 to compute the required steady-state probabilities $p[n, \alpha_1, \alpha_2]$ of states of internal form via (3.42) and those of the remaining boundary states from (3.43)-(3.50).
4. Determine performance measures. Determine the production rate from (3.1), in-process inventory from (3.6) and blocking and starvation probabilities from (3.7) and (3.8).

3.2.6 Determination of Performance Measures without Explicit Computation of all Steady-State Probabilities

When analyzing a two-machine line, we are not necessarily interested in the complete set of steady-state probabilities. It may be sufficient to determine aggregate performance measures such as expected production rate, average inventory level, and some selected steady-state probabilities. This is the case when a two-machine model is used in a decomposition approach.

The definition of the production rate PR and the average inventory $\bar{n}$ as given in (3.1) and (3.6) appears to require all steady-state probabilities $p[n, \alpha_1, \alpha_2]$. It is, however, possible to compute these quantities without explicitly determining the complete set of probabilities.[5] This is useful for two reasons:

1. Two-machine models are often solved repeatedly in decomposition approaches and the decomposition approaches are used repeatedly in optimization approaches. It may be necessary to solve hundreds of thousands two-machine models during a single optimization. Allocating the storage for all the internal steady-state probabilities may require a significant amount of time and computer storage, especially when the buffers are large. The numerical effort can be reduced significantly, if only the required probabilities and performance measures are computed explicitly.
2. Some algorithms to determine the optimal buffer allocation like those proposed in Schor (1995) use gradient techniques that treat the buffer

---

[5] This hint was due to Stanley B. Gershwin (private conversation). These ideas must have been used by Schor (1995, p. 34), but they are not clearly documented in Schor's thesis.

size as if it were continuous, i.e. they calculate performance measures for non-integer buffer sizes. Even though a non-integer buffer size has no physical meaning, the production rate as a function of the continuous buffer size appears to be a smooth, continuous function. In order to use these powerful gradient approaches, we need expressions for production rates and inventory levels as functions of continuous buffer sizes.

For this reason we show in this subsection how to derive the performance measures without prior determination of all the steady-state probabilities.

If we separate for the production rate boundary and internal states for (3.18) as in

$$
PR_1 = u_1 \left[\mathbf{p}[0,1,1] + \mathbf{p}[1,1,0] + \mathbf{p}[1,1,1] + \right.
$$
$$
\sum_{n=2}^{N-2} \left(\mathbf{p}[n,1,0] + \mathbf{p}[n,1,1] \right) +
$$
$$
\left. \mathbf{p}[N-1,1,0] + \mathbf{p}[N-1,1,1] \right]
\tag{3.57}
$$

and apply (3.42) for the internal states, we find after a small algebraic manipulation

$$
PR_1 = u_1 \left[\mathbf{p}[0,1,1] + \mathbf{p}[1,1,0] + \mathbf{p}[1,1,1] + \right.
$$
$$
\sum_{j=1}^{4} c_j \left(Y_{1j} + Y_{1j} Y_{2j} \right) \sum_{n=2}^{N-2} X_j^n +
$$
$$
\left. \mathbf{p}[N-1,1,0] + \mathbf{p}[N-1,1,1] \right].
\tag{3.58}
$$

From the equation for the geometric sum we know that for $X_j \neq 1$

$$
\sum_{n=0}^{N} X_j^n = \frac{1 - X_j^{N+1}}{1 - X_j}
\tag{3.59}
$$

or

$$
\sum_{n=2}^{N-2} X_j^n = \frac{1 - X_j^{N-1}}{1 - X_j} - 1 - X_j = \frac{X_j^2 - X_j^{N-1}}{1 - X_j}
\tag{3.60}
$$

and therefore for all $X_j \neq 1$

$$PR_1 = u_1 \Bigg[\mathbf{p}[0,1,1] + \mathbf{p}[1,1,0] + \mathbf{p}[1,1,1] +$$

$$\sum_{j=1}^{4} c_j \left(Y_{1j} + Y_{1j} Y_{2j} \right) \frac{X_j^2 - X_j^{N-1}}{1 - X_j} +$$

$$\mathbf{p}[N-1,1,0] + \mathbf{p}[N-1,1,1] \Bigg] \qquad (3.61)$$

which allows us to compute the production rate without computing the complete set of probabilities for the internal states. For $X_j = 1$ we find $\sum_{n=2}^{N-2} X_j = N - 3$ and replace $(X_j^2 - X_j^{N-1})/(1 - X_j)$ in (3.61) by $N - 3$.

A similar argument holds for the average inventory level $\bar{n}$ with

$$\bar{n} = \sum_{n=0}^{N} n \left(\mathbf{p}[n,0,0] + \mathbf{p}[n,0,1] + \mathbf{p}[n,1,0] + \mathbf{p}[n,1,1] \right)$$

$$= \mathbf{p}[1,0,0] + \mathbf{p}[1,0,1] + \mathbf{p}[1,1,0] + \mathbf{p}[1,1,1] +$$

$$\sum_{n=2}^{N-2} n \left(\mathbf{p}[n,0,0] + \mathbf{p}[n,0,1] + \mathbf{p}[n,1,0] + \mathbf{p}[n,1,1] \right) +$$

$$(N-1) \left(\mathbf{p}[N-1,0,0] + \mathbf{p}[N-1,0,1] + \right.$$

$$\left. \mathbf{p}[N-1,1,0] + \mathbf{p}[N-1,1,1] \right) +$$

$$N \left(\mathbf{p}[N-1,1,0] + \mathbf{p}[N,1,1] \right) \qquad (3.62)$$

Differentiating (3.60) we find

$$\sum_{n=2}^{N-2} n X_j^{n-1} = \frac{(2X_j - (N-1)X_j^{N-2})(1 - X_j) + (X_j^2 - X_j^{N-1})}{(1 - X_j)^2}$$

$$= \frac{2X_j - X_j^2 - (N-1)X_j^{N-2} + (N-2)X_j^{N-1}}{(1 - X_j)^2} \qquad (3.63)$$

and after multiplying by X_j

$$\sum_{n=2}^{N-2} n X_j^n = \frac{2X_j^2 - X_j^3 - (N-1)X_j^{N-1} + (N-2)X_j^N}{(1 - X_j)^2} \qquad (3.64)$$

which can be used to express the average inventory $\bar{n}$ in Equation (3.6) after inserting (3.42) as

$$
\begin{aligned}
\bar{n} &= \sum_{n=0}^{N} n \left(\mathbf{p}[n,0,0] + \mathbf{p}[n,0,1] + \mathbf{p}[n,1,0] + \mathbf{p}[n,1,1] \right) \\
&= \mathbf{p}[1,0,0] + \mathbf{p}[1,0,1] + \mathbf{p}[1,1,0] + \mathbf{p}[1,1,1] + \\
&\quad \sum_{j=1}^{4} c_j \left[\frac{2X_j^2 - X_j^3 - (N-1)X_j^{N-1} + (N-2)X_j^N}{(1-X_j)^2} \right. \\
&\quad \left. (1 + Y_{1j} + Y_{2j} + Y_{1j}Y_{2j}) \right] + \\
&\quad (N-1)\Big(\mathbf{p}[N-1,0,0] + \mathbf{p}[N-1,0,1] + \\
&\quad \mathbf{p}[N-1,1,0] + \mathbf{p}[N-1,1,1] \Big) + \\
&\quad N\Big(\mathbf{p}[N-1,1,0] + \mathbf{p}[N,1,1] \Big)
\end{aligned}
\tag{3.65}
$$

where probabilities related to boundary states are computed from (3.43)-(3.50).

For $X_j = 1$ we note that (3.64) becomes

$$
\sum_{n=2}^{N-2} n1^n = \frac{N(N+1)}{2} - 1 - (N-1) - N = \frac{N(N+1) - 4N}{2}
\tag{3.66}
$$

and make the appropriate substitutions in Equation (3.65).

A similar set of substitutions is required in the four equations (3.53)-(3.56) that are used to determine the coefficients c_j in (3.42). If we do not want to compute all steady-state probabilities explicitly or if we want to determine performance measures for non-integer buffer sizes, this fast approach can be used.

3.2.7 Numerical Results for the Two-Machine Systems

This section shows that the model behaves as expected for a variety of parameter sets. Following an example in Gershwin (1994, p. 197) of a so-called deterministic two-machine line with deterministic and identical processing times, we analyze a stochastic two-machine line with random processing times. As in Gershwin's example, the system parameters $p_1 = 0.1$, $p_2 = 0.1$, and $r_2 = 0.1$ are held constant and we vary buffer size N and Machine $M_1's$ repair probability r_1 from 0.06 to 0.14. However, while in Gershwin's example an operational machine that is neither blocked nor starved always completes processing a part at the end of a period ($u_i = 1.0$), we assume that this happens only with a probability $u_1 = u_2 = 0.9$. The isolated production rate of M_2 is thus $IPR_2 = u_2 \frac{r_2}{r_2+p_2} = 0.9 \cdot 0.1/0.2 = 0.45$, and it is an upper limit of the transfer line production rate. If we vary r_1, the isolated production rate of Machine M_1 raises from 0.3375 for $r_1 = 0.06$ up to 0.525 for $r_1 = 0.14$. Thus, for a low value of r_1 we expect Machine M_1 to limit the system performance, whereas for $r_1 > 0.1$ the isolated production rate of M_1 is higher than that of M_2, which is then the bottleneck.

Figure 3.9 confirms this expectation. No matter how many buffers we add, for $r_1 = 0.06$ there is an upper limit of 0.3375 for the production rate. Figure 3.10 shows that the average buffer level $\bar{n}$ is low. When the line is balanced, i.e. $r_1 = 0.1$, adding buffers has the strongest impact on the production rate and the average buffer level is always $N/2$. For $r_1 > 0.1$ the production rate is quickly limited by the isolated production rate of M_2 and the buffer is full or almost full most of the time. This behavior is completely analogous to those of the deterministic two-machine line.

In the next experiment we hold $p_1 = p_2 = r_1 = r_2 = 0.1$, constant and study the system behavior for values of the processing probabilities $u_1 = u_2$ ranging from 0.6 to 0.9. Using the Gershwin and Schick model in Gershwin (1994, p. 76-93) we also compute the production rates of the deterministic model, i.e. the $u_1 = u_2 = 1.0$ case. Figure 3.11 demonstrates that the production rate of the stochastic transfer line does indeed approach the limit of the Schick and Gershwin deterministic line as processing probabilities u_1 and u_2 get close to one.

In a last numerical experiment, we show that for small time units, i.e. $u_i, p_i, r_i << 1, i = 1, 2$, this discrete time model behaves very much like the continuous time model by Gershwin and Berman. Assume that in the continuous time model processing times, times to repair and times to failure are exponentially distributed with rates $\mu_1 = 0.007, p_1 = 0.0001, r_1 = 0.001, \mu_2 = 0.008, p_2 = 0.0002$ and $r_2 = 0.001$. For these values the first machine has an isolated production rate of approximately 0.006363, whereas the second machine is a bit more efficient with an isolated production rate of 0.006666. The (extended) buffer size $N = C + 2$ is 10. In the upper part of Table 3.2, we compare production rates and average buffer levels for increasingly smaller time units in the discrete time model. The last row contains the values of the Gershwin and Berman continuous time model and demonstrates again the expected limiting behavior. Note the different time scales in the three cases of the discrete time model. For small time units the discrete time model behaves very much like the continuous time model, whereas for big time units the production rate is relatively higher.

An explanation for this behavior is the lower coefficient of variation of all random variables in the discrete time model compared to the continuous time model: Processing times, times to failure and times to repair are assumed to be geometrically distributed in the discrete time model. A geometrically dis-

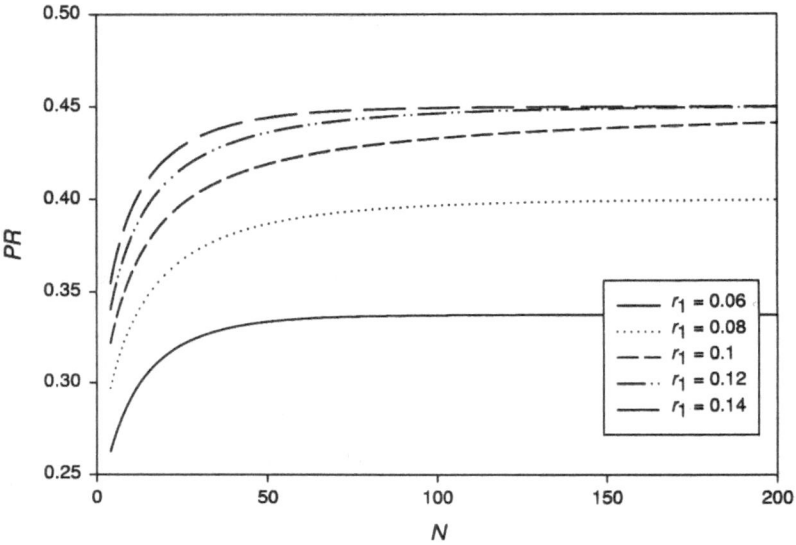

Fig. 3.9. Effect of Repair Probability and Buffer Size on Production Rate

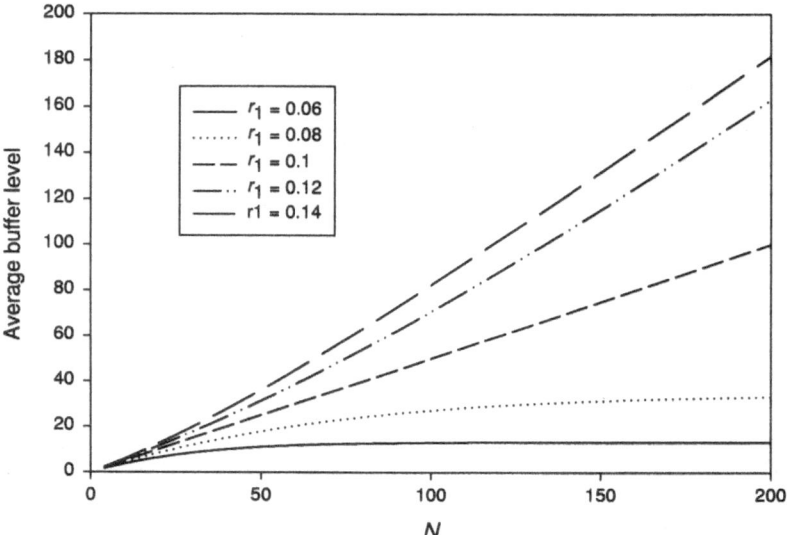

Fig. 3.10. Effect of Repair Probability and Buffer Size on Average Buffer Level

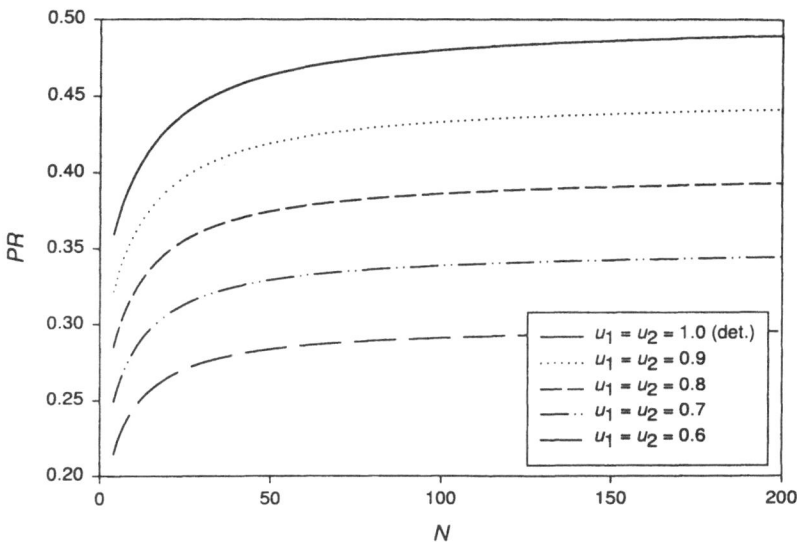

Fig. 3.11. Effect of Production Probability and Buffer Size on Production Rate

Discrete Time Model	PR	$\bar{n}$
$u_1 = 0.7,\ p_1 = 0.01,\ r_1 = 0.1$ $u_2 = 0.8,\ p_2 = 0.02,\ r_2 = 0.1$	0.584929	4.021
$u_1 = 0.07,\ p_1 = 0.001,\ r_1 = 0.01$ $u_2 = 0.08,\ p_2 = 0.002,\ r_2 = 0.01$	0.055707	4.481
$u_1 = 0.007,\ p_1 = 0.0001,\ r_1 = 0.001$ $u_2 = 0.008,\ p_2 = 0.0002,\ r_2 = 0.001$	0.005545	4.502
Continuous Time Model	PR	$\bar{n}$
$u_1 = 0.007,\ p_1 = 0.0001,\ r_1 = 0.001$ $u_2 = 0.008,\ p_2 = 0.0002,\ r_2 = 0.001$	0.005543	4.504

Table 3.2. Discrete vs. Continuous Time Model

tributed random variable x with a minimum value of one has a coefficient of variation of $\sqrt{1 - 1/E\{x\}}$ where $E\{x\}$ denotes the expected value of x. If time units are very small, the probabilities $u_i, p_i, r_i, i = 1, 2$ are also small. As a result, $E\{x\}$ is large and the coefficient of variation of the geometrically distributed random variables is only slightly smaller than one. Here, the continuous time model may well approximate the discrete time system. However, if time units are big the probabilities $u_i, p_i, r_i, i = 1, 2$ are not small. In this case, $E\{x\}$ is small and the coefficient of variation will be somewhere between zero and one but not as close to one as in the case of small time units. The lower variability compared to the coefficients of variation that equal one in the continuous time model leads to an increase in the production rate. That is, we may analyze a discrete time system by a continuous time model when all production, failure and repair probabilities are small. If they are not small, however, one should rather work with a discrete time model like the one developed in this section. This applies as well when we use this model as a building block in an approximate decomposition approach directed at the evaluation of a more complex transfer line or A/D system.

3.3 Decomposition Equations for Assembly/Disassembly Systems

The goal of the derivation below is to determine a set of equations for the parameters of the virtual machines arising in the decomposition of the A/D system.[6]

3.3.1 Conservation of Flow Equation

Define $PR_{i,q}$ as the production rate through Buffer $B_{i,q}$ in the real system. Since the A/D network does not contain cycles and there is no mechanism to create or destroy material, flow is conserved, and thus the production rates through all buffers are equal, i.e.

$$PR_{i,q} = PR_{j,k} \tag{3.67}$$

for all pairs $B_{i,q}$ and $B_{j,k}$ of buffers. The decomposition into a set of two-machine lines as described in Section 2.5 leads to estimated production rates $PR(i, q)$ for each virtual two-machine line $L(i, q)$. The decomposition must be performed in a way that the production rates of the two-machine lines are identical, i.e.

$$PR(i, q) = PR(j, k) \tag{3.68}$$

---

[6] See Section 2.5 on Page 25.

for all pairs $L(i, q)$ and $L(j, k)$ of two-machine lines in the decomposition.

This conservation of flow (COF) equation is used as a stopping criterion in the iterative algorithm that is used to solve the decomposition equations. Note that the COF equation for an A/D system implies that the production rates of all machines are identical.

Conservation of flow equations for the continuous time model. The conservation of flow equations for the discrete time model and the continuous time model have the same structure. However, in the discrete time model, PR_i is the fraction of periods during which Machine M_i processes a part, i.e. a *percentage* that is always lower than 100% due to random failures and repairs. In the continuous time model, however, PR_i is the average *rate* at which M_i processes parts, i.e. a number that can be larger than one given appropriate processing rates μ_i.

3.3.2 Flow Rate-Idle Time Equations

The goal of this section is to derive an expression for the production probabilities of the virtual machines in the decomposition. The production rate of a machine is determined by its speed, its failure and repair processes, and starving and blocking due to other machines in the network. The *flow rate-idle time* (FRIT) equation relates the production rate to these factors. If Machine M_i in the discrete time model operates in isolation, i.e. without preceding or succeeding machines, the fraction of time e_i the machine is up (the *isolated efficiency*) depends on its mean time to failure $MTTF_i = 1/p_i$ and its mean time to repair $MTTR_i = 1/r_i$ according to

$$e_i = \frac{MTTF_i}{MTTF_i + MTTR_i} = \frac{r_i}{r_i + p_i}. \tag{3.69}$$

If the machine operates, it completes an operation during a production cycle or period with probability u_i, i.e. the isolated production rate IPR_i of this machine is

$$IPR_i = u_i e_i = u_i \frac{r_i}{r_i + p_i}. \tag{3.70}$$

Machine M_i can only operate if none of its input buffers is empty and none of its output buffers is full. Define $n(l, i)$ as the buffer level related to Buffer $B_{l,i}$. The production rate PR_i related to Machine M_i in the real system is the product of the isolated production rate IRR_i times the probability that none of the input buffer is empty and none of the output buffers is full, i.e

$$PR_i = IPR_i \, \text{prob} \left[\begin{array}{ll} \{n(l,i) > 0, & \forall (l,i) \in U(i)\} \text{ and} \\ \{n(i,q) < N(i,q), & \forall (i,q) \in D(i)\} \end{array} \right] \quad (3.71)$$

where the set of all buffers immediately upstream of Machine M_i is denoted as $U(i)$ and the set of all buffers immediately downstream as $D(i)$.

A similar equation must hold for the production rate $PR(i)$ related to machine M_i in the decomposition. As an approximation, we assume that

- buffer levels are independent and that
- the probability of simultaneous starvation and blocking is negligible.

Using these assumptions, the FRIT equation can be approximated as

$$PR(i) \approx u_i e_i \left[1 - \sum_{(l,i) \in U(i)} p_s(l,i) - \sum_{(i,q) \in D(i)} p_b(i,q) \right] \quad (3.72)$$

for all M_i where $p_s(l,i)$ denotes the probability of starvation of Machine $M_d(l,i)$ due to an empty Buffer $B_{l,i}$ and $p_b(i,q)$ the probability of blocking due to a full Buffer $B(i,q)$.

From the two-machine models related to Lines $L(l,i)$ and $L(i,q)$ we find:

$$PR(l,i) = u_d(l,i)e_d(l,i)(1 - p_s(l,i)) \quad (3.73)$$
$$PR(i,q) = u_u(i,q)e_u(i,q)(1 - p_b(i,q)) \quad (3.74)$$

Equation (3.73) says that the production rate $PR(l,i)$ of Line $L(l,i)$ is the product of the isolated production rate $u_d(l,i)e_d(l,i)$ of its downstream machine $M_d(l,i)$ times the probability $(1 - p_s(l,i)$ that this machine is not starved. We can therefore express the FRIT equation as

$$PR(i) \approx u_i e_i \left[1 + \sum_{(l,i) \in U(i)} \left(\frac{PR(l,i)}{u_d(l,i)e_d(l,i)} - 1 \right) \right.$$
$$\left. + \sum_{(i,q) \in D(i)} \left(\frac{PR(i,q)}{u_u(i,q)e_u(i,q)} - 1 \right) \right] \quad (3.75)$$

Since the production rates $PR(i)$ of all Machines in the decomposition must be equal, we can cancel and rearrange for Machine M_i

$$\frac{1}{u_i e_i} = \frac{1}{PR(i)} + \sum_{(l,i) \in U(i)} \left(\frac{1}{u_d(l,i)e_d(l,i)} - \frac{1}{PR(l,i)} \right)$$
$$+ \sum_{(i,q) \in D(i)} \left(\frac{1}{u_u(i,q)e_u(i,q)} - \frac{1}{PR(i,q)} \right) \quad (3.76)$$

Since the production rate in an A/D system is identical for all machines and buffers,[7] i.e. $PR(i) = PR(j, i) = PR(i, q)$, we can solve (3.76) to derive expressions for production probabilities $u_u(i, m)$ and $u_d(j, i)$ of the virtual machines $M_u(i, m)$ and $M_d(l, i)$

$$u_u(i, m) = \frac{1}{e_u(i, m)} \frac{1}{K_1} = \frac{p_u(i, m) + r_u(i, m)}{r_u(i, m)} \frac{1}{K_1} \qquad (3.77)$$

$$u_d(j, i) = \frac{1}{e_d(j, i)} \frac{1}{K_2} = \frac{p_d(j, i) + r_d(j, i)}{r_d(j, i)} \frac{1}{K_2} \qquad (3.78)$$

with

$$K_1 = \frac{1}{u_i e_i} - \sum_{(l,i) \in U(i)} \left(\frac{1}{u_d(l, i) e_d(l, i)} - \frac{1}{PR(l, i)} \right)$$
$$- \sum_{\substack{(i,q) \in D(i) \\ q \neq m}} \left(\frac{1}{u_u(i, q) e_u(i, q)} - \frac{1}{PR(i, q)} \right) \qquad (3.79)$$

$$K_2 = \frac{1}{u_i e_i} - \sum_{\substack{(l,i) \in U(i) \\ l \neq j}} \left(\frac{1}{u_d(l, i) e_d(l, i)} - \frac{1}{PR(l, i)} \right)$$
$$- \sum_{(i,q) \in D(i)} \left(\frac{1}{u_u(i, q) e_u(i, q)} - \frac{1}{PR(i, q)} \right). \qquad (3.80)$$

This is a useful result as it relates, for example for Line $L(i, m)$ in (3.77), the production probability $u_u(i, m)$ of Machine $M_u(i, m)$ to its failure and repair probabilities $p_u(i, m)$ and $r_u(i, m)$, to the given isolated production rate $u_i e_i$ of Machine M_i in the real system, and to parameters of other two-machine lines in the decomposition. Whenever $u_u(i, m)$ has to be computed, all the other parameters are given or can be easily determined.

Note that the virtual upstream machine related to the two-machine line $L(i, m)$ is expressed in terms of its failure and repair probabilities $p_u(i, m)$ and $r_u(i, m)$, respectively. The coefficient K_1 in Equation (3.79), however, does not contain any parameters of the virtual line $L(i, m)$. The situation for Line $L(j, i)$ in Equations (3.78) and (3.80) is identical. Separating these quantities in K_1 from $u_u(i, m)$, $p_u(i, m)$, and $r_u(i, m)$ in (3.77) will later allow us to solve the complete set of decomposition equations simultaneously in a way proposed in Burman (1995) that leads to an improved convergence behavior of the algorithm.

---

[7] This substitution was introduced for flow lines in Dallery et al. (1988) to improve the propagation of production rate estimates in the iterative algorithm.

Flow rate-idle time equations for the continuous time model. The flow rate-idle time equations for the continuous time model with exponentially distributed processing times, times to failure and to repair have exactly the same structure as those for the discrete time model given above. The production probabilities u_i have to be replaced by processing rates μ_i, and the p_i and r_i have to be interpreted as failure and repair rates in continuous time. The equation is a generalization of the corresponding equation in Choong and Gershwin (1987).

3.3.3 Resumption of Flow Equations

A second set of equations is required to determine the repair probabilities $r_u(i, m)$ and $r_d(i, m)$ for each line $L(i, m)$ in the decomposition. Since the flow resumes after each of these repairs, the equations are frequently called *resumption of flow* equations (Gershwin (1994)). They are needed in a form similar to the flow rate-idle time equations, i.e. one that requires quantities that are either given system parameters or that can be computed from other two-machine models in the decomposition.

The virtual upstream machine $M_u(i, m)$ in Line $L(i, m)$ is considered to be *up* if and only if the following conditions hold simultaneously:

1. Machine M_i is up at time t. This event is denoted as $\{\alpha_i(t) = 1\}$.
2. The (input) buffer related to each line $L(l, i)$ is either non-empty ($\{n[(l, i), t - 1] > 0\}$), or the virtual upstream machine $M_u(l, i)$ is up. (A long processing time of the virtual upstream machine $M_u(l, i)$ is not treated as a failure of $M_u(i, m)$.) Otherwise, Machine M_i is starved.
3. The (output) buffer related to each line $L(i, q)$ other than $L(i, m)$ is either non-full ($\{n[(i, q), t - 1] < N(i, q)\}$), or the virtual downstream machine $M_d(i, q)$ is up. Otherwise, Machine M_i is blocked.

Define $\{\alpha_u[(i, m), t] = 1\}$ as the event of the virtual upstream machine $M_u(i, m)$ being up at time t:

$$\{\alpha_u[(i, m), t] = 1\} \quad \text{iff} \quad \{\alpha_i(t) = 1\} \text{ and}$$

$$\left\{ \begin{aligned} &\{n[(l, i), t - 1] > 0\} \text{ or} \\ &\{n[(l, i), t - 1] = 0\} \text{ and } \{\alpha_u[(l, i), t] = 1\}, \\ &\forall (l, i) \in U(i) \end{aligned} \right\} \text{ and}$$

$$\left\{ \begin{aligned} &\{n[(i, q), t - 1] < N(i, q)\} \text{ or} \\ &\{n[(i, q), t - 1] = N(i, q)\} \text{ and } \{\alpha_d[(i, q), t] = 1\}, \\ &\forall (i, q) \in D(i), q \neq m \end{aligned} \right\} \quad (3.81)$$

The first condition says that Machine M_i is up. The second demands that Machine M_i is not starved due to an upstream failure, and the third demands that M_i is not blocked due to a downstream failure in Line $L(i, q)$ other than $L(i, m)$.

Machine $M_u(i, m)$ is down if it is not up, i.e.

$$\{\alpha_u[(i,m), t] = 0\} \quad \text{iff} \quad \{\alpha_i(t) = 0\} \text{ or}$$
$$\{n[(l,i), t-1] = 0\} \text{ and } \{\alpha_u[(l,i), t] = 0\},$$
$$\text{for some } (l,i) \in U(i) \Big\} \text{ or}$$
$$\{n[(i,q), t-1] = N(i,q)\} \text{ and } \{\alpha_d[(i,q), t] = 0\},$$
$$\text{for some } (i,q) \in D(i), q \neq m \Big\} \qquad (3.82)$$

Using Equations (3.81) and (3.82), we can express the repair probability $r_u(i, m)$ as the conditional probability that Machine $M_u(i, m)$ is up at time $t + 1$ given that it was down at time t, i.e.

$$r_u(i,m) = \text{prob}\big[\ \{\alpha_u[(i,m), t+1] = 1\}\big|\{\alpha_u[(i,m), t] = 0\} \text{ and} \qquad (3.83)$$
$$\{n[(i,m), t-1] < N(i,m)\}\big]$$

Note that a failure of Machine $M_u(i, m)$ at time t implies that $M_u(i, m)$ is not blocked ($\{n[(i,m), t-1] < N(i,m)\}$), because a blocked machine cannot fail. We apply the definition of virtual machine states in Equations (3.81) and (3.82) to (3.83) and decompose the conditioning event as in Gershwin (1991). Using the same approach for $M_d(j, i)$ and rearranging the resulting terms, we find

$$r_u(i,m) = \left[r_i + K_3 \frac{r_u(i,m)u_u(i,m)}{p_u(i,m)} \right] \qquad (3.84)$$

$$r_d(j,i) = \left[r_i + K_4 \frac{r_d(j,i)u_d(j,i)}{p_d(j,i)} \right] \qquad (3.85)$$

where we define

$$K_3 = \sum_{(l,i)\in U(i)} (r_u(j,i) - r_i)\mathbf{p}[(j,i); 001]$$
$$+ \sum_{\substack{(i,q)\in D(i) \\ q\neq m}} (r_d(i,q) - r_i)\mathbf{p}[(i,q); N10] \qquad (3.86)$$

$$(3.87)$$

$$K_4 = \sum_{\substack{(l,i)\in U(i) \\ l\neq j}} (r_u(j,i) - r_i)\mathbf{p}[(j,i);001]$$

$$+ \sum_{(i,q)\in D(i)} (r_d(i,q) - r_i)\mathbf{p}[(i,q);N10] \qquad (3.88)$$

It should be noted that contrary to the two-machine, deterministic processing time model by Gershwin and Schick,[8] $\mathbf{p}[(j,i);001]$ is not the probability of starvation of Line $L(j,i)$ as $\mathbf{p}[(j,i);011]$ is not transient for $u_i < 1$. However, if all production probabilities u_i approach one, $\mathbf{p}[(j,i);011]$ becomes transient and in this case Equations (3.84) and (3.85) are identical to the corresponding equations in the deterministic processing time model.

Resumption of flow equations for the continuous time model. The resumption of flow equations for the continuous time model with exponentially distributed processing times, times to failure and to repair have exactly the same structure as those for the discrete time model given above. The production probabilities u_i have to be replaced by processing rates μ_i, and the parameters p_i and r_i have to be interpreted as failure and repair rates in continuous time instead of probabilities in discrete time.

3.3.4 Interruption of Flow Equations

A last set of equations needs to be derived for the probability of flow being interrupted due to machine failures. Define the auxiliary expression $U_u(i, m, t)$ for the event that Machine $M_u(i, m)$ is up at time t and Buffer $B(i, m)$ is not full at the end of the previous period $t - 1$:

$$U_u(i, m, t) := \{\alpha_u[(i, m), t] = 1\} \text{ and } \{n[(i, m), t - 1] < N(i, m)\} \qquad (3.89)$$

Using this auxiliary expression and the definition of virtual machine states in Equations (3.81) and (3.82), the failure probability

$$p_u(i, m) = \text{prob}\Big[\{\alpha_u[(i, m), t + 1] = 0\}\Big|U_u(i, m, t)\Big] \qquad (3.90)$$

can be expressed as the conditional probability that the virtual machine $M_u(i, m)$ is down at time $t + 1$ given that it was up at time t. Since the probability of more than one buffer being full or empty at the same time is small, we can use Equation (3.82) to approximate:

---

[8] Gershwin (1994, p. 76-93)

$$p_u(i,m) \tag{3.91}$$
$$= \operatorname{prob}\left[\{\alpha_i(t+1)=0\}\Big|U_u(i,m,t)\right]$$
$$+ \sum_{(l,i)\in U(i)} \operatorname{prob}\left[\{\{n[(l,i),t]=0\} \text{ and } \{\alpha_u[(l,i),t+1]=0\}\}\Big|\right.$$
$$\left. U_u(i,m,t)\right]$$
$$+ \sum_{\substack{(i,q)\in D(i)\\ q\neq m}} \operatorname{prob}\left[\{\{n[(i,q),t]=N(i,q)\} \text{ and } \{\alpha_d[(i,q),t+1]=0\}\}\Big|\right.$$
$$\left. U_u(i,m,t)\right]$$

The first term on the right hand side of Equation (3.91) is the probability p_i that Machine M_i in the real system fails:

$$\operatorname{prob}\left[\{\alpha_i(t+1)=0\}\Big|U_u(i,m,t)\right] = p_i \tag{3.92}$$

With respect to the second term on the right hand side of Equation (3.91), we conclude that due the dynamics of the two-machine model related to Line $L(l,i)$, the downstream machine $M_d(l,i)$ must be up if the upstream machine $M_u(l,i)$ is down and Buffer $B(l,i)$ is empty. The reason is that a machine cannot fail if it is starved. This leads to the following equation for the second term

$$\operatorname{prob}\left[\{\{n[(l,i),t]=0\} \text{ and } \{\alpha_u[(l,i),t+1]=0\}\}|U_u(i,m,t)\right]$$
$$= \operatorname{prob}\left[\{\{n[(l,i),t]=0\} \text{ and }\right.$$
$$\left.\{\alpha_u[(l,i),t+1]=0\} \text{ and } \{\alpha_d[(l,i),t+1]=1\}\}|U_u(i,m,t)\right]$$
$$= \mathbf{p}\left[((l,i);001) \text{ at } t+1\big|U_u(i,m,t)\right] \tag{3.93}$$

where $\mathbf{p}\left[((l,i);n\alpha_u\alpha_d) \text{ at } t+1\big|U_u(i,m,t)\right]$ denotes the conditional probability that Line $L(l,i)$ is in state (n,α_u,α_d) at time $t+1$ given that (event $U_u(i,m,t)$) Machine $M_u(i,m)$ is up and not blocked at time t.

Using the transition equations of the corresponding two-machine model, the state of Line $L(l,i)$ at time $t+1$ can be related to the possible states at time t:

$$\mathbf{p}\left[((l,i);001) \text{ at } t+1 \mid U_u(i,m,t)\right] = \tag{3.94}$$
$$\mathbf{p}\left[((l,i);001) \text{ at } t \mid U_u(i,m,t)\right] (1-r_u(l,i)) +$$
$$\mathbf{p}\left[((l,i);011) \text{ at } t \mid U_u(i,m,t)\right] p_u(l,i) +$$
$$\mathbf{p}\left[((l,i);100) \text{ at } t \mid U_u(i,m,t)\right] (1-r_u(l,i)) r_d(l,i) u_d(l,i) +$$
$$\mathbf{p}\left[((l,i);101) \text{ at } t \mid U_u(i,m,t)\right] (1-r_u(l,i)) (1-p_d(l,i)) u_d(l,i) +$$
$$\mathbf{p}\left[((l,i);110) \text{ at } t \mid U_u(i,m,t)\right] p_u(l,i) r_d(l,i) u_d(l,i) +$$
$$\mathbf{p}\left[((l,i);111) \text{ at } t \mid U_u(i,m,t)\right] p_u(l,i) (1-p_d(l,i)) u_d(l,i)$$

Some of the terms on the right hand side of Equation (3.94) are zero due to the conditioning event $U_u(i, m, t)$. Consider the first term: if Machine $M_u(i, m)$ is up at time t (which follows from event $U_u(i, m, t)$) and Buffer $B(l, i)$ is empty, then Machine $M_u(l, i)$ cannot be down at time t and the corresponding probability $\mathbf{p}[((l, i); 001)$ at $t|U_u(i, m, t)]$ is zero.[9]

The third and fifth term on the right hand side of (3.94) are zero since the downstream machine in Line $L(l, i)$, Machine $M_d(l, i)$, cannot be down at time t given that $M_u(i, m)$ is up and not blocked. For all the remaining terms that are not zero, Machine $M_d(l, i)$ is up which implies event $U_u(i, m, t)$. We may thus write:

$$\mathbf{p}[((l, i); 001) \text{ at } t + 1 \mid U_u(i, m, t)] = \tag{3.95}$$
$$\frac{\mathbf{p}[(l, i); 011)]}{\text{prob}[U_u(i, m, t)]} p_u(l, i) +$$
$$\frac{\mathbf{p}[(l, i); 101)]}{\text{prob}[U_u(i, m, t)]} (1 - r_u(l, i))(1 - p_d(l, i)) u_d(l, i) +$$
$$\frac{\mathbf{p}[(l, i); 111)]}{\text{prob}[U_u(i, m, t)]} p_u(l, i)(1 - p_d(l, i)) u_d(l, i)$$

Since $U_u(i, m, t)$ is the event that Machine $M_u(i, m)$ is up and not blocked, the probability of event $U_u(i, m, t)$ times the production probability $u_u(i, m)$ of Machine $M_u(i, m)$ is the production rate $PR(i, m)$:

$$PR(i, m) = \text{prob}[U_u(i, m, t)]u_u(i, m) \tag{3.96}$$

Using this equation, we can reformulate Equation (3.95)

$$\mathbf{p}[((l, i); 001) \text{ at } t + 1 \mid U_u(i, m, t)] = \tag{3.97}$$
$$\Big[\mathbf{p}[(l, i); 011)] \, p_u(l, i) +$$
$$\mathbf{p}[(l, i); 101)] \, (1 - r_u(l, i))(1 - p_d(l, i)) u_d(l, i) +$$
$$\mathbf{p}[(l, i); 111)] \, p_u(l, i)(1 - p_d(l, i)) u_d(l, i)\Big] \frac{u_u(i, m)}{PR(i, m)}$$

After a similar analysis for the third term on the right hand side of (3.91) and applying the conservation of flow equation, we finally see for $p_u(i, m)$, and, in exactly the same manner, for $p_d(j, i)$

$$p_u(i, m) \approx p_i + u_u(i, m) K_5 \tag{3.98}$$
$$p_d(j, i) \approx p_i + u_d(j, i) K_6 \tag{3.99}$$

---

[9] If Machine $M_u(l, i)$ were down while $B(l, i)$ is empty, $M_u(i, m)$ would also be down.

with

$$K_5 \tag{3.100}$$

$$
\begin{aligned}
= & \sum_{(l,i)\in U(i)} \Big[\mathbf{p}\,[(l,i);011)]\ p_u(l,i) + \\
& \mathbf{p}\,[(l,i);101)]\ (1 - r_u(l,i))\,(1 - p_d(l,i))\,u_d(l,i) + \\
& \mathbf{p}\,[(l,i);111)]\ p_u(l,i)\,(1 - p_d(l,i))\,u_d(l,i) \Big] \cdot \tfrac{1}{PR(l,i)} \\[1em]
+ & \sum_{\substack{(i,q)\in D(i) \\ q\neq m}} \Big[\mathbf{p}\,[(i,q);N11)]\ p_d(i,q) + \\
& \mathbf{p}\,[(i,q);N-1,10)]\ (1 - p_u(i,q))\,(1 - r_d(i,q))\,u_u(i,q) + \\
& \mathbf{p}\,[(i,q);N-1,11)]\ (1 - p_u(i,q))\,p_d(i,q)\,u_u(i,q) \Big] \cdot \tfrac{1}{PR(i,q)}
\end{aligned}
$$

$$K_6 \tag{3.101}$$

$$
\begin{aligned}
= & \sum_{\substack{(l,i)\in U(i) \\ l\neq j}} \Big[\mathbf{p}\,[(l,i);011)]\ p_u(l,i) + \\
& \mathbf{p}\,[(l,i);101)]\ (1 - r_u(l,i))\,(1 - p_d(l,i))\,u_d(l,i) + \\
& \mathbf{p}\,[(l,i);111)]\ p_u(l,i)\,(1 - p_d(l,i))\,u_d(l,i) \Big] \cdot \tfrac{1}{PR(l,i)} \\[1em]
+ & \sum_{(i,q)\in D(i)} \Big[\mathbf{p}\,[(i,q);N11)]\ p_d(i,q) + \\
& \mathbf{p}\,[(i,q);N-1,10)]\ (1 - p_u(i,q))\,(1 - r_d(i,q))\,u_u(i,q) + \\
& \mathbf{p}\,[(i,q);N-1,11)]\ (1 - p_u(i,q))\,p_d(i,q)\,u_u(i,q) \Big] \cdot \tfrac{1}{PR(i,q)}
\end{aligned}
$$

Note that these equations differ from those for Gershwin's model of an A/D network with deterministic processing times.[10] This is because we derived the interruption of flow equations directly via Equation (3.90), whereas Gershwin derived them indirectly as a reformulation of his approximation of the resumption of flow equation. One can also directly derive a slightly different interruption of flow equation in the deterministic processing time model using the equivalent of (3.90). The result is then a special case of (3.98)-(3.101) where all production probabilities equal one and the then transient states ((l, i);011) and ((i, q);N(i,q),1,1) are omitted. Since we have to determine three parameters for each machine in the two-machine transfer lines, we will need the interruption of flow relationships to compute the third parameter, whereas they are not needed in Gershwin's model.

Interruption of flow equations for the continuous time model. To derive an interruption of flow equation for the continuous time model, we first define virtual machine states for Machine $M_u(i, m)$ in a way similar to the one in Equation (3.81):[11]

---

[10] Gershwin (1994, p. 193-194)

[11] These two equations are **not** identical. In Equation (3.81) of the discrete time model, we refer to buffer levels at the end of the discrete period $t - 1$, whereas here we refer to buffer levels at time t in continuous time.

$\{\alpha_u[(i,m),t] = 1\}$ iff $\{\alpha_i(t) = 1\}$ and

$$\left\{ \begin{array}{l} \{n[(l,i),t] > 0\} \text{ or} \\ \quad \{n[(l,i),t] = 0\} \text{ and } \{\alpha_u[(l,i),t] = 1\}, \\ \quad \forall (l,i) \in U(i) \end{array} \right\} \text{ and}$$

$$\left\{ \begin{array}{l} \{n[(i,q),t] < N(i,q)\} \text{ or} \\ \quad \{n[(i,q),t] = N(i,q)\} \text{ and } \{\alpha_d[(i,q),t] = 1\}, \\ \quad \forall (i,q) \in D(i), q \neq m \end{array} \right\} \qquad (3.102)$$

Machine $M_u(i,m)$ is down if it is not up and the definition of event $\{\alpha_u[(i,m),t] = 0\}$ is, therefore, the negation of Equation (3.102). The definition of virtual machine states for the downstream machine is analogous. Note that (3.102) differs from (3.81) since in the continuous time model, machine states and buffer levels can change at any moment in time.

The previous decomposition equations (flow rate-idle time and resumption of flow), as well as the conservation of flow equation are identical for the discrete and the continuous time model. However, the interruption of flow equations

$$p_u(i,m) \approx p_i + u_u(i,m) K_5^{cont} \qquad (3.103)$$
$$p_d(j,i) \approx p_i + u_d(j,i) K_6^{cont} \qquad (3.104)$$

with

$$K_5^{cont} = \sum_{(l,i) \in U(i)} \frac{\mathbf{p}[(l,i); 001)] \; r_u(l,i)}{PR(l,i)}$$

$$+ \sum_{\substack{(i,q) \in D(i) \\ q \neq m}} \frac{\mathbf{p}[(i,q); N10)] \; r_d(i,q)}{PR(i,q)} \qquad (3.105)$$

$$K_6^{cont} = \sum_{\substack{(l,i) \in U(i) \\ l \neq j}} \frac{\mathbf{p}[(l,i); 001)] \; r_u(l,i)}{PR(l,i)}$$

$$+ \sum_{(i,q) \in D(i)} \frac{\mathbf{p}[(i,q); N10)] \; r_d(i,q)}{PR(i,q)} \qquad (3.106)$$

are different in the continuous time case. They are a generalization of the
equations given in Choong and Gershwin (1987) for the n-machine transfer
line with exponentially distributed processing times, times to failure and to
repair. Note that this is the only decomposition equation that differs for the
discrete and the continuous time model. The structure of Equations (3.98)
and (3.103), as well as (3.99) and (3.104) is identical, but the expressions K_5
and K_5^{cont} as well as K_6 and K_6^{cont} differ.

It is also possible to derive this equation indirectly from the interruption
of flow equations for the discrete time model if the time steps approach zero.
In this case K_5 in (3.100) approaches $\tilde{K}_5$ with

$$
\begin{aligned}
\tilde{K}_5 \;=\; & \sum_{(l,i)\in U(i)} \frac{\mathbf{p}\,[(l,i);011]\; p_u(l,i) + \mathbf{p}\,[(l,i);101]\; u_d(l,i)}{PR(l,i)} \\
& + \sum_{\substack{(i,q)\in D(i) \\ q\neq m}} \frac{\mathbf{p}\,[(i,q);N11]\; p_d(i,q) + \mathbf{p}\,[(i,q);N10]\; u_u(i,q)}{PR(i,q)}
\end{aligned} \tag{3.107}
$$

and we have $\tilde{K}_5 = K_5^{cont}$ due to the transition equations of the two-machine
models in Gershwin and Berman (1981) and in Section 3.2. The result for
the downstream machine is analogous.

3.3.5 Simultaneous Solution of the Decomposition Equations

We can use the decomposition equations (3.77), (3.78), (3.84), (3.85), (3.98),
and (3.99) to update the parameters of the two-machine pseudo transfer lines
in a modified version of the iterative DDX algorithm proposed by Dallery,
David and Xie,[12] to be described in detail in Section 3.4. In this algorithm,
to analyze Machine $M_u(i,m)$, we first determine a new value $p_u(i,m)$ using
(3.98):

$$
p_u(i,m) \;=\; p_i + K_5 u_u(i,m)
$$

However, the new value $p_u(i,m)$ depends on the old value of $u_u(i,m)$ from
the previous iteration. When we next update $r_u(i,m)$ using (3.84)

$$
r_u(i,m) \;=\; r_i + K_3 \frac{u_u(i,m)r_u(i,m)}{p_u(i,m)}
$$

and finally $u_u(i,m)$ using (3.77)

---

[12] Dallery et al. (1988)

$$u_u(i,m) \;=\; \frac{1}{K_1}\frac{r_u(i,m)+p_u(i,m)}{r_u(i,m)},$$

we again use the parameters of Line $L(i,m)$ from the previous and/or current iteration. Unfortunately, the DDX algorithm with this updating procedure does often not converge for large A/D networks or transfer lines consisting of inhomogeneous machines. However, in a recent Ph.D. thesis directed at a continuous material model of a transfer line, Burman has realized that the decomposition equations in his problem constitute two systems of equations that can be solved simultaneously.[13] This is possible in our case as well. Equations (3.77), (3.84), and (3.98) are a non-linear system of equations with the following solution:

$$u_u(i,m) \;=\; \frac{p_i+r_i}{K_1 r_i + K_3 - K_5} \tag{3.108}$$

$$r_u(i,m) \;=\; \frac{K_1 p_i r_i + K_3 p_i + K_5 r_i}{K_1 p_i - K_3 + K_5} \tag{3.109}$$

$$p_u(i,m) \;=\; \frac{K_1 p_i r_i + K_3 p_i + K_5 r_i}{K_1 r_i + K_3 - K_5} \tag{3.110}$$

We can use (3.108), (3.109), and (3.110) to update $u_u(i,m)$, $r_u(i,m)$, and $p_u(i,m)$ without falling back on any parameter of machine $M_u(i,m)$ from the current or previous iteration since these parameters are not contained in K_1, K_3, or K_5. Equations (3.78), (3.85), and (3.99) can be solved in a similar way:

$$u_d(j,i) \;=\; \frac{p_i+r_i}{K_2 r_i + K_4 - K_6} \tag{3.111}$$

$$r_d(j,i) \;=\; \frac{K_2 p_i r_i + K_4 p_i + K_6 r_i}{K_2 p_i - K_4 + K_6} \tag{3.112}$$

$$p_d(j,i) \;=\; \frac{K_2 p_i r_i + K_4 p_i + K_6 r_i}{K_2 r_i + K_4 - K_6} \tag{3.113}$$

The numerical results presented in Section 3.5 indicate that updating parameters using (3.108)-(3.113) leads to both a dramatic improvement of convergence reliability and a strong acceleration of the algorithm. Following Burman (1995), we call this procedure the accelerated DDX or **ADDX** algorithm.

3.3.6 Comparison with Jeong and Kim's Decomposition Equations for the Continuous Time Case

Jeong and Kim (1998) presented a decomposition approach for *exactly* the same model of an A/D system with exponentially distributed processing

---

[13] Burman (1995, p. 84-87)

times, times to failure and to repair shortly before the submission of this thesis. (They did not treat the discrete time case.)

Their elegant derivation of decomposition equations (flow rate-idle time, interruption of flow, resumption of flow) differs from ours in several ways:

1. Jeong and Kim (1998, p. 42) assume that a machine is starved if any one of its input buffers is empty when the machine discharges a processed unit. This differs from the definition of starvation in Choong and Gershwin (1987) and in this thesis where a machine is assumed to be starved only if the input buffer is empty *and* the virtual upstream machine is down. These different definitions of virtual machine states lead to slightly different resumption-of-flow and interruption-of-flow equations.

2. In the decomposition presented in this thesis, the processing rates $\mu_u(i, m)$ and $\mu_d(i, m)$ in the two-machine line $L(i, m)$ can have values that differ from μ_i and μ_m in the real system. Jeong and Kim (1998), however, set $\mu_u(i, m) = \mu_i$ and $\mu_d(i, m) = \mu_m$ to determine processing rates for the two-machine subsystems. This should work well if all processing rates are almost identical, but it is not clear that convergence reliability and accuracy of the algorithm are as high as in our approach if the system is very inhomogeneous. In the numerical results that they present, processing rates of all machines differ no more than $\pm 10\%$. This may be sufficiently inhomogeneous for many practical purposes.

3. Jeong and Kim do not actually modify the virtual processing rates, but derive *three* decomposition equations (flow rate-idle time (FRIT), interruption of flow (IOF), and resumption of flow (ROF)) to determine the remaining *two* parameters for failure and repair rates of the virtual machines, i.e. they have one equation more than needed. In one algorithm called IF they use their IOF- and FRIT-equations and in a second one called RF they use their ROF- and FRIT-equations to determine the remaining two parameters.

4. Jeong and Kim do not use a simultaneous solution of the decomposition equations as in Section 3.3.5, but determine parameter updates sequentially. Furthermore, they appear to propagate updated production rate estimates in way that differs from the approach proposed by Dallery et al. (1988) and Gershwin (1994, p. 195-196) to solve the decomposition equations.

5. While in our derivation we use the *extended buffer level*, which includes the work space at the machines in the calculation of the buffer size N, Jeong and Kim count exactly those parts that are actually stored in the buffer between the machines.

Given these differences, one should not be surprised to see at least slightly different estimates from these two different decomposition approaches. We come back to this topic in the section on numerical experiments for A/D systems.

3.4 Two Algorithms to Determine Performance Measures

Two algorithms are available to approximate the production rate and in-process inventory of the A/D systems modeled above. The first algorithm is called DDX algorithm since it is based on the work of Dallery, David, and Xie.[14] It evaluates the decomposition equations in the sequence Gershwin proposed for a transfer line consisting of machines with exponentially distributed processing times, times to failure and times to repair.[15] The second procedure named ADDX algorithm[16] uses the reformulation of the decomposition equations developed in Section 3.3.5. Both algorithms consist of three parts.[17]

In the first part, an evaluation sequence of the pseudo machines in the two-machine transfer lines is determined. This sequence is necessary to make sure that while updating any parameter, we always use the most current estimates of other machine parameters that are available so far. Second, the parameters of the pseudo machines are initialized. In the third step, they are modified in an iterative manner until the production rates of all two-machine pseudo transfer lines are sufficiently close.

The evaluation sequence. The following procedure is used to determine two evaluation sequences denoted as S_1 and S_2. We call an input (output) machine with exactly one succeeding (preceding) machine a *pure* input (output) machine, respectively.

1. Set $S_1 = \{\}$ and $S_2 = \{\}$ (both sequences are empty).
2. Determine all buffers $B_{i,q}$ connected to pure input machines in the A/D system and append the corresponding machines $M_u(i,q)$ to sequence S_1. Remove these pure input machines M_i and the connected buffers $B_{i,q}$ from the (remaining) A/D system.
3. Determine all buffers $B_{j,i}$ connected to pure output machines in the A/D system and append the corresponding machines $M_d(j,i)$ to sequence S_1. Remove these pure output machines M_i and the connected buffers $B_{j,i}$ from the (remaining) A/D system.
4. If there are buffers left in the A/D system, go to Step 2, otherwise go to Step 5.
5. Set $S_1 = S_2$. Reverse the sequence of machines in S_2. Replace upstream machines in S_2 by downstream machines and *vice versa*.

For the system in Figure 3.1, we find the following sequences:

---

[14] Dallery et al. (1988)
[15] Gershwin (1989)
[16] Burman (1995)
[17] Gershwin (1991)

$$\begin{aligned}
S_1 \quad = \quad & \{M_u(1,4), M_u(3,6), M_d(7,11), M_d(8,12), M_u(6,10), M_d(4,7), \\
& M_d(4,8), M_u(10,13), M_d(2,4), M_d(9,13), M_u(2,5), M_d(5,9)\} \\
S_2 \quad = \quad & \{M_u(5,9), M_d(2,5), M_u(9,13), M_u(2,4), M_d(10,13), M_u(4,8), \\
& M_u(4,7), M_d(6,10), M_u(8,12), M_u(7,11), M_d(3,6), M_d(1,4)\}
\end{aligned}$$

This procedure appears to be easier to describe and program than those proposed in Gershwin (1991) and Di Mascolo et al. (1991).

Initialization. The virtual machines in the two-machine lines $L(i,q)$ corresponding to each buffer $B_{i,q}$ are initialized with the parameters of the real machines M_i and M_q:

$$\begin{aligned}
u_u(i,q) = u_i; \quad p_u(i,q) = p_i; \quad r_u(i,q) = r_i; \\
u_d(i,q) = u_q; \quad p_d(i,q) = p_q; \quad r_d(i,q) = r_q;
\end{aligned}$$

Consider the situation that Machine M_i is a pure input machine, i.e. it has just one immediate successor as it does not perform disassembly operations. In this case, the virtual upstream machine $M_u(i,q)$ as seen by an observer in Buffer $B(i,q)$ has to behave exactly like the real machine M_i in the original A/D system. The same holds for $M_d(i,q)$ if M_q is a pure output machine. In both cases, the initial parameters of $M_u(i,q)$ and $M_d(i,q)$ do not have to be modified. However, for all other virtual machines, the virtual machine parameters have to be modified using the decomposition equations in order to reflect random events up- and downstream of Machines M_i and M_q.

If the physical buffer between Machines M_i and M_q can hold $C_{i,q}$ parts, set the extended buffer size of the virtual line $L(i,q)$

$$N(i,q) = C_{i,q} + 2 \tag{3.114}$$

to include the workspace at the two machines M_i and M_q.[18]

Iteration. Consider sequentially all machines in Sequences S_1 and S_2. If the current machine is an upstream machine $M_u(i,m)$, update $u_u(i,m)$, $r_u(i,m)$, and $p_u(i,m)$ using (3.77), (3.84), and (3.98) in case of the DDX algorithm and using (3.108), (3.109), and (3.110) in case of the ADDX algorithm. If it is a downstream machine $M_d(j,i)$, update $u_d(j,i)$, $r_d(j,i)$, and $p_d(j,i)$ using (3.78), (3.85), and (3.99) in case of the DDX algorithm, and using (3.111), (3.112), and (3.113) in case of the ADDX algorithm. Solve the two-machine model in Section 3.2 (the model in Gershwin and Berman (1981) in the continuous time case) to update steady state probabilities and performance measures for the current line $L(i,m)$ or $L(j,i)$.

---

[18] See Equation (2.13) on Page 25 of Section 2.5.

Termination. Stop the procedure if the production rates $PR(i, q)$ of the different two-machine lines in the decomposition meet the conservation of flow equation (3.68) to a sufficient degree of accuracy.

3.5 Numerical Results

The first part of the numerical results concentrates on the behavior of the algorithm. In this part, we are primarily interested in the convergence reliability and speed of the iterative algorithm and in the accuracy of the production rate and inventory level estimates. A second set of numerical results is directed at the manufacturing system behavior of A/D systems. Here, we ask how the behavior of an A/D system differs from those of a transfer line and how assembly and disassembly operations relate to each other.

3.5.1 Behavior of the Algorithm

We analyzed 11 systems based on two different A/D networks to investigate the accuracy and the computational effort of the two algorithms. The first network introduced in Gershwin (1991) is depicted in Figure 3.12.

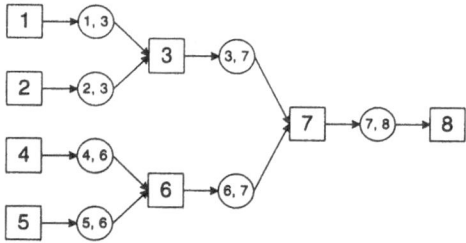

Fig. 3.12. Eight-Machine Assembly Network

Cases 1 to 5 described in Table 3.3 are based on this eight-machine network. In Gershwin's deterministic processing time version of this system, an operational machine that is neither blocked nor starved always completes a part at the end of a period. In Case 1, the processing probabilities u_i in our geometric processing time model are very close to one, our decomposition should hence give results close to those for the deterministic processing time case. Table 3.4 gives the approximated production rate and buffer levels from our geometric processing time decomposition and from Gershwin's deterministic processing time decomposition.[19]

It is assuring to note that the relative differences are smaller than one percent. In a Visual Basic program running on a 60 Mhz Pentium PC, the

---

[19] Gershwin (1991)

DDX algorithm required 14 iterations to reach a maximum relative difference of 0.00001 % between the production rates of the seven two-machine pseudo transfer lines, whereas the ADDX version needed 12 iterations (or 3 seconds).

In Table 3.5, approximated and simulated production rates are shown for Cases 2 to 5. These cases differ with respect to buffer sizes and selected machine failure probabilities. The last column in this table gives the average of the absolute values of the relative deviation between the seven simulated and approximated buffer levels for each case. As in other decomposition approaches,[20] the approximated extended storage levels $\bar{n}_{i,q}$ are less accurate in relative terms than the approximated production rates. The production rate estimates appear accurate. All simulation results reported in this section are based on 20 independent runs over 110,000 time units where the first 10,000 time units of each replication were truncated as a warm-up period. We used Siman V to perform the simulations and the Arena Output Processor to compute mean values and confidence intervals.

In the next six cases (Cases 6 to 11 in Table 3.3), we analyzed the 40-machine A/D system depicted in Figure 3.13. The numbers of the machines are above the squares. The buffers are omitted in the graphical representation. We compared the DDX algorithm and the ADDX algorithm with respect to accuracy, reliability, and speed of convergence. To study systems with different degrees of inhomogeneity with respect to machines, we introduced a

---

[20] Examples are Di Mascolo et al. (1991), Gershwin (1989).

Case	Processing probability u_i	Failure probability p_i	Repair probability r_i	Extended buffer size $N(i, q)$
1	.99	.01	.1	10
2	.8	.01	.1	5
3	.8	$p_2 = 0.1, p_6 = 0.2,$ $p_i = 0.01$, all other i	.1 .1	5
4	.8	.01	.1	20
5	.8	$p_2 = 0.1, p_6 = 0.2,$ $p_i = 0.01$, all other i	.1 .1	20
6-11	$u_7 = 0.9\gamma,$ $u_{21} = 0.9\gamma,$ $u_{32} = 0.9\gamma,$ $u_i = 0.9,$ all other i	$p_7 = 0.01, \gamma,$ $p_{21} = 0.01\gamma,$ $p_{32} = 0.01\gamma,$ $p_i = 0.01,$ all other i	$r_7 = 0.1\gamma,$ $r_{21} = 0.1\gamma,$ $r_{32} = 0.1\gamma,$ $r_i = 0.1,$ all other i	20

Table 3.3. System Parameters for Cases 1 to 11

Quantity	Geom.	Det.	Quantity	Geom.	Det.
PR	0.29179	0.29354	$\overline{n}_{5,6}$	7.3417	7.3551
$\overline{n}_{1,3}$	7.3417	7.3551	$\overline{n}_{3,7}$	5.6452	5.6516
$\overline{n}_{2,3}$	7.3417	7.3551	$\overline{n}_{6,7}$	5.6452	5.6516
$\overline{n}_{4,6}$	7.3417	7.3551	$\overline{n}_{7,8}$	2.6578	2.6449

Table 3.4. Results for Case 1

Case	Production rate				Buffer level
	Appr.	Sim. Mean	95% Interval	Dev. [%]	Dev. [%]
2	0.4924	0.4921	(0.4902, 0.4941)	0.06	1.81
3	0.2419	0.24	(0.239, 0.2409)	0.79	4.4
4	0.6357	0.6294	(0.6285, 0.6303)	1.00	1.4
5	0.2663	0.2662	(0.2649, 0.2675)	0.04	2.3

Table 3.5. Numerical Results in Cases 2 to 5

new parameter γ, $0 < \gamma < 1$, and set $u_i = 0.9\gamma$, $p_i = 0.01\gamma$, and $r_i = 0.1\gamma$ for the three machines M_7, M_{21}, and M_{32}. The results are presented in Table 3.6.

In Case 6 we set $\gamma = 1$, i.e. it is a homogeneous system with $u_i = 0.9$, $p_i = 0.01$, and $r_i = 0.1$, for all machines $M_i, i = 1, ..., 40$. The isolated production rate of any machine M_i is therefore $IPR_i = 0.9 \cdot 0.1/(0.01 + 0.1) \approx 0.818$. It is an upper bound of the overall production rate in this case. Note that the isolated efficiency $e_i = r_i/(r_i + p_i)$ of Machines M_7, M_{21}, and M_{32} does not depend on γ. However, since decreasing γ leads to a decrease of u_7, u_{21}, and u_{32}, we now have three bottlenecks in the system, which limit the overall

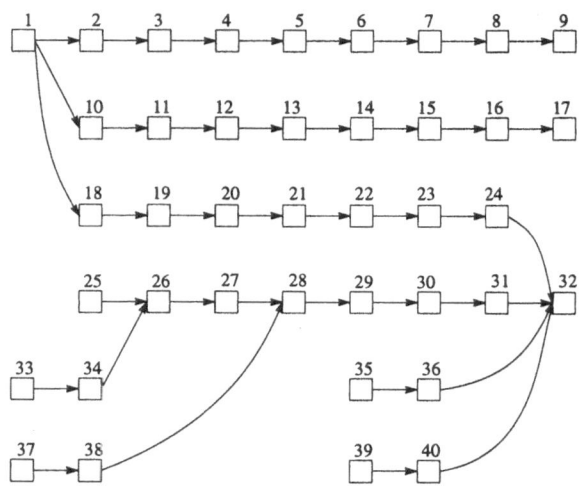

Fig. 3.13. 40-Machine Assembly/Disassembly Network

production rate. In Cases 7 to 11, we evaluated the production rate for values of γ of 0.8, 0.6, 0.4, 0.2, and 0.05.

When a system is both very large and inhomogeneous, many iterations are necessary to reach a maximum relative difference of the production rates in the two-machine lines of less than 0.5%. For this reason, we stopped the algorithm when the relative deviation of the production rates was less than 0.5% in Cases 6 to 11. The last column in Table 3.6 gives the relative deviation of the production rate approximated by the ADDX algorithm from the simulated mean production rate. The simulation for each case took more than 12 hours, whereas the seven iterations of the ADDX approximation in Case 6 required about 25 seconds using Visual Basic.

	DDX		ADDX			Simulation	
Case	PR	# It.	PR	# It.	Mean	95% Interval	Dev. [%]
6	0.696	19	0.699	7	0.6688	(0.6676, 0.67)	4.52
7	0.626	64	0.629	33	0.6123	(0.6112, 0.6133)	2.73
8	0.486	46	0.489	84	0.474	(0.473, 0.475)	3.16
9	-	-	0.327	23	0.3181	(0.3172,0.3189)	2.80
10	-	-	0.164	21	0.1594	(0.1585, 0.1603)	2.89
11	-	-	0.0409	15	0.03979	(0.0329, 0.04028)	2.79

Table 3.6. Numerical Results in Cases 6 to 11

Note that the ADDX algorithm converged even for the most inhomogeneous system, whereas the DDX algorithm failed in Cases 9, 10, and 11. Therefore, the ADDX algorithm appears to be a very important improvement. Furthermore, the ADDX algorithm required fewer iterations than the DDX algorithm to reach the same level of accuracy in two of the three cases where the DDX algorithm did not fail. The relative differences between approximated and simulated production rates are larger than in Cases 1 to 5. However, their order of magnitude is comparable to those of other decomposition approaches for A/D systems[21] or longer transfer lines.[22] The ADDX algorithm based on Burman's reformulation of the decomposition equations appears therefore to be efficient, quite accurate, and reliable even for large and inhomogeneous systems.

When using the corresponding decomposition equations for the continuous time model, we observed a very similar accuracy and algorithmic behavior, i.e. the decomposition is more precise for small than for large networks and convergence is much more reliable for the ADDX algorithm than for the DDX algorithm. We also observed that for each parameter set the production rate in the continuous time model is lower than in the corresponding discrete time model. The reason is that the geometrically distributed random variables in

---

[21] Di Mascolo et al. (1991)

[22] Burman (1995)

the discrete time model have a lower coefficient of variation than the exponentially distributed random variables in the continuous time model. Higher variation leads to a lower production rate. The largest relative deviation of the simulated production rate between the continuous time model and its discrete time counterpart occurred in Case 2 (identical machines, small buffers) where the continuous time model resulted in a production rate of 0.4157 as opposed to 0.4921 for the discrete time model. This high relative deviation indicates the usefulness of having both models. However, if the probabilities/rates of the random events are very small, both approaches yield very similar results as our analytical results for small time steps suggest.[23]

In their paper on the continuous time version of the A/D model, Jeong and Kim (1998, p. 48) publish a completely documented numerical case. They apply their method to a body assembly shop of a automobile manufacturing company. The tree-structured system with 21 work stations and 20 buffers is depicted in Figure 3.14. The data for the machines or work stations and the buffer capacities is repeated in Tables 3.7 and 3.8.

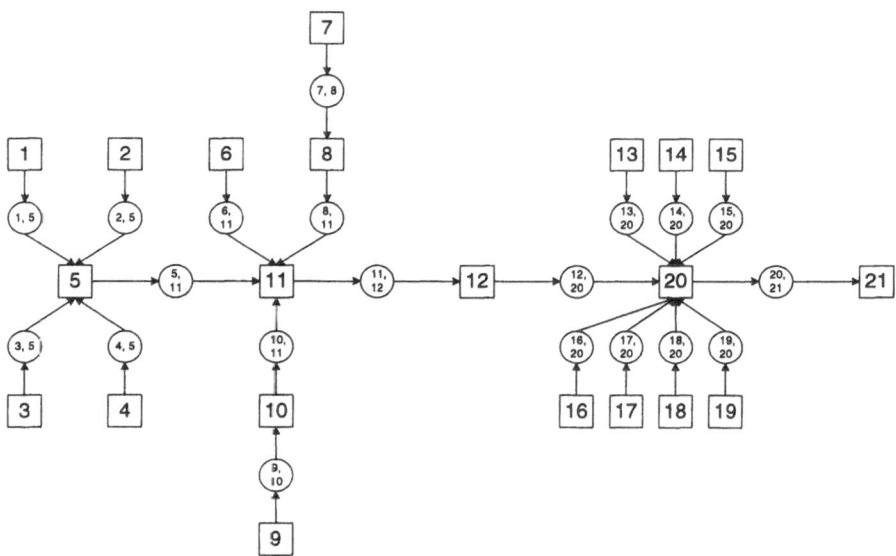

Fig. 3.14. Schematic View of the Body Assembly Shop Studied by Jeong and Kim

Our ADDX-type algorithm needed seven iterations to meet the conservation of flow equation with a maximum relative difference of the production rates estimates in the two-machine lines of less than 0.01%, the DDX-type algorithm required 14 iterations. In both cases, the estimated production rate was 0.739. This is a deviation of -0.94% from the simulation-based produc-

---

[23] See Equation (3.107) on Page 71.

Work station	Failure rate (per minute)	Repair rate (per minute)	Processing rate (per minute)
1	0.00190	0.05722	1.22449
2	0.00123	0.05769	1.22449
3	0.00123	0.05630	1.22449
4	0.00323	0.05681	1.22449
5	0.00611	0.05642	1.17647
6	0.00001	0.05000	1.22449
7	0.00105	0.05159	1.22449
8	0.00347	0.05672	1.17647
9	0.00105	0.05159	1.22449
10	0.00347	0.05672	1.17647
11	0.00290	0.05687	1.15385
12	0.00492	0.05644	1.15385
13	0.00028	0.06960	1.30435
14	0.00025	0.06600	1.30435
15	0.00018	0.06781	1.30435
16	0.00030	0.07053	1.39535
17	0.00022	0.07044	1.39535
18	0.00025	0.06600	1.30435
19	0.00018	0.06781	1.30435
20	0.00122	0.05732	1.11111
21	0.00006	0.08404	1.11111

Table 3.7. Machine Data for the Jeong and Kim Example

Buffer	(1, 5)	(2, 5)	(3, 5)	(4, 5)	(5, 11)	(6, 11)	(7, 8)
Size	10	11	10	11	10	10	3
Buffer	(8, 11)	(9, 10)	(10, 11)	(11, 12)	(12, 20)	(13, 20)	(14, 20)
Size	12	10	12	10	11	8	5
Buffer	(15, 20)	(16, 20)	(17, 20)	(18, 20)	(19, 20)	(20, 21)	
Size	7	14	5	5	7	10	

Table 3.8. Buffer Sizes C for the Jeong and Kim Example

tion rate estimate of 0.746 given in Jeong and Kim (1998, p. 49). Jeong and Kim report deviations of their methods of 5.6% and do not give the required number of iterations.

In the auto body example, processing rates range from 1.11111 to 1.30435 for the real machines. Our decomposition, however, computed processing rates of virtual machines ranging from 0.89632 to 1.39535. In our decomposition method, the speed of the *virtual* machines can differ from those of the corresponding machines in the real system, whereas this is not the case in Jeong and Kim's methods, see Page 73. We interpret this as an additional 'degree of freedom' of our method and conjecture that this flexibility might lead to more accurate results. However, no serious conclusion can be drawn from a single experiment.

In order to compare the two methods thoroughly, the hundreds of random problems studied by Jeong and Kim should be analyzed using our method as well. Furthermore, one should look for problems and parameter constellations for which both methods fail in order to explore their respective limitations. This is a task for further research.

3.5.2 Behavior of Assembly/Disassembly Systems

Transfer lines and A/D systems have several properties in common. The production rate appears to decrease in both cases as machines get less reliable, i.e. as the isolated efficiency $e_i = r_i/(r_i + p_i)$ decreases. It also decreases for a constant isolated efficiency as the average duration $1/r_i$ of a repair increases. Since buffers reduce the propagation of disruptions in the flow of material, increasing any buffer appears to increase the production rate, even though the increase may be very small and may not be economically justifiable.[24]

Another property, which appears to hold for A/D systems as well as for transfer lines, is *equivalence*. The equivalence theorem[25] states that

- the production rate remains unchanged if the direction of the flow of material through any buffer of an A/D system is reversed to create an equivalent system and that
- the average number of parts in any buffer of the initial system equals the the average number of empty spaces or 'holes' in any reversed buffer of the equivalent system. If the direction of flow through a buffer in both systems is identical, so is the average buffer level.

Consider the different five-machine systems in Figure 3.15. Assume that the vector of processing probabilities u_i is $(0.91, 0.92, 0.93, 0.94, 0.95)$, the vector of failure probabilities p_i is $(0.011, 0.012, 0.013, 0.014, 0.015)$, all repair probabilities are $r_i = 0.1, i = 1, ..., 5$, and the physical space between any two adjacent machines can hold ten parts. In this case, the decomposition method developed in this section yields a production rate estimate of an average of 0.689 parts per period or production cycle for all seven systems.

An intuitive way to understand this behavior is to imagine that for each part moving downstream the system, an empty space or 'hole' is moving upstream.[26] The production rate of parts is thus equal to the corresponding rate of holes. On the other hand, the production rate of parts of System 6 in Figure 3.15 is equal to the rate at which holes move upstream in System 1, which is in turn equal to the production rate of System 1.

---

[24] Note that, from a mathematical point of view, these are *conjectures* that should actually be proved from the assumptions of the mathematical model. Due to the large state space of the model, this should be rather difficult. However, these conjectures appear to be highly plausible and we are not aware of any case where they did not hold.

[25] Gershwin (1994, p. 203), see also Ammar and Gershwin (1989) and Dallery et al. (1994).

[26] See Ammar and Gershwin (1989, p. 239).

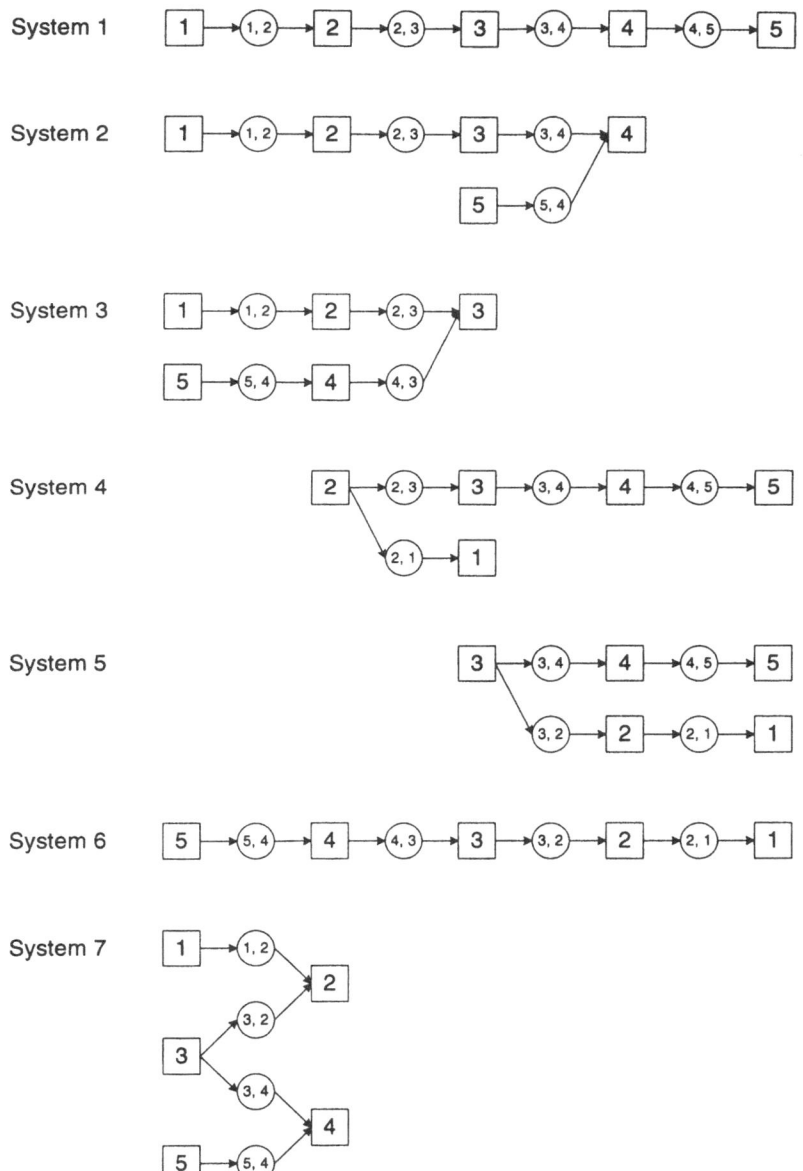

Fig. 3.15. Transfer Line and Equivalent Assembly/Disassembly Networks

Machine M_4 in System 1 can be considered to 'assemble' parts from Buffer $B_{3,4}$ and holes from Buffer $B_{4,5}$. Its production rate is therefore equal to the situation in System 2 where it actually assembles parts coming from Machines M_3 and M_5. The reasoning for the other systems in Figure 3.15 is similar.

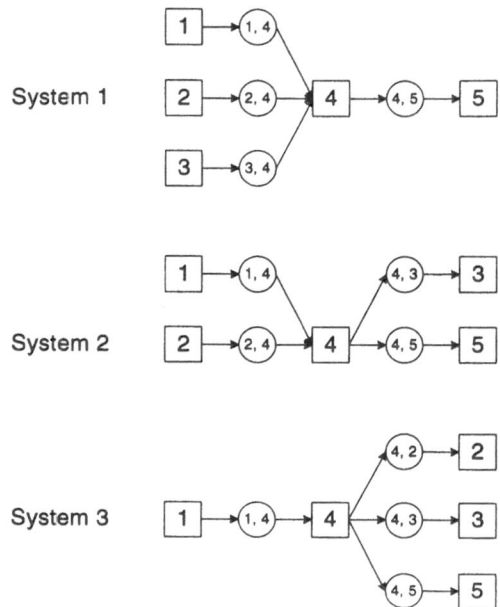

Fig. 3.16. Multiple Equivalent Assembly/Disassembly Networks

Now consider the systems in Figure 3.16. These three systems are equivalent to each other as well. Another intuitive way to see that the assembly operation in System 1 in Figure 3.16 is equivalent to the disassembly in System 3 is that what is happening in the latter is basically that the manufacturing process is "running backwards." The estimated average production rate for all three systems in Figure 3.16 is 0.677. Note that this is approximately 1.7% lower than for the systems in Figure 3.15. This is not a coincidence or an error of the analytical method. The reason for the lower production rate of the systems in Figure 3.16 is that, due to the assembly operation performed by Machine M_4, the machines are more closely coupled than, for example, in System 1 of Figure 3.15.

This effect gets stronger as the number of input buffers at the assembly machine increases and the buffer sizes decreases. Consider the two 40-machine systems depicted in Figure 3.17. Assume that all machines M_i have processing probabilities $u_i = 0.9$, failure probabilities $p_i = 0.01$, and repair probabilities $r_i = 0.1$ while between any two adjacent machines there are two buffer spaces. The isolated production rate of each of these machines is therefore $u_i e_i = 0.8 \cdot 10/11 \approx 0.818$.

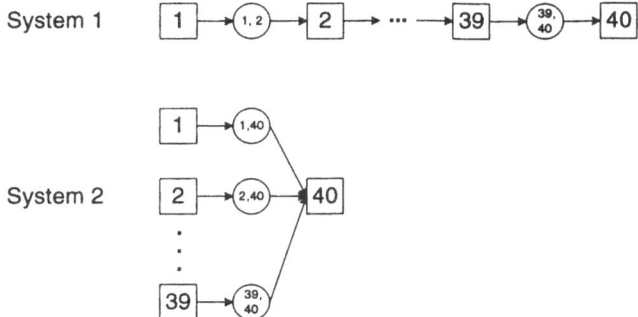

Fig. 3.17. Transfer Line and Non-Equivalent Assembly Network

For the serial arrangement in System 1 of Figure 3.17, the decomposition method predicts a production rate of 0.454 parts per period, i.e. in this system each machine is starved and/or blocked approximately $(1 - 0.454/0.818) = 45\%$ of the time it is operational. Now consider the assembly structure in System 2. The estimated production rate for this case is only 0.273, i.e. each machine is starved or blocked $(1 - 0.273/0.818) = 67\%$ of the time it is operational.[27] This is $(1 - 0.273/0.454) = 40\%$ below the production rate of the serial system and indicates that the decrease in the production rate due to the assembly structure can be significant.

An intuitive way to understand this phenomenon is to ask how many buffer spaces separate any two machines in the network. Consider Machines M_1 and M_{40}. In the transfer line in System 1, they are separated by 39 buffers with two buffer spaces each and 38 Machines with one workspace each. It takes some time for a failure of M_1 to starve M_{40}. In System 2 in Figure 3.17, however, only two buffer spaces separate M_1 and M_{40} and no more than $2 \cdot 2 = 4$ buffer spaces plus the workspace at M_{40} separate *any* two of the machines M_1 to M_{40}. Thus, *all* machines are tightly coupled in the second system and each disruption in the flow of material propagates quickly through the network, yielding a lower production rate.

This insight into the behavior of A/D systems relative to transfer lines leads to the following rule of thumb for manufacturing system design: Whenever assembly patterns in the flow of material can be avoided by a serial arrangement of work stations, this leads—*ceteris paribus*—to a (possibly small) increase in the production rate. However, this increase in the production rate does not come for free. If Machines M_1 to M_{39} in System 2 of Figure 3.17 perform operations on subassemblies, which can be physically separated from each other, the buffers, as well as the value of the work-in-process, differs from

---

[27] A simulation over 100,000 time units after an initial phase of 10,000 time units resulted in an estimated production rate of 0.423 parts per time unit for the 40-machine transfer line and of 0.276 for the 40-machine assembly network, i.e. the difference is *not* due the analytical method.

those in the serial arrangement in System 1. The reason is that a subassembly from, for example Machine M_1, has to travel through all the buffers in System 1, whereas it passes through only one buffer in System 2. That is, an average part spends less time in System 2.

3.6 Optimal Design of Assembly/Disassembly Systems

The problem of finding an optimal design has been analyzed from a cash-flow perspective in Section 2.3 and a design was considered to be optimal if it leads to the maximum possible expected net present value (NPV) of the corresponding cash flow. For machines and buffers with identical performance and cost parameters, the buffer space was allocated according to the shape of a symmetrical bowl turned upside down, i.e. buffers in the middle of the line were larger than those at the beginning or end as machines in this part of the system are particularly likely to be starved or blocked.[28] In this section, we ask how this picture changes if the system performs assembly operations.

Consider the eight-machine assembly system depicted in Figure 3.18, which was derived from the system in Figure 2.10 on Page 15 by adding two additional machines, M_7 and M_8, immediately upstream of M_2 (which is now performing assembly operations). Assume that all processing probabilities u_i are 0.95, all failure probabilities $p_i = 0.01$, all repair probabilities $r_i = 0.1$, and the initial buffer capacity is $C_{i,j} = 2$ for all buffers $B_{i,j}$. The isolated production rate of each machine is thus $0.95 \cdot 10/11 \approx 0.864$ parts per period or production cycle. Each machine costs \$160,000 and each buffer \$2,000, the scrap value is 10% in each case. The raw material per product unit costs \$200, each unit is sold for \$245, and the annual fixed cost is \$220,000. The system is expected to operate for 4 years with 24,000 production cycles each and the interest rate on a perfect capital market is 10% per year.

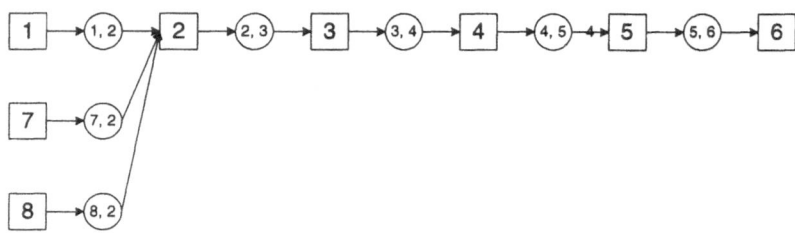

Fig. 3.18. Eight-Machine Assembly System

In the initial situation with two buffer spaces at each buffer, the decomposition method predicts a production rate of 0.585 parts per production cycle.

---

[28] See the example on Page 16.

The corresponding cash flow has an NPV of \$137,468, i.e. this investment is profitable.

An optimization algorithm based on gradient techniques[29] shows that the investment can be made much more profitable by adding buffer spaces. The optimal buffer allocation for the initial situation is depicted in Figure 3.19, where the numbers between the machines represent the capacity of the corresponding buffer. Thus, Buffer $B_{2,3}$ can hold up to 28 parts etc.

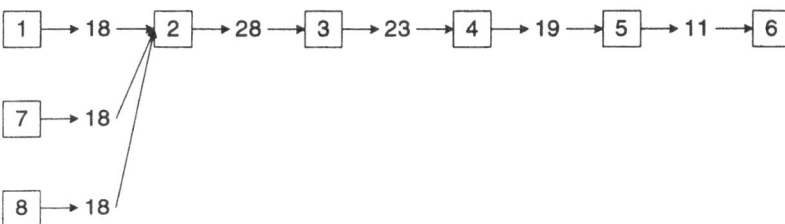

Fig. 3.19. Optimal Buffer Allocation for Identical Buffer Costs

Note that the majority of the buffers is now allocated in front of and behind the assembly machine M_2. The total number of additional buffer spaces is 121. This leads to an increase of the expected production rate from 0.5851 to 0.7651 parts per production cycle that results in an increase of the expected NPV from \$137,468 to \$551,323. Adding buffer spaces makes this particular system *much* more profitable, and the buffer spaces are attracted by the assembly machine.

In a last experiment we ask how this picture changes if buffer costs are no longer identical and assume that buffers immediately up- and downstream of the assembly machine are twice as expensive as the other buffers, i.e. cost \$4,000 each and have again a scrap value of 10%.

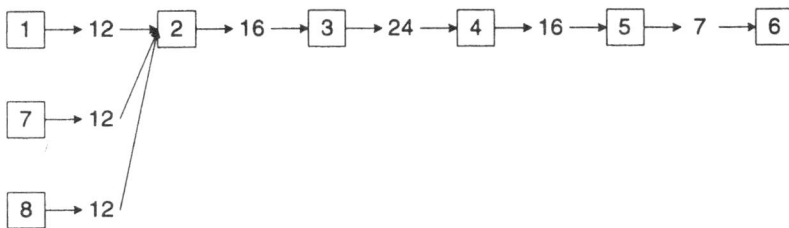

Fig. 3.20. Optimal Buffer Allocation for Non-Identical Buffer Costs

It is not surprising that the NPV corresponding to the initial two-spaces-per-buffer situation is now lower, with a value of \$122,540. The optimization

---

[29] Schor (1995)

algorithm proposes the buffer allocation depicted in Figure 3.20, which corresponds to an NPV of $430,000. This is lower than the $551,323 for the situation in Figure 3.19, but still much higher than the $122,540 for the two-spaces-per-buffer situation.

If we compare the results of the two experiments, the following observations can be made: In both cases, the assembly machine attracts many buffer spaces, even if they are much more expensive than buffer spaces elsewhere in the system. In the design in Figure 3.19, 82 out of 135 buffer spaces (or about 61%) are allocated immediately in front or behind the assembly machine. Even if these buffer spaces are made twice as expensive, we still find 52 out of 99 buffer spaces (i.e. 53%) at the assembly machine, see Figure 3.20. Note that not only the NPV of the investment decrease as buffers get more expensive, but also the total number of buffer spaces decreases for an optimal or close to optimal solution.

As a result we can state that—other things being equal—assembly operations attract buffer spaces, even if these should be much more costly than spaces elsewhere in the system. The reason is that assembly operation couple more than two adjacent machines and thus intensify the propagation of disruptions in the flow of material. As opposed to the examples studied above, machines and buffers in real systems often do not have identical parameters, and in this case it is not at all obvious how to allocate buffer space in an economically sound way. In this situation, powerful analytical methods for performance evaluation and optimization can be useful to achieve a system design that leads to an efficient use of resources.

4. Flow Lines with Rework Loops and Identical Processing Times

4.1 Discrete-Material Flow Line Model with Identical Deterministic Processing Times

To analyze production lines with scrapping and rework, we extend an existing model of a transfer line[1] by allowing for two additional phenomena concerning the flow of material. These two phenomena are split and merge operations.[2] The previously existing model as well as the numerical technique to determine performance measures assumed a purely linear flow of material. This situation is depicted in Figure 4.1.

Fig. 4.1. Production Line with Linear Flow of Material

The system produces discrete parts. Processing times are assumed to be deterministic and identical for all machines and are taken as the time unit. In a system with a linear flow of material, a machine processes a part during a time period if it is not starved (its input buffer is not empty), not blocked (its output buffer is not full), and it does not fail. We assume geometrically distributed operation dependent failures (ODFs) at the machines, i.e. a machine M_i that is is neither starved nor blocked and could thus process a part fails at the beginning of a period with probability p_i. If it is either starved or blocked, it cannot fail. Machine M_i is repaired at the beginning of a period with probability r_i if it was down during the previous period, i.e. times to repair are also geometrically distributed.

We also assume blocking after service (BAS): Machine M_i is blocked if it has processed a part and finds its output buffer full. In this case, the processed part remains at the workspace of Machine M_i until a space in the downstream buffer becomes available. We further assume that machine states change (due to failures and repairs) at the beginning of periods, whereas buffer levels change (due to completion of processing) at the end of periods.

---

[1] Gershwin (1987, 1994)
[2] Helber (1997a,b)

Travel times within buffers are zero. A more detailed and formal description of the model is given in Gershwin (1994, pages 71–74).

The first new phenomenon that we model is the *split operation* depicted in Figure 4.2. Machine M_2 has *multiple alternative* immediate successors. Each part processed at M_2 is randomly sent to one of M_2's immediate successors.

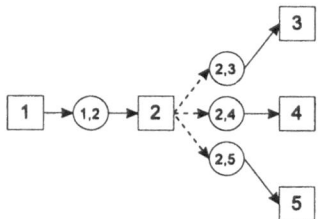

Fig. 4.2. System with a Split Operation

The routing decision is made *after* the part has been processed and it is modeled as the flip of a multi-sided coin. Formally, if $D(i)$ is the set of buffers immediately downstream of Machine M_i, a part processed by Machine M_i is sent to M_q with probability $d_{i,q}$ where $\sum_{(i,q)\in D(i)} d_{i,q} = 1$. The broken arcs between Machine M_2 and its immediately succeeding buffers in Figure 4.2 represent *alternative* routings.[3]

We furthermore model the *merge operation* shown in Figure 4.3. Machine M_3 has two immediate predecessors, M_1 and M_2, which produce the same type of parts. To perform an operation, Machine M_3 takes a part out of either Buffer $B_{1,3}$ or $B_{2,3}$. Machine M_3 is starved if both buffers are empty.

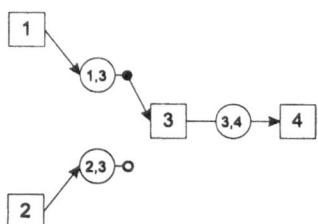

Fig. 4.3. System with a Merge Operation

We assume that a merging machine uses a *priority rule* to select an input buffer. In Figure 4.3, the priority one buffer $B_{1,3}$ is always chosen unless it is empty. It is only in this case that a part is taken from the priority two

---

[3] See Section 2.2.2, especially Page 8, for a more detailed discussion of split operations.

buffer $B_{2,3}$. In Figure 4.3, the priority one buffer $B_{1,3}$ is connected to its downstream machine, whereas the priority two buffer $B_{2,3}$ is not.[4]

If we allow for machines where the flow of material splits or merges, we can model systems with rework loops like the one shown in Figure 2.8 on Page 10.

We finally state the assumptions that each splitting machine has exactly one immediate predecessor, that each merging machine has exactly one successor, that all *input machines* without preceding machines are never starved, that all *output machines* without succeeding machines are never blocked, and that there is at least one input and one output machine in the system. If each machine has no more than one predecessor and successor, the transfer line model in Gershwin (1987) results.

4.2 Decomposition Equations for Loops and Identical Processing Times

4.2.1 Conservation of Flow Equation

The conservation of flow (COF) property for systems with split and merge operations differs from those for transfer lines and A/D networks. In A/D systems, the rate of flow of material through each buffer is the same. In a split or merge system, however, the flow to each machine M_i over all immediate upstream buffers $U(i)$ equals the flow from each machine over all immediate downstream buffers $D(i)$. Define E_i as the production rate of Machine M_i in the real system and $E_{j,i}$ as the production rate passing through Buffer $B_{j,i}$.

The conservation of flow equation for the real system

$$E_i = \sum_{(j,i)\in U(i)} E_{j,i} = \sum_{(i,q)\in D(i)} E_{i,q}, \ \forall i \tag{4.1}$$

states that the flow into a machine equals the flow out of the machine. The decomposition must be performed in a way that a similar condition is met by the production rates in all the virtual two-machine lines.

Define $E(j,i)$ and $E(i,q)$ as the production rate in the decomposed two-machine lines $L(j,i)$ and $L(i,q)$, respectively. The decomposition must satisfy the following conservation of flow equation

$$\sum_{(j,i)\in U(i)} E(j,i) = \sum_{(i,q)\in D(i)} E(i,q), \ \forall i \tag{4.2}$$

---

[4] See Section 2.2.2, especially Page 8, for a more detailed discussion of merge operations.

which couples the solutions for the different two-machine models.

$E(i)$ is the total production rate related to Machine M_i in the two-machine lines $L(i, q)$ or $L(j, i)$, i.e.

$$E(i) = \sum_{(i,q) \in D(i)} E(i, q) = \sum_{(j,i) \in U(i)} E(j, i), \; \forall i. \tag{4.3}$$

The following condition must hold *in addition* if we assume that a part that has been processed by Machine M_i is routed to Machine M_m with a routing probability $d_{i,m}$, irrespective of buffer levels or machine states:

$$d_{i,m} = \frac{E(i, m)}{\sum_{(i,q) \in D(i)} E(i, q)}, \; (i, m) \in D(i), \forall i \tag{4.4}$$

It says that the ratio of the flow rates through the different output buffers of Machine M_i is determined by the routing probabilities. This is because the routing decision is made *after* the part has been processed. If the selected buffer, for example for good parts, is full, the part just processed remains at the workspace of the machine, which is now blocked until a space in the buffer becomes available. The COF equation for the transfer line model in Gershwin (1987) is a special case of (4.2) where each machine has no more than one input buffer and one output buffer, respectively.

The COF equation serves as a stopping criterion in the iterative algorithm that is based on the decomposition. The decomposition must lead to a set of failure and repair parameters for the virtual machines such that production rates in the two-machine lines satisfy the conservation of flow equations for all the machines to a pre-specified accuracy.

4.2.2 Flow Rate-Idle Time Equations

In this section, a set of equations is derived that can be used to determine the failure probabilities $p_u(j, i)$ and $p_d(j, i)$ of the virtual machines $M_u(j, i)$ and $M_d(j, i)$ in each line $L(j, i)$ of the decomposition. The equations are needed in a form that depends on parameters of real and/or virtual machines and performance measures of two-machine lines as these are the only available quantities in a two-machine decomposition. An additional set of equations is needed to compute the repair probabilities.

The flow rate through a machine is determined by the failures and repairs of the machine and by the probability that a machine is starved or blocked due to events that happen up- or downstream in the system. The *flow rate-idle time* (FRIT) equation relates the production rate of a machine to these two factors.

The effect of failures of Machine M_i can be analyzed under the assumption that M_i operates in isolation. In isolation, a Machine M_i is always either up and working (and waiting for the next failure), or it is down (and waiting for the next repair).

The mean time to failure $MTTF_i$ of a Machine M_i that is never starved nor blocked is the inverse of the failure probability, i.e.

$$MTTF_i = \frac{1}{p_i}, \tag{4.5}$$

and a similar equation holds for the mean time to repair $MTTR_i$ with

$$MTTR_i = \frac{1}{r_i}. \tag{4.6}$$

Define e_i as the isolated production rate of the real machine M_i if it is never starved nor blocked Gershwin (1994, p. 75), i.e. the fraction of time Machine M_i is up. Since a machine operating in isolation can only be up or down, the following holds:

$$e_i = \frac{MTTF_i}{MTTF_i + MTTR_i} = \frac{r_i}{r_i + p_i} \tag{4.7}$$

Machine M_i can only fail if it is neither starved nor blocked. All input buffers must be empty for Machine M_i to be starved, whereas one full output buffer is sufficient to block it. The flow rate-idle time (FRIT) relationship is therefore

$$E_i = e_i \text{ prob} \left[\begin{array}{ll} \{n(l,i) > 0, & \text{some } (l,i) \in U(i)\} \text{ and} \\ \{n(i,q) < N(i,q), & \forall (i,q) \in D(i)\} \end{array} \right] \tag{4.8}$$

where $n(l,i)$ denotes the buffer level in Buffer $B_{l,i}$. It says that the production rate E_i is the probability that Machine M_i is up and neither blocked nor starved.

We assume that the probability of a machine being blocked and starved simultaneously is negligible. This is a common assumption in the analysis of *linear transfer lines* which helps to approximate the probability that M_i is neither starved nor blocked. A similar assumption appears to be reasonable in the analysis of *split and merge systems*.

The reason is that, compared to a purely linear transfer line, having two instead of one input buffer makes starvation of Machine M_i *ceteris paribus* less likely. Similarly, having several output buffers makes blocking less likely. Thus, for a machine with multiple predecessors or successors, the probability of it being starved and blocked simultaneously is smaller than for a machine in a linear transfer line.

If a machine has multiple output buffers due to a split operation, only one of these buffers can be full at any time. However, the two input buffers of a merge machine can be empty simultaneously. The probability of having an empty priority two buffer depends on the level of the corresponding priority one buffer: The priority two buffer is more likely to be empty if the priority one buffer is empty as well. As an approximation, however, we assume that input buffer levels are independent to find

$$
E(i) \approx e_i \left[\left(1 - \prod_{(l,i) \in U(i)} \text{prob}[\{n(l,i) = 0\}] \right) \left(1 - \sum_{(i,q) \in D(i)} \text{prob}[\{n(i,q) = N(i,q)\}] \right) \right]. \tag{4.9}
$$

Given that the probability of a machine being blocked and starved simultaneously is negligible, we can further approximate:

$$
E(i) \approx e_i \left[1 - \prod_{(l,i) \in U(i)} \text{prob}[\{n(l,i) = 0\}] - \sum_{(i,q) \in D(i)} \text{prob}[\{n(i,q) = N(i,q)\}] \right] \tag{4.10}
$$

Define $e_u(i,q) = r_u(i,q)/(r_u(i,q) + p_u(i,q))$ as the isolated production rate of the virtual upstream machine in the two-machine line $L(i,q)$ and $e_d(l,i) = r_d(l,i)/(p_d(l,i) + r_d(l,i))$ as the isolated production rate of the virtual downstream machine in Line $L(l,i)$. The two-machine flow rate-idle time equations Gershwin (1994, p. 81-82) can be expressed as

$$
\text{prob}[\{n(i,q) = N(i,q)\}] = 1 - \frac{E(i,q)}{e_u(i,q)} \tag{4.11}
$$

and

$$
\text{prob}[\{n(l,i) = 0\}] = 1 - \frac{E(l,i)}{e_d(l,i)} \tag{4.12}
$$

yielding

$$
E(i) = e_i \left[1 - \prod_{(l,i) \in U(i)} \left(1 - \frac{E(l,i)}{e_d(l,i)} \right) - \sum_{(i,q) \in D(i)} \left(1 - \frac{E(i,q)}{e_u(i,q)} \right) \right]. \tag{4.13}
$$

This leads to two sets of equations used to determine the parameters of the two-machine lines:

$$\frac{r_u(i,m) + p_u(i,m)}{r_u(i,m)} = \frac{1}{e_u(i,m)} = K_1 \tag{4.14}$$

$$\frac{r_d(j,i) + p_d(j,i)}{r_d(j,i)} = \frac{1}{e_d(j,i)} = K_2 \tag{4.15}$$

where

$$K_1 = \frac{\frac{E(i)}{e_i} + \prod_{(l,i) \in U(i)}(1 - \frac{E(l,i)}{e_d(l,i)}) + \sum_{\substack{(i,q) \in D(i) \\ q \neq m}}(1 - \frac{E(i,q)}{e_u(i,q)})}{E(i,m)} \tag{4.16}$$

$$K_2 = [\frac{\frac{E(i)}{e_i} + \sum_{(i,q) \in D(i)}(1 - \frac{E(i,q)}{e_u(i,q)}) - 1}{\prod_{\substack{(l,i) \in U(i) \\ l \neq j}}(1 - \frac{E(l,i)}{e_d(l,i)})} + 1]\frac{1}{E(j,i)} \tag{4.17}$$

This is a useful result as it relates, for example for Line $L(i,m)$ in (4.14), failure and repair probabilities $r_u(i,m)$ and $p_u(i,m)$ to the given isolated efficiency e_i of Machine M_i, to the isolated efficiencies $e_d(l,i)$ and $e_u(i,q)$ of other virtual machines, and to production rates $E(l,i)$ and $E(i,q)$ in other two- machine models of the decomposition. Whenever the parameters $r_u(i,m)$ and $p_u(i,m)$ for the upstream machine $M_u(i,m)$ of Line $L(i,m)$ are updated, all these other quantities are given or can be easily computed.

Note that the term K_1 in (4.16) related to Line $L(i,m)$ contains parameters and performance measures of adjacent two-machine systems *other* than $L(i,m)$. Separating these parameters from $p_u(i,m)$ and $r_u(i,m)$ in (4.14) will later allow us to solve the complete set of decomposition equations simultaneously in a way proposed in Burman (1995) that leads to an improved convergence behavior of the algorithm. The same holds for K_2 and the downstream parameters $p_d(j,i)$ and $r_d(j,i)$. The FRIT equation for the transfer line model in Gershwin (1987, 1994) is again a special case of (4.14) and (4.15).

4.2.3 Resumption of Flow Equations I: Split Operations

A second set of equations is required to determine the repair probabilities $r_u(i,m)$ and $r_d(i,m)$ for each line $L(i,m)$ in the decomposition. Since the flow resumes after each of these repairs, the equations are frequently called *resumption of flow* equations (Gershwin (1994)). They are required in a form similar to the flow rate-idle time equations, i.e. one that requires quantities that are either given system parameters or that can be computed from other two-machine models in the decomposition.

These resumption of flow equations reflect the perspective of an observer in a buffer. His perspective depends on whether he is up- or downstream of a split or merge system. For this reason, two different sets of resumption of flow equations have to be derived. However, the approach is the same in all cases and the results show a common structure. In this subsection, split operations are analyzed. Merge operations are studied in the next subsection.

The resumption of flow equations for a purely linear transfer line as depicted in Figure 4.1 are given in Gershwin (1987, 1994). The first new component that we model in Figure 4.4 consists of a machine M_i that has exactly one upstream machine denoted as M_j and multiple downstream machines.

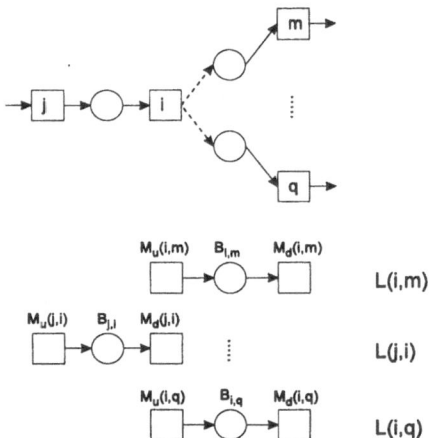

Fig. 4.4. Split System and its Decomposition

Each part processed at Machine M_i in Figure 4.4 is randomly routed to one of Machine M_i's downstream machines. To produce a part at time t and send it to Machine M_m with $(i,m) \in D(i)$, several conditions must hold simultaneously. First, Machine M_i must be up at time t. Second, it must not be starved, i.e. the level $n[(j,i), t-1]$ of its one and only upstream buffer $B_{j,i}$ must be positive. Third, it must not be blocked. Since we assume blocking after service this means that there must not be an already processed part waiting at the workspace of Machine M_i for its selected output buffer to become non-full. In the two-machine model used for the decomposition, the workspaces at the two machines are included in the extended buffer size: $N(i,m) = C_{i,m} + 2$. Thus, if Machine M_i is blocked due to an already processed part that is waiting for a space in Buffer $B_{i,m}$, we have $n(i,m) = N(i,m)$ for Line $L(i,m)$. For this reason, Machine M_i is not blocked if we have $n[(i,q), t-1] < N(i,q), \forall (i,q) \in D(i)$. Finally, the part produced at time t must be sent to Machine M_m instead of being sent to one of the other machines M_q with $(i,q) \in D(i), q \neq m$.

4.2.3.1 Upstream Machine. To an observer in Buffer $B_{i,m}$, the virtual upstream machine $M_u(i,m)$ is up at time t when a part enters Buffer $B_{i,m}$ at time t or when Buffer $B_{i,m}$ is full and $M_u(i,m)$ is blocked (since a blocked machine cannot fail). For this to happen, Machine M_i must be up, it must not be starved or blocked due to a full buffer $B_{i,q}, q \neq m$, and the processed part must be sent to Machine M_m.

Define $\{\alpha_u[(i,m),t] = 1\}$ as the event that the virtual upstream machine $M_u(i,m)$ is up at time t:

$$\{\alpha_u[(i,m),t] = 1\} \quad \text{iff} \quad \{\alpha_i(t) = 1\} \text{ and} \qquad\qquad (4.18)$$
$$\{n[(j,i),t-1] > 0\} \text{ and}$$
$$\{n[(i,q),t-1] < N(i,q), \ \forall (i,q) \in D(i), q \neq m\} \text{ and}$$
$$\{\beta_i(t) = m\}.$$

The first condition on the right hand side of (4.18) says that Machine M_i must be up. Since we assume in our model that a split machine has exactly one input buffer denoted as $B_{j,i}$, the second condition demands that this input buffer is not empty. The third condition says that Machine M_i must not be blocked due to a part that is waiting for a space in Buffer $B_{i,q}$ ($q \neq m$), i.e. one of the *other* buffers downstream of Machine M_i. (If Machine M_i is blocked due to a part that is waiting for a space in Buffer $B_{i,m}$, then Machine $M_u(i,m)$ as seen from an observer in Buffer $B_{i,m}$ is **up** as it is trying to deliver a part.) Finally, the part produced at time t must be randomly routed to Machine M_m. This is denoted as the event $\{\beta_i(t) = m\}$.

Machine $M_u(i,m)$ is down if it is not up, i.e:

$$\{\alpha_u[(i,m),t] = 0\} \quad \text{iff} \quad \{\alpha_i(t) = 0\} \text{ or} \qquad\qquad (4.19)$$
$$\{n[(j,i),t-1] = 0\} \text{ or}$$
$$\{n[(i,q),t-1] = N(i,q),$$
$$\text{for some } (i,q) \in D(i), q \neq m\} \text{ or}$$
$$\{\beta_i(t) = q, \text{ for some } (i,q) \in D(i), q \neq m\}$$

In (4.19), the different events that can force Machine $M_u(i,m)$ down are approximately mutually disjoint: If Machine M_i is either starved or blocked it can neither fail nor send a part to a M_q with $(i,q) \in D(i), q \neq m$. If a part is sent to Machine M_q, M_i cannot be down. Finally, we have already assumed in the derivation of the flow rate-idle time equation on Page 93 that the probability of a machine being blocked and starved simultaneously is very small and can be neglected in an approximation.

The repair probability $r_u(i,m)$ of the virtual upstream machine $M_u(i,m)$ is the probability of seeing a part being sent into Buffer $B_{i,m}$ at time $t+1$ given that no part was sent into Buffer $B_{i,m}$ at time t and that Machine M_i

was not blocked due to a part for Buffer $B_{i,m}$, i.e. $n[(i, m), t - 1] < N(i, m)$. This can be written as

$$
\begin{aligned}
r_u(i, m) &= \text{prob}\Big[\{\alpha_u[(i, m), t + 1] = 1\} \mid \{\alpha_u[(i, m), t] = 0\} \text{ and} \\
&\quad \{n[(i, m), t - 1] < N(i, m)\}\Big] \quad (4.20) \\
&= \text{prob}\Big[\{\alpha_u[(i, m), t + 1] = 1\} \mid \Big\{\{\alpha_i(t) = 0\} \text{ or} \\
&\quad \{n[(j, i), t - 1] = 0\} \text{ or} \\
&\quad \{n[(i, q), t - 1] = N(i, q), \text{ for some } (i, q) \in D(i), q \neq m\} \text{ or} \\
&\quad \{\beta_i(t) = q, \text{ for some } (i, q) \in D(i), q \neq m\}\Big\} \text{ and} \\
&\quad \{n[(i, m), t - 1] < N(i, m)\}\Big] \quad (4.21)
\end{aligned}
$$

if we use the definition of an upstream machine being down (4.19).

In Appendix A.1.1 on Page 199 this expression is approximated by decomposing the conditioning event to find

$$
r_u(i, m) = \left[r_i + K_3 \frac{r_u(i, m)}{p_u(i, m)} \right] d_{i,m} \quad (4.22)
$$

where we define

$$
\begin{aligned}
K_3 &= \Big[(r_u(j, i) - r_i)\mathbf{p}[(j, i); 001] \quad (4.23) \\
&\quad + \sum_{(i,q) \in D(i), q \neq m} (r_d(i, q) - r_i)\mathbf{p}[(i, q); N(i, q)10] \\
&\quad + \sum_{(i,q) \in D(i), q \neq m} (1 - p_i)F(i, q) - r_i \, \text{prob}[\{\beta_i(t) = q\}] \Big] \frac{1}{E(i, m)}
\end{aligned}
$$

with the auxiliary expression

$$\text{prob}[\{\beta_i(t) = q\}]$$

$$\approx \left[\sum_{n=0}^{N(i,q)-1} \mathbf{p}[(i,q); n11] + \mathbf{p}[(i,q); n10]\right] \cdot$$

$$\left[\prod_{\substack{(i,k)\in D(i) \\ k\neq m \\ k\neq q}} \sum_{n=0}^{N(i,k)-1} \mathbf{p}[(i,k); n00] + \mathbf{p}[(i,k); n01]\right] \cdot$$

$$\left[\sum_{n=1}^{N(j,i)} \mathbf{p}[(j,i); n11] + \mathbf{p}[(j,i); n01]\right] \tag{4.24}$$

and

$$F(i,q) = \left[\sum_{n=0}^{N(i,q)-2} \Big(\mathbf{p}[(i,q); n11] + \mathbf{p}[(i,q); n10]\Big)\right.$$

$$+ (1 - p_d(i,q))\mathbf{p}[(i,q); N(i,q) - 1, 11]$$

$$\left. + r_d(i,q)\mathbf{p}[(i,q); N(i,q) - 1, 10]\right] \cdot$$

$$\left[\prod_{\substack{(i,k)\in D(i) \\ k\neq m \\ k\neq q}} \sum_{n=0}^{N(i,k)-1} \Big(\mathbf{p}[(i,k); n00] + \mathbf{p}[(i,k); n01]\Big)\right] \cdot$$

$$\left[\sum_{n=2}^{N(j,i)} \Big(\mathbf{p}[(j,i); n11] + \mathbf{p}[(j,i); n01]\Big)\right.$$

$$+ (1 - p_u(j,i))\mathbf{p}[(j,i); 111]$$

$$\left. + r_u(j,i)\mathbf{p}[(j,i); 101]\right] \tag{4.25}$$

where $d_{i,m}$ is the fraction of parts sent from Machine M_i to M_m. The factor K_3 in (4.23) contains parameters of two-machine lines other than Line $L(i,m)$. This is again a generalization of the corresponding equation for the transfer line model in Gershwin (1987).

4.2.3.2 Downstream Machine. In this section, the resumption of flow probability $r_d(j,i)$ for Line $L(j,i)$ upstream of Machine M_i in Figure 4.4 on Page 96 is derived.

To an observer in Buffer $B_{j,i}$, the virtual downstream machine $M_d(j,i)$ is up at time t when a part leaves the buffer at time t or when Buffer $B_{j,i}$ is empty and Machine $M_d(j,i)$ is starved (since a starved machine cannot fail).

Define $\{\alpha_d[(j,i),t]=1\}$ as the event of the virtual downstream machine $M_d(j,i)$ being up at time t. Machine $M_d(j,i)$ is up if Machine M_i is up and not blocked, i.e:

$$\{\alpha_d[(j,i),t]=1\} \quad \text{iff} \quad \{\alpha_i(t)=1\} \text{ and} \tag{4.26}$$
$$\{n[(i,q),t-1] < N(i,q), \ \forall(i,q) \in D(i)\}$$

Machine $M_d(j,i)$ is down if it is not up, i.e:

$$\{\alpha_d[(j,i),t]=0\} \quad \text{iff} \quad \{\alpha_i(t)=0\} \text{ or} \tag{4.27}$$
$$\{n[(i,q),t-1] = N(i,q), \text{ for some } (i,q) \in D(i)\}$$

Note that this definition *also* describes the perspective of an observer upstream of a *disassembly machine* (Gershwin (1991)). In both cases, i.e. split as well as disassembly operations, Machine M_i must be up and not blocked in order for $M_d(j,i)$ to be up and one full downstream buffer $B_{i,q}$ is sufficient to block Machine M_i.

The resumption of flow probability $r_d(j,i)$ is defined as

$$r_d(j,i) = \text{prob} \ \Big[\{\alpha_d[(j,i),t+1]=1\} \ | \tag{4.28}$$
$$\{\alpha_d[(j,i),t]=0\} \text{ and } \{n[(j,i),t-1]>0\}\Big]$$

and is evaluated in Appendix A.1.2 on Page 211 by decomposing the conditioning event to find

$$r_d(j,i) \quad = \quad r_i F + K_4 \frac{r_d(j,i)}{p_d(j,i)} \tag{4.29}$$

with

$$F \quad = \quad 1 \tag{4.30}$$

$$K_4 \quad = \quad \sum_{(i,q)\in D(i)} (r_d(i,q) - r_i)\mathbf{p}[(i,q); N(i,q)10]\frac{1}{E(j,i)} \tag{4.31}$$

This general form involving the auxiliary parameter F allows to formulate the resumption of flow equation for merge systems (to be derived below) in a very similar way. As in (4.23), we have in (4.31) a factor (K_4) containing parameters of two-machine lines other than Line $L(j,i)$. This is exactly the same equation as for a disassembly machine in Gershwin (1991), Gershwin (1994).

4.2.4 Resumption of Flow Equations II: Merge Operations

The second type of subsystem depicted in Figure 4.5 for which we derive resumption of flow equations consists of Machine M_i that has two immediately preceding machines denoted as Machines M_{j_1} and M_{j_2}, and one immediately succeeding machine denoted as Machine M_q.

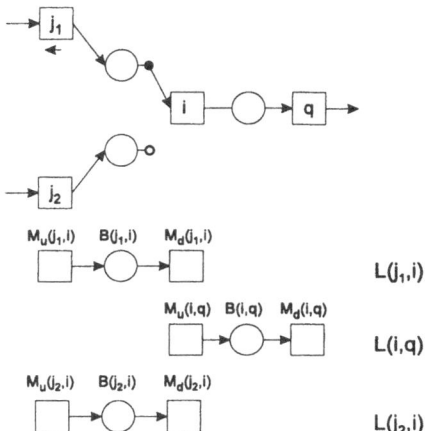

Fig. 4.5. Merge System and its Decomposition

Buffer $B_{j_1,i}$ has priority one and $B_{j_2,i}$ has priority two. From the perspective of Machine M_i, both Machine M_{j_1} and Machine M_{j_2} produce the same type of parts. Thus, Machine M_i is starved if both of its upstream buffers are empty.

Note that this differs from the situation in an assembly system in two ways: First, in an assembly system, different types of parts are arriving from the different machines upstream of the assembly machine. In a merge system, however, the parts arriving from different upstream machines are identical. Second, an assembly machine is starved if at least *one* upstream buffer is empty as all the different components have to be assembled together. In a merge system, a machine is starved if *all* upstream buffers are empty. If one of them is non-empty, the machine can serve this buffer and is therefore not starved.

There are three types of two-machine lines that arise in the decomposition of the system depicted in Figure 4.5. The perspective of an observer in Buffer $B_{i,q}$ is analyzed first.

4.2.4.1 Upstream Machine. In this section, the repair probability of the virtual machine $M_u(i,q)$ as seen by an observer downstream of Machine M_i is determined. To an observer in the buffer corresponding to Line $L(i,q)$ in

Figure 4.5, the virtual machine $M_u(i,q)$ is up if Machine M_i is up and not starved, i.e.

$$\{\alpha_u[(i,q),t] = 1\} \quad \text{iff} \quad \{\alpha_i(t) = 1\} \text{ and} \qquad (4.32)$$
$$\{n[(j,i),t-1] > 0, \text{ for some } j \in \{j_1,j_2\}\}.$$

Machine $M_u(i,q)$ is down if it is not up, i.e

$$\{\alpha_u[(i,q),t] = 0\} \quad \text{iff} \quad \{\alpha_i(t) = 0\} \text{ or} \qquad (4.33)$$
$$\{n[(j,i),t-1] = 0, \forall j \in \{j_1,j_2\}\}.$$

The different events that can force Machine $M_u(i,q)$ down are mutually disjoint since Machine M_i cannot fail if it is starved.

We derive the resumption of flow equation in the usual way using the definition of virtual machine states:

$$
\begin{aligned}
r_u(i,q) \quad = \quad & \text{prob} \left[\{\alpha_u[(i,q),t+1] = 1\} \bigg| \right. \qquad (4.34) \\
& \left. \{\alpha_u[(i,q),t] = 0\} \text{ and } \{n[(i,q),t-1] < N(i,q)\} \right] \\
= \quad & \text{prob} \left[\{\alpha_u[(i,q),t+1] = 1\} \bigg| \right. \qquad (4.35) \\
& \left\{ \{\alpha_i(t) = 0\} \text{ or} \right. \\
& \left. \{n[(j,i),t-1] = 0, \forall j \in \{j_1,j_2\}\} \right\} \text{ and} \\
& \left. \{n[(i,q),t-1] < N(i,q)\} \right]
\end{aligned}
$$

Since the different events that force Machine $M_u(i,q)$ down are mutually disjoint, we can break Equation (4.35) down in Appendix A.2.1 on Page 213 by decomposing the conditioning event to find

$$r_u(i,q) = \left[r_i + K_3 \frac{r_u(i,q)}{p_u(i,q)} \right] d_{i,q} \qquad (4.36)$$

with

$$d_{i,q} \quad = \quad 1 \qquad (4.37)$$

$$K_3 \quad = \quad \left(1 - \prod_{\forall j \in \{j_1,j_2\}} (1 - r_u(j,i)) - r_i \right) \prod_{\forall j \in \{j_1,j_2\}} \text{p}[(j,i);001] \frac{1}{E(i,q)}$$
$$(4.38)$$

Using the routing probability $d_{i,q} = 1$ in (4.36) leads to a form of the resumption of flow equation that is identical to those for an upstream machine related to a split machine (see (4.22) on Page 98). If there is just one line $L(j,i)$ with $(j,i) \in U(i)$, equation (4.36) reduces to the corresponding equation of the purely linear transfer line in Gershwin (1987).

4.2.4.2 Downstream Machine. *Priority One Line.* To an observer in the priority one buffer $B_{j_1,i}$ of Line $L(j_1,i)$ in Figure 4.5 on Page 101, the virtual downstream machine $M_d(j_1,i)$ is down if Machine M_i is either down or blocked. Since we assume that a machine with two upstream buffers has exactly one downstream buffer, there is exactly one virtual downstream machine $M_d(i,q)$ that can block Machine M_i.

However, as there are two input buffers, Machine M_i can fail even if the priority one buffer is empty (and the priority two buffer is not empty). Thus, to an observer in Buffer $B_{j_1,i}$, the failures of Machine M_i do not appear to be completely operation dependent. Repairs, however, are independent of buffer levels and the repair behavior of Machine $M_d(j_1,i)$ is hence exactly as in a transfer line for which the resumption of flow equation

$$r_d(j_1,i)) = r_iF + K_4\frac{r_d(j_1,i)}{p_d(j_1,i)} \tag{4.39}$$

with

$$F = 1 \tag{4.40}$$

$$K_4 = \left(r_d(i,q) - r_i\right)\mathbf{p}[(i,q); N(i,q),10]\frac{1}{E(j_1,i)} \tag{4.41}$$

is given in Gershwin (1987). The resumption of flow equation (4.39) has the same structure as (4.29) on Page 100.

Priority Two Line. To an observer in the priority two buffer $B_{j_2,i}$, there is an additional reason to see Machine $M_d(j_2,i)$ down: The buffer of the priority one line $L(j_1,i)$ may be non-empty. In this case, the next part processed at Machine M_i is always taken from the priority one buffer and the virtual machine $M_d(j_2,i)$ is down since nothing can be taken from the buffer of Line $L(j_2,i)$. Thus, three conditions must hold for Machine $M_d(j_2,i)$ to be up: Machine M_i must be up, it must not be blocked by its virtual downstream machine $M_d(i,q)$, and its priority one upstream buffer $B_{j_1,i}$ must be empty. Formally,

$$\{\alpha_d[(j_2,i),t] = 1\} \quad \text{iff} \quad \{\alpha_i(t) = 1\} \text{ and}$$
$$\{n[(i,q),t-1] < N(i,q)\} \text{ and}$$
$$\{n[(j_1,i),t-1] = 0\}.$$

Machine $M_d(j_2, i)$ is down if it is not up:

$$\{\alpha_d[(j_2, i), t] = 0\} \quad \text{iff} \quad \{\alpha_i(t) = 0\} \text{ or}$$
$$\{n[(i, q), t-1] = N(i, q)\} \text{ or}$$
$$\{n[(j_1, i), t-1] > 0\}. \tag{4.42}$$

The resumption of flow probability $r_d(j_2, i)$ describing repairs of Machine $M_d(j_2, i)$ is defined as

$$r_d(j_2, i) = \text{prob}\Big[\{\alpha_d[(j_2, i), t+1] = 1\} \mid$$
$$\{\alpha_d[(j_2, i), t] = 0\} \text{ and } \{n[(j_2, i), t-1] > 0\}\Big] \tag{4.43}$$

and is approximated in Section A.2.2 on Page 215 as

$$r_d(j_2, i) = r_i F + K_4 \frac{r_d(j_2, i)}{p_d(j_2, i)} \tag{4.44}$$

where we define

$$F = (1 - r_u(j_1, i)) \tag{4.45}$$

$$K_4 = \Big[r_d(i, q) F \mathbf{p}[(i, q); N(i, q)10]\mathbf{p}[(j_1, i); 001] +$$
$$(1 - p_i) \cdot$$
$$\Big[\sum_{n=0}^{N(i,q)-2} \mathbf{p}[(i, q); n10] + \mathbf{p}[(i, q); n11]$$
$$+(1 - p_d(i, q))\mathbf{p}[(i, q); N(i, q) - 1, 11]$$
$$+r_d(i, q)\mathbf{p}[(i, q); N(i, q) - 1, 10]\Big]$$
$$\Big[p_u(j_1, i)\mathbf{p}[(j_1, i); 111] + (1 - r_u(j_1, i))\mathbf{p}[(j_1, i); 101]\Big]$$
$$-r_i F\Big[(1 - \mathbf{p}[(i, q); N(i, q)10])(1 - \mathbf{p}[(j_1, i); 001])$$
$$+\mathbf{p}[(i, q); N(i, q)10]\Big]\Big] \frac{1}{E(j_2, i)]} \tag{4.46}$$

Note that the term K_4 in (4.46) does not contain parameters of Line $L(j_2, i)$. Now all resumption of flow equations for split and merge systems have been determined.

4.2.5 Boundary Equations

In the definition of the model, we assume that machines without preceding machines do not perform split operations. Thus, such an input machine M_i with $U(i) = \{\}$ has exactly one downstream buffer denoted as $B_{i,l}$. The observer in this buffer sees the original behavior of this input machine, so

$$p_u(i, l) = p_i, \qquad \forall i \text{ with } U(i) = \{\} \qquad (4.47)$$
$$r_u(i, l) = r_i, \qquad \forall i \text{ with } U(i) = \{\} \qquad (4.48)$$

i.e. the parameters of Machine $M_u(i, l)$ are those of M_i if Machine M_i is an input machine.

We further assume that machines without succeeding machines do not perform merge operations. Thus, such an output machine M_l with $D(l) = \{\}$ has exactly one upstream buffer denoted as $B_{i,l}$. The observer in this buffer sees the original behavior of the output machine, so

$$p_d(i, l) = p_l, \qquad \forall l \text{ with } D(l) = \{\} \qquad (4.49)$$
$$r_d(i, l) = r_l, \qquad \forall l \text{ with } D(l) = \{\} \qquad (4.50)$$

i.e. the parameters of Machine $M_d(i, l)$ are those of M_l if Machine M_l is an output machine. Thus, the parameters of virtual machines that correspond to input or output machines in the real system are determined by the boundary conditions. The decomposition equations are required to determine the parameters for the other virtual machines not corresponding to input or output machines.

If a system has B buffers, there are B two-machine lines in the decomposition and $4B$ failure and repair probabilities of virtual machines have to be determined. This requires $4B$ equations that are given by the boundary equations in this subsection, the flow rate-idle time equations (4.14), (4.15), and the resumption of flow equations (4.22), (4.29), (4.36), (4.39), and (4.44).

4.2.6 Simultaneous Solution of the Decomposition Equations

The decomposition equations (4.14), (4.15),

$$\frac{r_u(i, m) + p_u(i, m)}{r_u(i, m)} = \frac{1}{e_u(i, m)} = K_1$$

$$\frac{r_d(j, i) + p_d(j, i)}{r_d(j, i)} = \frac{1}{e_d(j, i)} = K_2$$

as well as (4.22), (4.29), (4.36), (4.39), and (4.44)

$$r_u(i,m) = \left[r_i + K_3 \frac{r_u(i,m)}{p_u(i,m)} \right] d_{i,m}$$

$$r_d(j,i) = r_i F + K_4 \frac{r_d(j,i)}{p_d(j,i)}$$

can be solved for the parameters of Lines $L(i,m)$ and $L(j,i)$, respectively, to find:

$$r_u(i,m) = d_{i,m} \frac{K_3 + r_i(K_1 - 1)}{K_1 - 1} \tag{4.51}$$

$$p_u(i,m) = d_{i,m} \left(K_3 + r_i(K_1 - 1) \right) \tag{4.52}$$

$$r_d(j,i) = \frac{K_4 + r_i F(K_2 - 1)}{K_2 - 1} \tag{4.53}$$

$$p_d(j,i) = K_4 + r_i F(K_2 - 1) \tag{4.54}$$

This type of reformulation has been proposed in Burman (1995), in a Ph.D. thesis on the analysis of flow lines.

4.3 The Algorithm to Determine Performance Measures

4.3.1 Purpose, Background, and Basic Structure of the Algorithm

The purpose of the algorithm is to determine production rates and inventory levels for the model of a transfer line with split and merge operations. Since the real system is decomposed into a set of virtual two-machine lines, these quantities are approximated from the analysis of the virtual two-machine lines. The failure and repair probabilities for the virtual machines in these two-machine lines must satisfy the decomposition equations derived in the previous section. The basic idea of the algorithm is to solve the decomposition equations in an iterative way and to hope that the algorithm converges to a set of parameters for the two-machine lines that meets all the decomposition equations to some degree of accuracy. The performance measures of the two-machine lines are an approximation of those seen by an observer in the respective part of the real system.

The algorithm is based on the DDX-algorithm[5] for linear transfer lines. It uses a reformulation of the decomposition similar to those proposed in Burman (1995) and evaluates the two-machine lines in a sequence similar to those proposed in Gershwin (1991) for assembly/disassembly systems.

The basic structure of the algorithm is as follows:

---

[5] Dallery et al. (1988)

1. Determine an evaluation sequence for the pseudo-machines in the two-machine lines (Section 4.3.2).
2. Initialize the parameters of the two-machine models for each line (Section 4.3.3).
3. For each virtual upstream machine $M_u(i,q)$ (Section 4.3.4.1):
   a) Update failure and repair probabilities $p_u(i,q)$ and $r_u(i,q)$.
   b) Compute new steady-state probabilities and performance measures for the two-machine line of the virtual upstream machine.
4. For each virtual downstream machine $M_d(j,i)$ (Section 4.3.4.2):
   a) Update failure and repair probabilities $p_d(j,i)$ and $r_d(j,i)$.
   b) Compute new steady-state probabilities and performance measures for the two-machine line of the virtual downstream machine.
5. Go to Step 3 if the production rates of the two-machine lines do not satisfy the conservation of flow equation and an upper limit on the number of iterations is not exceeded.
6. Stop.

In the remainder of this section, the details of the algorithm are discussed.

4.3.2 Determination of the Evaluation Sequence

The evaluation sequence consists of two parts. The first part denoted as S_U describes the order in which virtual upstream machines in Phase 3 of the algorithm are updated. The second part denoted as S_D describes the order for virtual downstream machines in Phase 4.

Consider the structure in Figure 4.6 where an index $I_{B_{i,q}}$ is assigned to each buffer $B_{i,q}$. The indices are depicted above the respective buffers. In systems with loops, there does not appear to be an obvious way how to assign these indices. As a general rule, we tried to follow the flow of material, i.e. to assign lower indices to buffers at earlier production stages.

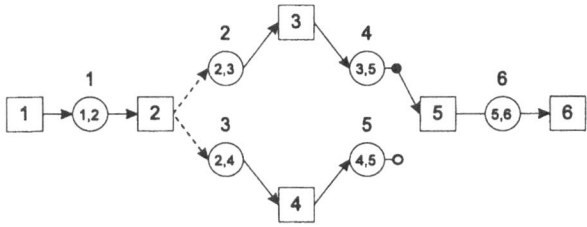

Fig. 4.6. Structure with Indices Assigned to Buffers

We first construct S_U by starting with the pseudo-machines that are *immediately upstream* of priority two buffers. Priority two buffers are considered according to ascending indices. In the next step, the remaining pseudo-machine upstream of priority one or single buffers are added, again according to ascending indices.

In the structure in Figure 4.6, there is one priority two buffer: $B_{4,5}$. Machine $M_u(4,5)$ is therefore placed at the head of Sequence S_U. All other virtual upstream machines are related to priority one or single input buffers. They are appended to the sequence according to ascending indices, starting with Machine $M_u(1,2)$ related Buffer $B_{1,2}$ that has the lowest index $I_{B_{1,2}} = 1$. Using this simple rule for the given assignment of buffer indices, we find the following upstream sequence:

$$S_U = \{M_u(4,5), M_u(1,2), M_u(2,3), M_u(2,4), M_u(3,5), M_u(5,6)\}$$

That is, we first determine new parameters for Machine $M_u(4,5)$, then for $M_u(1,2)$, and so on. (Note that the parameters for Machine $M_u(1,2)$ are actually determined by the boundary conditions as M_1 is an input machine.)

To construct the downstream sequence S_D, we set $S_D = S_U$, reverse the order of machines in S_D, and replace upstream machines by downstream machines to find:

$$S_D = \{M_d(5,6), M_d(3,5), M_d(2,4), M_d(2,3), M_d(1,2), M_d(4,5)\}$$

We hence start with Machine $M_d(5,6)$ and go through the sequence until we finally update the parameters for Machine $M_d(4,5)$.

Due to this sequence, the parameters of downstream machines related to priority two buffers (in this case, $M_d(4,5)$) are always updated *after* the parameters of the downstream machine of the corresponding priority one buffer (in the example $M_d(3,5)$). Numerical test that are not reported her indicate that this helps to improve the convergence reliability especially for merge systems.

4.3.3 Initialization

The failure and repair parameters of the two-machine lines are initialized with the respective values of the corresponding machines in the real system.

For each line $L(i,m)$, set

$$
\begin{align}
p_u(i,m) &= p_i, & (4.55)\\
r_u(i,m) &= r_i, & (4.56)\\
p_d(i,m) &= p_m, & (4.57)\\
r_d(i,m) &= r_m. & (4.58)
\end{align}
$$

The two-machine model by Gershwin and Schick used in the decomposition includes the workspace at the two machines in the extended buffer space. For this reason, set

$$N(i, m) = C_{i,m} + 2 \tag{4.59}$$

where $C_{i,m}$ is the number of physical buffer spaces between Machines M_i and M_m and $N(i, m)$ is the extended buffer size. Solve all the two-machine lines to determine initial steady state probabilities $\mathbf{p}[(i, m)n\alpha_u\alpha_d]$ and production rates $E(i, m)$ (see also Gershwin (1994, p. 76-93)).

4.3.4 Iterative Solution of the Decomposition Equations

The iterative procedure is used to update the parameters of virtual machines that do not correspond to input or output machines. The parameters of those virtual machines that correspond to input or output machines are given by the boundary equations (4.47)-(4.50) and stay at their initial values.

4.3.4.1 Upstream Phase. Consider sequentially all machines $M_u(i, m)$ in Sequence S_U except for *input machines*, i.e. machines without preceding machines. For each machine $M_u(i, m)$, determine $E(i)$ from (4.60) and $E(i, m)$ from (4.61):

$$E(i) \quad = \sum_{(j,i)\in U(i)} E(j, i) \tag{4.60}$$

$$E(i, m) \quad = \quad d_{i,m} E(i) \tag{4.61}$$

Equation (4.61) reflects the random routing of the parts leaving Machine M_i.

Determine K_1 from (4.16):

$$K_1 \quad = \quad \frac{\frac{E(i)}{e_i} + \prod_{(l,i)\in U(i)}(1 - \frac{E(l,i)}{e_d(l,i)}) + \sum_{(i,q)\in D(i), q\neq m}(1 - \frac{E(i,q)}{e_u(i,q)})}{E(i, m)}$$

If Machine M_i performs split operations, determine K_3 from (4.23)

$$
\begin{aligned}
K_3 \quad = \quad & \Big[(r_u(j, i) - r_i)\mathbf{p}[(j, i); 001] \\
+ \quad & \sum_{(i,q)\in D(i), q\neq m} (r_d(i, q) - r_i)\mathbf{p}[(i, q); N(i, q)10] \\
+ \quad & \sum_{(i,q)\in D(i), q\neq m} (1 - p_i)F(i, q) - r_i \, \text{prob}[\{\beta_i(t) = q\}] \Big] \frac{1}{E(i, m)}
\end{aligned}
$$

where $F(i, q)$ is given in (A.35) and $\text{prob}[\{\beta_i(t) = q\}]$ in (A.26).

If Machine M_i performs merge operations, determine K_3 from (4.38):

$$K_3 = \left(1 - \prod_{j \in \{j_1, j_2\}} (1 - r_u(j, i)) - r_i\right) \prod_{j \in \{j_1, j_2\}} \mathbf{p}[(j, i); 001] \frac{1}{E(i, q)}$$

If Machine M_i performs neither split nor merge operations, either equation for K_3 can be used. Determine preliminary values for upstream parameters $r_u^*(i, m)$ from (4.51) and $p_u^*(i, m)$ from (4.52):

$$r_u^*(i, m) = d_{i,m} \frac{K_3 + r_i(K_1 - 1)}{K_1 - 1}$$
$$p_u^*(i, m) = d_{i,m} (K_3 + r_i(K_1 - 1))$$

Especially during the first iterations of the algorithm, we sometimes observed estimates $p_u^*(i, m)$ and $r_u^*(i, m)$ that were either negative or larger than 1. Since these are meaningless values for probabilities, we imposed a set of hard constraints on the parameter updates to be described below. We encountered a few cases where the algorithm appeared to oscillate. To avoid this behavior whenever the algorithm did not converge within 100 iterations, we exponentially smoothed all parameters updates in iteration k in the following way

$$p_u^{**}(i, m) = \epsilon \, p_u^*(i, m) + (1 - \epsilon) \, p_u^{k-1}(i, m) \qquad (4.62)$$
$$r_u^{**}(i, m) = \epsilon \, r_u^*(i, m) + (1 - \epsilon) \, r_u^{k-1}(i, m) \qquad (4.63)$$

where $p_u^{k-1}(i, m)$ and $r_u^{k-1}(i, m)$ denote the parameters from the previous iteration.

During early iterations, even smoothed parameters guesses $p_u^{**}(i, m)$ were occasionally larger than 1 or smaller than p_i, the failure probability of Machine M_i. Since the virtual machine $M_u(i, m)$ cannot possibly fail less often than Machine M_i in the real system, $p_u(i, m)$ cannot be smaller than p_i.

We also occasionally observed smoothed parameters guesses $r_u^{**}(i, m)$ larger than 1 or smaller than 0. In order to update parameters within the possible range of values only, we used the following updating scheme:

$$p_u^k(i, m) = \begin{cases} p_u^{**}(i, m) & \text{if } p_i \leq p_u^{**}(i, m) < 1 \\ p_i + 0.5 \, (p_u^{k-1}(i, m) - p_i) & \text{if } p_u^{**}(i, m) < p_i \quad (4.64) \\ p_u^{k-1}(i, m) + 0.5 \, (1 - p_u^{k-1}(i, m)) & \text{if } p_u^{**}(i, m) \geq 1 \end{cases}$$

$$r_u^k(i, m) = \begin{cases} r_u^{**}(i, m) & \text{if } 0 < r_u^{**}(i, m) < 1 \\ 0.5 \, r_u^{k-1}(i, m) & \text{if } r_u^{**}(i, m) \leq 0 \quad (4.65) \\ r_u^{k-1}(i, m) + 0.5 \, (1 - r_u^{k-1}(i, m)) & \text{if } r_u^{**}(i, m) \geq 1 \end{cases}$$

If $p_u^{**}(i,m)$ is smaller than 1 and larger than p_i, we use $p_u^{**}(i,m)$ in (4.64) to update $p_u^k(i,m)$. However, if $p_u^{**}(i,m)$ is below p_i or above 1, we reduce or increase the previous estimates $p_u^{k-1}(i,m)$ or $r_u^{k-1}(i,m)$ within these bounds. The procedure for the repair parameters (4.65) is similar.

When the virtual failure and repair probabilities $p_u^k(i,m)$ and $r_u^k(i,m)$ for Machine $M_u(i,m)$ in Line $L(i,m)$ have been determined, the new steady-state probabilities $p[(i,m)n\alpha_u\alpha_d]$ and the production rate $E(i,m)$ are computed using the approach by Gershwin and Schick in Gershwin (1994, p. 76-93).

4.3.4.2 Downstream Phase. Consider all machines $M_d(j,i)$ in the downstream part S_D of the evaluation sequence except for *output machines*, i.e. machines without succeeding machines. For each machine $M_d(j,i)$, determine $E(i)$ from (4.66).

$$E(i) \;=\; \text{Min}\left\{e_i, \sum_{(i,q)\in D(i)} E(i,q)\right\} \tag{4.66}$$

During early iterations of the algorithm, the sum of the production rates $E(i,q)$ may be higher than the isolated efficiency e_i of Machine M_i or even higher than 1, the isolated production rate of a perfectly reliable machine. However, the flow out of Machine M_i in the real system, $\sum_{(i,q)\in D(i)} E_{i,q}$, cannot possibly be higher than the isolated production rate e_i of Machine M_i. Furthermore, a production rate higher than 1 is meaningless in the context of the underlying model. We therefore impose in (4.66) an upper limit e_i on the approximated throughput $E(i)$ of Machine M_i. To determine $E(j,i)$, we have to consider three different cases:

- Case A: Machine M_i does not perform merge operations.
- Case B: Machine M_i performs merge operations and we are considering the production rate $E(j_1,i)$ related to the priority one buffer.
- Case C: Machine M_i performs merge operations and we are considering the production rate $E(j_2,i)$ related to the priority two buffer.

In Case A, we set $E(j,i) = E(i)$, as proposed in Dallery et al. (1988) (see also Gershwin (1994, p. 152)) for the simple transfer line. In Cases B and C, we have to enforce that $E(j_1,i)$ and $E(j_2,i)$ for the priority one and priority two buffers add up to $E(i)$ as determined in (4.66). We found that the following procedure leads to a reasonable convergence behavior: During the first five iterations, update

$$E(j_1,i)^k \;=\; \frac{E(j_1,i)^{k-1}}{E(j_1,i)^{k-1} + E(j_2,i)^{k-1}} E(i)^k \tag{4.67}$$

$$E(j_2,i)^k \;=\; \frac{E(j_2,i)^{k-1}}{E(j_1,i)^k + E(j_2,i)^{k-1}} E(i)^k \tag{4.68}$$

where superscripts k and $k-1$ denote the respective iteration. During the first few iterations, the sum of $E(j_1,i)^{k-1}$ and $E(j_2,i)^{k-1}$ may be larger than $E(i)$, or even larger than 1. In order to propagate positive new values for $E(j_1,i)^k$ and $E(j_2,i)^k$ that add up to $E(i)^k$, we use the updating schemes in (4.67) and (4.68).

After five iterations, we usually find

$$E(j_1,i)^{k-1} < E(i)^k \tag{4.69}$$

and

$$E(j_2,i)^{k-1} < E(i)^k \tag{4.70}$$

We therefore switch to the following updating scheme:

$$E(j_1,i)^k = E(i)^k - E(j_2,i)^{k-1} \tag{4.71}$$
$$E(j_2,i)^k = E(i)^k - E(j_1,i)^k \tag{4.72}$$

Using $E(i)$ and $E(j,i)$, determine K_2 from (4.17):

$$K_2 = [\frac{\frac{E(i)}{e_i} + \sum_{(i,q)\in D(i)}(1 - \frac{E(i,q)}{e_u(i,q)}) - 1}{\prod_{(l,i)\in U(i),l\neq j}(1 - \frac{E(l,i)}{e_d(l,i)})} + 1]\frac{1}{E(j,i)}$$

In Cases A and B, set $F = 1$. If Machine M_i in Case A does not perform a merge operation, it either performs a split operation or it sends material to exactly one succeeding machine. In both cases, determine K_4 from (4.31):

$$K_4 = \sum_{(i,q)\in D(i)} (r_d(i,q) - r_i)\mathbf{p}[(i,q); N(i,q)10]\frac{1}{E(j,i)}$$

If in Case B, Machine M_i performs a merge operation and we are considering the priority one buffer, determine K_4 from (4.41):

$$K_4 = \left(r_d(i,q) - r_i\right)\mathbf{p}[(i,q); N(i,q), 10]\frac{1}{E(j_1,i)}$$

In Case C, set $F = (1 - r_u(j_1,i))$ and determine K_4 from (4.46):

$$K_4 = \left[r_d(i,q)F\mathbf{p}[(i,q);N(i,q)10]\mathbf{p}[(j_1,i);001] + \right.$$

$$(1-p_i) \cdot$$

$$\left[\sum_{n=0}^{N(i,q)-2} \mathbf{p}[(i,q);n10] + \mathbf{p}[(i,q);n11] \right.$$

$$+(1-p_d(i,q)\mathbf{p}[(i,q);N(i,q)-1,11]$$

$$\left. +r_d(i,q)\mathbf{p}[(i,q);N(i,q)-1,10] \right]$$

$$\left[p_u(j_1,i)\mathbf{p}[(j_1,i);111] + (1-r_u(j_1,i))\mathbf{p}[(j_1,i);101] \right]$$

$$-r_iF\left[(1-\mathbf{p}[(i,q);N(i,q)10])(1-\mathbf{p}[(j_1,i);001]) \right.$$

$$\left. \left. +\mathbf{p}[(i,q);N(i,q)10] \right] \right] \frac{1}{E(j_2,i)]}$$

Compute preliminary values for downstream parameters $r_d^*(j,i)$ from (4.53) and $p_d^*(j,i)$ from (4.54):

$$r_d^*(j,i) = \frac{K_4 + r_iF(K_2-1)}{K_2-1}$$

$$p_d^*(j,i) = K_4 + r_iF(K_2-1)$$

As for the upstream machines, compute smoothed updates $r_d^{**}(j,i)$ from (4.74) and $p_d^{**}(j,i)$ from (4.73).

$$p_d^{**}(j,i) = \epsilon p_d^*(j,i) + (1-\epsilon)p_d^{k-1}(j,i) \qquad (4.73)$$

$$r_d^{**}(j,i) = \epsilon r_d^*(j,i) + (1-\epsilon)r_d^{k-1}(j,i) \qquad (4.74)$$

Finally, impose the hard constraints in (4.75) and (4.76) to determine the updated failure and repair probabilities for Machine $M_d(j,i)$ in iteration k:

$$p_d^k(j,i) = \begin{cases} p_d^{**}(j,i) & \text{if } p_i \le p_d^{**}(j,i) < 1 \\ p_i + 0.5\,(p_d^{k-1}(j,i) - p_i) & \text{if } p_d^{**}(j,i) < p_i \qquad (4.75) \\ p_d^{k-1}(j,i) + 0.5\,(1 - p_d^{k-1}(j,i)) & \text{if } p_d^{**}(j,i) \ge 1 \end{cases}$$

$$r_d^k(j,i) = \begin{cases} r_d^{**}(j,i) & \text{if } 0 < r_d^{**}(j,i) < 1 \\ 0.5\,r_d^{k-1}(j,i) & \text{if } r_d^{**}(j,i) \le 0 \qquad (4.76) \\ r_d^{k-1}(j,i) + 0.5\,(1 - r_d^{k-1}(j,i)) & \text{if } r_d^{**}(j,i) \ge 1 \end{cases}$$

When the virtual failure and repair probabilities $p_d^k(j,i)$ and $r_d^k(j,i)$ for Machine $M_d(j,i)$ in Line $L(j,i)$ have been computed, the new steady-state probabilities $\mathbf{p}[(j,i)n\alpha_u\alpha_d]$ and the production rate $E(j,i)$ are determined.

Termination. Stop the procedure if the conservation of flow (4.2) equations for all machines are met to a sufficient degree of accuracy for a number of successive iterations.

A proof of convergence for this type of algorithm is not available, but numerical experiments to be reported below show a high convergence reliability for a wide range of system parameters.

4.3.5 General Comments on Implementation and Algorithm Behavior

We always start the algorithm with a smoothing parameter of $\epsilon = 1.0$, i.e. without exponential smoothing. In the very few cases where the algorithm does not converge within 100 iterations, we reduce ϵ to 0.5 for an additional 100 iterations and finally to 0.25 for a last 100 iterations before we abort the evaluation. In almost all cases, reducing ϵ to 0.5 or 0.25 is not necessary. Furthermore, we encountered cases where the estimated performance measures appeared to be a function of the smoothing parameter. For this reason, we suggest not to use the exponential smoothing unless it is necessary to achieve convergence.

In our numerical study to be described below, we terminated the algorithm when the conservation of flow equation (4.2) was met within 0.01% for ten successive iterations. In most cases, these ten iterations lead to an even higher accuracy than 0.01%. Our numerical results show a few cases where the algorithm failed to converge.

4.4 Numerical Results: Algorithm and Flow Line Behavior

4.4.1 Introduction into the Numerical Study

4.4.1.1 Overview. In order to evaluate both the behavior of manufacturing systems with split and merge operations and the performance of our decomposition method, we performed a numerical study based on artificial problems of different size and structure. In this study, we compared the results of our decomposition approach to results from a discrete-event simulation. The simulation was programmed in C. For each structure, i.e. each arrangement of machines and buffers, we studied a set of 100 randomly generated problems. The purpose of this first part of the study was to evaluate the algorithm with respect to convergence reliability and accuracy over a wide range of machine and buffer parameters. For each of the random problems, we ran a simulation of 20 independent runs over 31,000 time units where the first 1,000 time units of each run were omitted as a warm-up period.

In the second part of the study, we varied in a systematic way

- the probabilities of failures and repairs,
- the size of the buffers, and
- the routing probabilities

in order to discover systematic effects. Here, the focus was primarily on the manufacturing system behavior. We started with smaller pure split or merge systems and later considered larger structures with loops where both split and merge operations were performed. Due to the systematic variation especially of buffer sizes, some of the systems were rather similar and, therefore, hard to compare by simulation results. In order to get production rate estimates with tight 95% confidence intervals, we increased the simulation time by running 20 independent runs over 110,000 time units. For each run, we omitted the first 10,000 time units as a warm-up period.

4.4.1.2 Generating Random Problems. To generate the random problems, we used parts of a procedure proposed in Burman (1995, p. 92-94) in the context of flow line analysis. In what follows, let RAN denote a pseudo random number generated by a call to the rnd() function in Visual Basic. This random number generator returns a number uniformly distributed between 0 and 1. Each reference to RAN_h in the formulas below represents a different value of this variable that is obtained by a separate call h of rnd(). Following Burman's approach,[6] we first generated a single random number

$$x = 1 + (9RAN_1) \tag{4.77}$$

for a given structure by a single call to the random number generator. In the next step, a set of different random numbers

$$y_i = -(1 + RAN_{2,i}), \tag{4.78}$$

one for each machine M_i, was generated. We then computed repair probabilities

$$r_i = x^{y_i} \tag{4.79}$$

for each machine M_i. Due to this approach, the repair probabilities are between 0.01 and 1. They are similar in their order of magnitude for all machines in a structure. Given these repair probabilities, an average repair takes between 1 and 100 periods. This covers a wide range of possible repair times.

The failure probabilities p_i are generated in a way that results in isolated efficiencies between 50% and 99%:

---

[6] Burman (1995, p. 93)

$$p_i = r_i * 10^{-(0.66RAN_{3,i}+0.66RAN_{4,i}+0.66RAN_{5,i})} \qquad (4.80)$$

The isolated efficiency cannot be higher than 100%. An isolated efficiency below 50% says that a machine is down more often than it is up. We do not consider this to be a very realistic assumption.

In general, this results in systems that are roughly balanced with respect to the isolated efficiencies $e_i = \frac{r_i}{r_i+p_i}$ of the machines. This is desirable to avoid random cases with slow bottleneck machines that are usually easy to analyze but not very realistic. However, due to split and merge operations, the bottleneck issue gets more complicated than it was in Burman's study of purely linear flow lines. In two of the cases to be described below, additional adjustments were necessary to generate random cases that were not extremely unbalanced.

The physical buffer sizes $C_{i,j}$ in a well-designed system are more or less proportional to the number of parts produced during the average repair time of the machines immediately up- or downstream of the buffer. If the buffer sizes are much smaller, an average failure propagates quickly through the system as machines are blocked and starved long before the broken machine is repaired.

The numerical solution technique used for the two-machine model in the decomposition is based on the assumption that there are at least two physical buffer spaces $C_{i,j}$ *between* any two machines.

To this minimal buffer size $C_{i,j} = 2$ we randomly added up to three times the amount of space required for the parts produced during a failure of average duration at the adjacent machines:

$$C_{i,j} = 2 + MAX[\frac{1}{r_i}, \frac{1}{r_j}]3RAN_{6,i,j} \qquad (4.81)$$

Since the two-machine model by Gershwin and Schick includes the workspaces at the machines in the calculation of the available storage, we have an extended buffer size $N(i,j) = C_{i,j} + 2$ of at least 4 in the two-machine lines used in the decomposition.

4.4.2 Pure Split Structures

4.4.2.1 Structure S1. We first studied Structure S1 depicted in Figure 4.7 for 100 random cases using the approach described above. The numbers above the buffers are the buffer indices used to determine the evaluation sequence. The routing probability $d_{2,3}$ was set to 0.9. We asked for the production rate in the branch between Machines M_2 and M_3.

The 100 random cases were analyzed using both simulation and our approximation technique and they were sorted according to ascending simulated production rates. Figure 4.8 shows the simulated production rates for the 100 random cases. We do not display the approximated production rate estimates

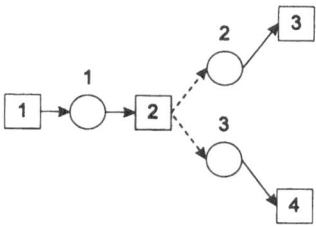

Fig. 4.7. Structure S1

in Figure 4.8 as they are very close to the simulation results. Instead, we display in Figure 4.9 the percentage error $(PR_{Ap} - PR_{Si})/PR_{Si}$ for each case. The algorithm converged in all 100 cases and the average over the absolute values of the relative errors was 0.81%. The probably somewhat less important buffer levels estimates are typically less accurate as we discuss in more detail below. Usually, the algorithm terminated after less then 20 iterations.

In the second part of the study for Structure S1, we analyzed 44 different sets of parameters. These 44 problems were arranged in 8 different problem classes. The performance measure was again the production rate in the branch between Machines M_2 and M_3 in Figure 4.7. In the first problem class, referred to as S1C1, we used the parameters in the upper part (first line) of Table 4.1. The results for different values of the routing probability $d_{2,3}$ are given in the lower part of Table 4.1 where BL denotes the relative buffer level deviation (described below) and PR_{Si} and PR_{Ap} denote simulated and approximated production rates, respectively.

Class S1C1: $C_{i,j} = 2, \forall(i,j); \; p_i = 0.01, r_i = 0.1, \forall i$						
Case	$d_{2,3}$	Iter.	BL [%]	PR_{Si}	PR_{Ap}	%
S1C1S1	0.95	13	6.809	.7375 ± .0024	.7532	2.1
S1C1S2	0.9	13	7.452	.7014 ± .0027	.7165	2.2
S1C1S3	0.7	13	8.172	.5505 ± .0016	.5620	2.1
S1C1S4	0.5	13	8.271	.3951 ± .0012	.4023	1.8
S1C1S5	0.3	12	8.163	.2357 ± .0011	.2409	2.2
S1C1S6	0.1	12	7.553	.0779 ± .0006	.0796	2.2
S1C1S7	0.05	12	6.838	.0388 ± .0003	.0396	2.1

Table 4.1. Results for Class S1C1 (Small Buffers)

In this problem class, the routing probability $d_{2,3}$ changed from 0.95 to 0.05. The last column in Table 4.1 reports the relative difference between the approximated and the simulated production rate for the branch between Machines M_2 and M_3. (It is not necessary to report results for more than one branch of the network since the *ratio* of the respective production rates in different branches is determined by the routing probabilities $d_{i,q}$ in the underlying model. For this reason, the accuracy of the production rate ap-

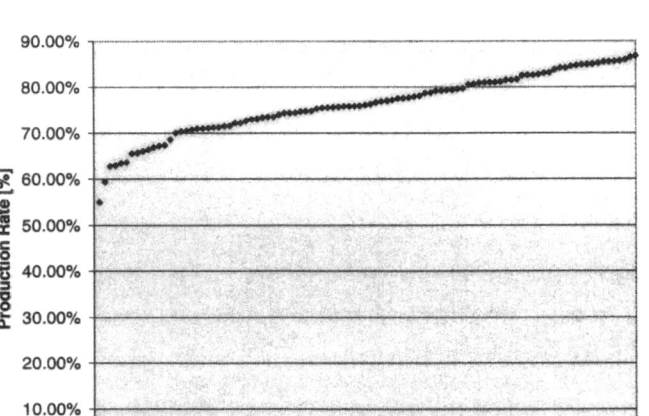

Fig. 4.8. Structure S1 - Simulated Production Rates for Random Problems

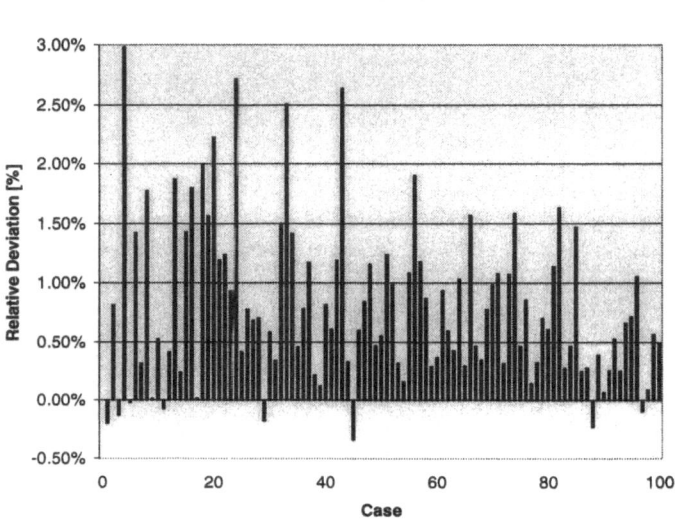

Fig. 4.9. Structure S1 - Percentage Errors for Random Problems

Class S1C2: $C_{i,j} = 8, \forall (i,j); p_i = 0.01, r_i = 0.1, \forall i$						
Case	$d_{2,3}$	Iter.	BL [%]	PR_{Si}	PR_{Ap}	%
S1C2S1	0.95	13	2.926	.7790 ± .0022	.7878	1.1
S1C2S2	0.9	13	2.944	.7441 ± .0024	.7502	0.8
S1C2S3	0.7	13	2.860	.5851 ± .0013	.5903	0.9
S1C2S4	0.5	13	2.939	.4201 ± .0012	.4232	0.7
S1C2S5	0.3	12	2.874	.2499 ± .0010	.2530	1.2
S1C2S6	0.1	12	2.884	.0820 ± .0004	.0834	1.6
S1C2S7	0.05	12	2.912	.0410 ± .0004	.0415	1.1

Table 4.2. Results for Class S1C2 (Larger Buffers)

proximation is identical for all branches of pure split structures.) We stopped the iterative approximation when the conservation of flow condition (4.2) was met to a relative accuracy of 0.0001 for each machine in the system for 10 consecutive iterations. That is, the approximation needed three iterations for Case S1C1S1 in Table 4.1 to meet the conservation of flow equation at an accuracy of 0.01% and then stayed within this level of accuracy for the next 10 iterations. (In most of the cases this led to an even higher accuracy with respect to the conservation of flow equation.)

The fourth column is a measure of the accuracy of the buffer level estimate. We determined for each buffer b the difference between the average buffer level from the decomposition approach $\bar{n}_{App}(b)$ and the the average buffer level from the simulation $\bar{n}_{Sim}(b)$. The absolute value of this difference is related to the extended buffer size $N(b)$. The average over the absolute values of these relative quantities is computed as follows

$$BL = \frac{1}{B} \sum_{b=1}^{B} \frac{|\bar{n}_{App}(b) - \bar{n}_{Sim}(b)|}{N(b)} \qquad (4.82)$$

where B is the number of buffers in the system.

The fifth column gives the average simulated production rate and the half-width of the respective 95% confidence interval.

For the given set of parameters, Cases S1C1S1 and S1C1S7 with routing probabilities $d_{2,3}$ of 0.95 and 0.05, respectively, are equivalent. The reason is that the perspective of an observer in Buffer $B_{2,3}$ of Case S1C1S1 is identical to those of an observer in Buffer $B_{2,4}$ of Case S1C1S7. Each of them sees 95% of the material entering 'its' buffer. For this reason, the decomposition method provided perfectly equivalent results. Now note the almost perfect symmetry of the production rate *errors* in the fifth row of Table 4.1. They suggest that the simulation did *also* provide almost perfectly equivalent results, yielding symmetrical deviations between the analytical and the simulation method. Since the simulated production rates are the result of a (pseudo-)random process, this is a basically a result of running the simulations for a very long time such that independent simulation runs for Cases

S1C1S1 and S1C1S7 produce almost identical results. However, note that for the equivalent Cases S1C1S3 and S1C1S5, the production rate errors differ slightly.

The results in Table 4.1 show that the buffer level estimate is less accurate than the production rate estimate, that the production rate estimate itself is very accurate and that the algorithm converges quickly for this problem class.

Class S1C3: $C_{i,j} = 2, \forall(i,j); p_i = 0.01, i = 1,2,4; r_i = 0.1, \forall i; d_{2,3} = 0.9$						
Case	p_3	Iter.	BL %	PR_{Si}	PR_{Ap}	%
S1C3S1	0.01	13	7.493	.7027 ± .0016	.7165	2.0
S1C3S2	0.04	13	6.285	.5874 ± .0023	.6100	3.8
S1C3S3	0.07	12	10.011	.5038 ± .0016	.5251	4.2
S1C3S4	0.1	13	12.140	.4415 ± .0015	.4586	3.9
S1C3S5	0.2	13	15.018	.3134 ± .0018	.3191	1.8

Table 4.3. Results for Class S1C3 (Small Buffers)

Class S1C4: $C_{i,j} = 8, \forall(i,j); p_i = 0.01, i = 1,2,4; r_i = 0.1, \forall i; d_{2,3} = 0.9$						
Case	p_3	Iter.	BL %	PR_{Si}	PR_{Ap}	%
S1C4S1	0.01	13	2.916	.7404 ± .0019	.7502	1.3
S1C4S2	0.04	14	2.359	.6473 ± .0027	.6613	2.2
S1C4S3	0.07	13	4.536	.5534 ± .0014	.5679	2.6
S1C4S4	0.1	13	5.601	.4817 ± .0017	.4907	1.9
S1C4S5	0.2	14	6.364	.3293 ± .0019	.3318	0.7

Table 4.4. Results for Class S1C4 (Larger Buffers)

Class S1C5: $C_{i,j} = 2, \forall(i,j); p_i = 0.01, i = 1,2,4; r_i = 0.1, \forall i; d_{2,3} = 0.1$						
Case	p_3	Iter.	BL %	PR_{Si}	PR_{Ap}	%
S1C5S1	0.01	12	7.453	.0779 ± .0005	.0796	2.2
S1C5S2	0.04	12	7.521	.0783 ± .0004	.0794	1.4
S1C5S3	0.07	12	7.506	.0774 ± .0004	.0791	2.3
S1C5S4	0.1	12	7.528	.0770 ± .0003	.0789	2.4
S1C5S5	0.2	12	7.585	.0764 ± .0004	.0780	2.1

Table 4.5. Results for Class S1C5 (Small Buffers)

In Class S1C2 we studied the impact of increasing all physical buffer sizes $C_{i,j}$ between machines from 2 to 8 and again varied the routing probabilities. The results in Table 4.2 are more accurate than those for smaller buffer sizes in Table 4.1. This is a typical pattern for decomposition approaches.

Class S1C6: $C_{i,j} = 8, \forall(i,j); p_i = 0.01, i = 1, 2, 4; r_i = 0.1, \forall i; d_{2,3} = 0.1$						
Case	p_3	Iter.	BL %	PR_{Si}	PR_{Ap}	%
S1C6S1	0.01	12	2.915	.0828 ± .0005	.0834	0.7
S1C6S2	0.04	12	2.996	.0826 ± .0004	.0834	0.9
S1C6S3	0.07	12	2.970	.0820 ± .0005	.0834	1.6
S1C6S4	0.1	12	2.948	.0823 ± .0004	.0833	1.2
S1C6S5	0.4	12	2.877	.0819 ± .0005	.0833	1.7

Table 4.6. Results for Class S1C6 (Larger Buffers)

Class S1C7: $C_{i,j} = 2, \forall(i,j); p_i = 0.01, i = 2, 3, 4; r_i = 0.1, \forall i; d_{2,3} = 0.9$						
Case	p_1	Iter.	BL %	PR_{Si}	PR_{Ap}	%
S1C7S1	0.01	13	7.473	.7010 ± .0021	.7165	2.2
S1C7S2	0.04	13	5.533	.5720 ± .0019	.5861	2.5
S1C7S3	0.07	13	5.927	.4841 ± .0019	.4935	1.9
S1C7S4	0.1	13	5.807	.4204 ± .0014	.4257	1.2
S1C7S5	0.2	13	4.504	.2888 ± .0015	.2913	0.9

Table 4.7. Results for Class S1C7 (Small Buffers)

Class S1C8: $C_{i,j} = 8, \forall(i,j); p_i = 0.01, i = 2, 3, 4; r_i = 0.1, \forall i; d_{2,3} = 0.9$						
Case	p_1	Iter.	BL %	PR_{Si}	PR_{Ap}	%
S1C8S1	0.01	13	2.967	.7431 ± .0020	.7502	1.0
S1C8S2	0.04	13	3.652	.6092 ± .0024	.6158	1.1
S1C8S3	0.01	13	3.391	.5104 ± .0024	.5153	0.9
S1C8S4	0.1	13	3.021	.4393 ± .0017	.4418	0.6
S1C8S5	0.2	13	2.160	.2979 ± .0015	.2981	0.1

Table 4.8. Results for Class S1C8 (Larger Buffers)

We next varied the failure probability of Machine M_3 from 0.01 to 0.2 for physical buffer sizes $C_{i,j}$ of 2 and 8 and the fixed routing probability $d_{2,3} = 0.9$. The results in Tables 4.3 and 4.4 indicate that the production rate estimate is still quite precise. The buffer level estimate deteriorates as the system becomes more and more unbalanced.

Problem Classes S1C5 and S1C6 are like Problem Classes S1C3 and S1C4 except for a different routing probability $d_{2,3} = 0.1$ instead of 0.9. The results in Tables 4.5 and 4.6 indicate that the algorithmic behavior is very similar with respect to accuracy. However, the manufacturing system behavior is completely different: In Classes S1C3 and S1C4, Machine M_3 becomes a bottleneck as its failure probability increases up to 0.2. This is because Machine M_3 has to process 90% of the parts that leave Machine M_2. When we reduce the routing probability $d_{2,3}$ to 0.1, Machine M_3 does not become a bottleneck for failure probabilities up to 0.2 and the production rate in Tables 4.5 and 4.6 remains almost unchanged.

In Problem Classes S1C7 and S1C8, we varied the failure probability of Machine M_1 and set the routing probability $d_{2,3}$ back to 0.9. Since all parts have to be processed by machine M_1 and we start with a balanced system, we observe in Tables 4.7 and 4.8 a strong reduction of the production rate that is precisely predicted by the approximation approach.

4.4.2.2 Structure S2. In a next step we added another machine downstream of Machine M_2. The resulting Structure S2 depicted in Figure 4.10 was evaluated for different routing probabilities. We asked for the production rate in the branch between Machines M_2 and M_3. We again generated 100 random problems that were sorted according to simulated production rates (see Figure 4.11). The routing probabilities were $d_{2,3} = 0.9$, $d_{2,4} = 0.05$, and $d_{2,4} = 0.05$. The percentage errors for each case are given in Figure 4.12. The algorithm converged in 99 out 100 cases and the average absolute value of the percentage error over the 99 random cases was 0.60%. In the one case where the algorithm did not converge, all repair probabilities were very close to 1.

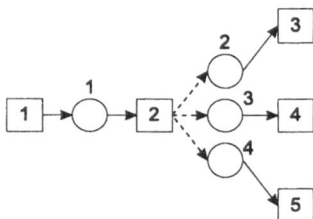

Fig. 4.10. Structure S2

Class S2C1: $C_{i,j} = 2, \forall(i,j)$; $p_i = 0.01, r_i = 0.1, \forall i$							
Case	$d_{2,3}$	$d_{2,4}$	$d_{2,5}$	Iter.	BL %	PR_{Si}	PR_{Ap}	%
S2C1S1	0.9	0.05	0.05	13	5.752	.7016 ± .0022	.7168	2.2
S2C1S2	0.8	0.1	0.1	13	6.035	.6283 ± .0016	.6414	2.1
S2C1S3	0.6	0.3	0.1	13	6.146	.4754 ± .0023	.4843	1.9

Table 4.9. Results for Class S2C1 (Small Buffers)

Class S2C2: $C_{i,j} = 8, \forall(i,j)$; $p_i = 0.01, r_i = 0.1, \forall i$							
Case	$d_{2,3}$	$d_{2,4}$	$d_{2,5}$	Iter.	BL %	PR_{Si}	PR_{Ap}	%
S2C2S1	0.9	0.05	0.05	13	2.488	.7430 ± .0015	.7502	1.0
S2C2S2	0.8	0.1	0.1	13	2.175	.6651 ± .0019	.6715	1.0
S2C2S3	0.6	0.3	0.1	12	2.184	.5048 ± .0012	.5081	0.7

Table 4.10. Results for Class S2C2 (Larger Buffers)

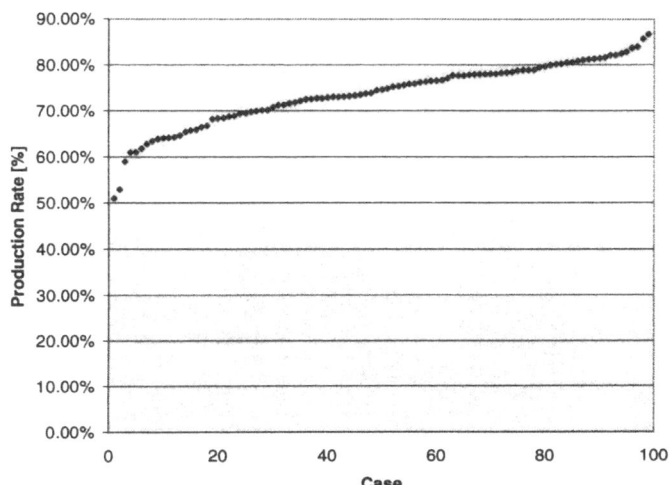

Fig. 4.11. Structure S2 - Simulated Production Rates for Random Problems

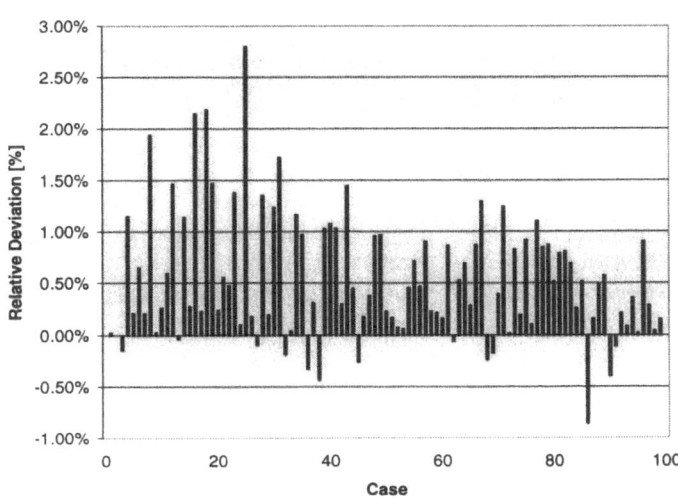

Fig. 4.12. Structure S2 - Percentage Errors for Random Problems

In the second part of the experiments for Structure S2, we briefly evaluated the effect of different routing probabilities. The results reported in Tables 4.9 and 4.10 show that the accuracy again increases as the buffer sizes increase. The decrease of the approximated production rate of Machine M_3 in Table 4.9 from 0.7168 to 0.4843 is almost proportional to the decrease of the routing probability $d_{2,3}$ from 0.9 to 0.6.

4.4.2.3 Structure S3. We next studied System S3 depicted in Figure 4.13, a network with multiple split operations. In the first part of the study, we again generated 100 random problems. In all of these problems, we set the routing probabilities to $d_{2,3} = 0.9, d_{2,5} = 0.1, d_{5,6} = 0.5, d_{5,7} = 0.5$ and asked for the production rate in the branch between Machines M_3 and M_4.

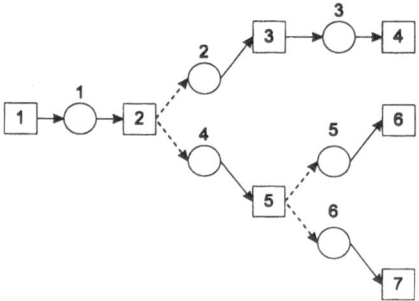

Fig. 4.13. Structure S3

Since the procedure to generate the failure and repair probabilities leads to machines with roughly similar isolated efficiencies, the random cases for Structure S3 are extremely unbalanced. Due to the routing probabilities, the average workload of Machine M_4 is 18 times as high as the workload of Machines M_6 or M_7. Since we assume identical processing times at all machines and generate similar isolated efficiencies, Machines M_6 and M_7 are idle most of the time. We created these artificial problems merely to show the limits of the numerical method. Out of the 100 random problems, the decomposition approach was able to analyze 83. It failed to converge in 17 cases. In Figure 4.14 we display the simulated production rate estimates for these 83 cases, sorted according to the simulated production rate. Figure 4.15 gives the respective percentage errors of the decomposition approach. The average over the 83 absolute values of the percentage error was 1.26%. That is, the algorithm was less reliable for this larger and unbalanced system, but when it converged the results were still accurate.

We briefly studied different routing probabilities with a more evenly distributed routing of the parts, for smaller as well as for larger buffer sizes. The parameters and the respective results are given in Tables 4.11 and 4.12 for the production rate in the branch between Machines M_3 and M_4.

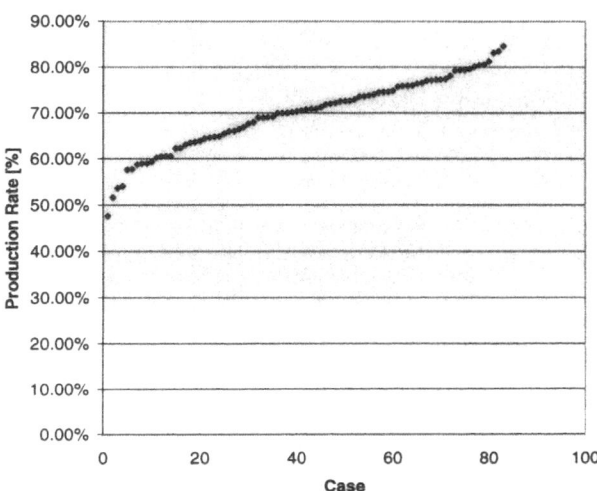

Fig. 4.14. Structure S3 - Simulated Production Rates for Random Problems

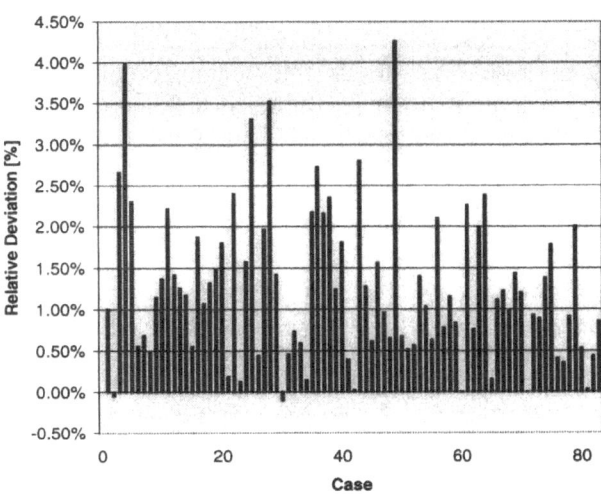

Fig. 4.15. Structure S3 - Percentage Errors for Random Problems

Class S3C1:
$C_{i,j} = 2, \forall(i,j)$
$p_i = 0.01, i = 1, 2, 3, 4; p_i = 0.1, i = 5, 6, 7; r_i = 0.1, \forall i$

Case	$d_{2,3}$	$d_{2,5}$	$d_{5,6}$	$d_{5,7}$	Iter.	BL %	PR_{Si}	PR_{Ap}	%
S3C1S1	0.9	0.1	0.5	0.5	14	5.616	.6578 ± .0025	.6813	3.6
S3C1S2	0.5	0.5	0.5	0.5	14	6.213	.2876 ± .001	.3022	5.1

Table 4.11. Results for Class S3C1 (Small Buffers)

Class S3C2:
$C_{i,j} = 8, \forall(i,j)$
$p_i = 0.01, i = 1, 2, 3, 4; p_i = 0.1, i = 5, 6, 7; r_i = 0.1, \forall i$

Case	$d_{2,3}$	$d_{2,5}$	$d_{5,6}$	$d_{5,7}$	Iter.	BL %	PR_{Si}	PR_{Ap}	%
S3C2S1	0.9	0.1	0.5	0.5	14	3.041	.7254 ± .0016	.7398	2.0
S3C2S2	0.5	0.5	0.5	0.5	14	2.611	.3694 ± .0016	.3766	2.0

Table 4.12. Results for Class S3C2 (Larger Buffers)

Note that the isolated efficiencies of the machines are not identical. The results suggest that the deviation between simulated and approximated production rates increases slightly as we add another split operation, but the decomposition approach is still fairly accurate.

4.4.3 Pure Merge Networks

4.4.3.1 Structure M1. To study networks with merge operations, we started with Structure M1 in Figure 4.16. In this structure, Machine M_3 always tries to take the next part from its priority one input buffer between Machines M_1 and M_3. We asked for the production rate in the branch between Machines M_3 and M_4. Out of the 100 random problems generated for Structure M1, the decomposition algorithm was able to solve 91. A general observation is that the algorithm may fail to converge if the machine performing the merge operation is almost never starved and the starvation probability hence approaches zero. A possible explanation is that this leads to numerical problems in the flow rate-idle time equation (4.15) where a division by this starvation probability is performed.

The simulated production rates for these 91 cases are displayed in Figure 4.17, sorted according to simulated production rates. Figure 4.18 gives the respective percentage errors.

The average absolute value of the percentage error was 0.29% in the 91 cases where the algorithm converged. The convergence reliability for merge systems is in general smaller than for split systems. However, the production rate estimate for the merge machine is usually relatively accurate whenever the algorithm converges.

We then asked for systematic effects with respect to the manufacturing system behavior. In the first two classes of systems, we analyzed the impact

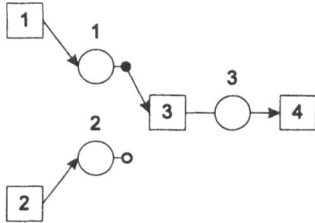

Fig. 4.16. Structure M1

of the failure probability p_1 and the buffer size $C_{i,j}$. The results in Tables 4.13 and 4.14 indicate the production rate estimate for the branch between Machines M_3 and M_4 is still quite accurate. The accuracy increases as the buffer sizes increase. We also see that there is only a modest decrease in the production rates of Machines M_3 and M_4 as the failure probability p_1 increases from 0.001 to 0.7. This is because of the high isolated efficiency of Machine M_2: As Machine M_1 fails more frequently, Machine M_2 is blocked less often, produces more and eventually carries the major part of the workload at this production stage.

Class M1C1: $C_{i,j} = 2, \forall (i,j)$; $p_i = 0.01, i = 2,3,4$; $r_i = 0.1, \forall i$						
Case	p_1	Iter.	BL %	PR_{Si}	PR_{Ap}	%
M1C1S1	0.001	21	24.771	.8347 ± .0022	.8409	0.7
M1C1S2	0.01	18	20.318	.8326 ± .0026	.8407	1.0
M1C1S3	0.04	17	16.085	.8271 ± .0025	.8380	1.3
M1C1S4	0.07	17	14.382	.8229 ± .0023	.8351	1.5
M1C1S5	0.1	17	14.094	.8160 ± .0025	.8323	2.0
M1C1S6	0.4	15	13.029	.7977 ± .0025	.8211	2.9
M1C1S7	0.7	15	10.758	.7898 ± .0026	.8143	3.1

Table 4.13. Results for Class M1C1 (Small Buffers)

Class M1C2: $C_{i,j} = 8, \forall (i,j)$; $p_i = 0.01, i = 2,3,4$; $r_i = 0.1, \forall i$						
Case	p_1	Iter.	BL %	PR_{Si}	PR_{Ap}	%
M1C2S1	0.001	22	9.872	.8502 ± .0016	.8562	0.7
M1C2S2	0.01	19	8.033	.8535 ± .0015	.8562	0.3
M1C2S3	0.04	18	8.215	.8500 ± .0024	.8557	0.7
M1C2S4	0.07	17	6.417	.8488 ± .0018	.8545	0.7
M1C2S5	0.1	17	8.163	.8448 ± .0021	.8538	1.1
M1C2S6	0.4	15	6.425	.8371 ± .0023	.8473	1.2
M1C2S7	0.7	15	4.118	.8294 ± .0019	.8436	1.7

Table 4.14. Results for Class M1C2 (Larger Buffers)

M1: Random Cases

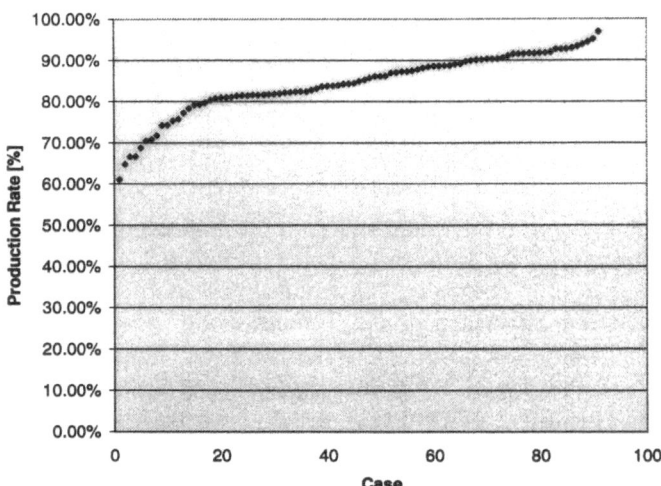

Fig. 4.17. Structure M1 - Simulated Production Rates for Random Problems

M1: Random Cases

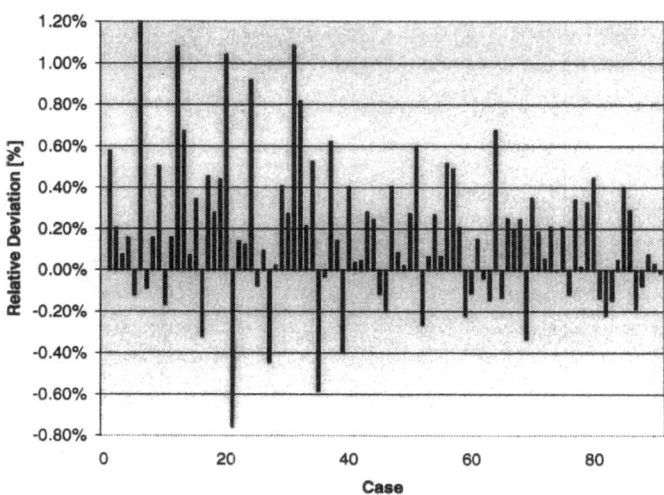

Fig. 4.18. Structure M1 - Percentage Errors for Random Problems

Line	BL_{Si}	BL_{Ap}	%	PR_{Si}	PR_{Ap}	%
(1 , 3)	2.15 ± .002	3.06	22.8	.8277 ± .0024	.8313	0.4
(2 , 3)	2.00 ± .000	3.98	49.5	.0070 ± .0005	.0097	38.9
(3 , 4)	1.92 ± .004	2.00	2.0	.8347 ± .0022	.8409	0.7

Table 4.15. Results for Case M1C1S1 (Failure Probability $p_1 = 0.001$)

Line	BL_{Si}	BL_{Ap}	%	PR_{Si}	PR_{Ap}	%
(1 , 3)	1.39 ± .005	1.68	7.2	.5173 ± .0025	.4822	-6.8
(2 , 3)	2.01 ± .002	3.41	35.2	.3056 ± .0024	.3529	15.5
(3 , 4)	1.89 ± .004	1.92	0.8	.8229 ± .0023	.8351	1.5

Table 4.16. Results for Case M1C1S4 (Failure Probability $p_1 = 0.07$)

Line	BL_{Si}	BL_{Ap}	%	PR_{Si}	PR_{Ap}	%
(1 , 3)	0.40 ± .003	.71	7.8	.1214 ± .0006	.1177	-3.1
(2 , 3)	2.00 ± .005	2.87	21.7	.6684 ± .0027	.6966	4.2
(3 , 4)	1.82 ± .007	1.70	-2.8	.7898 ± .0026	.8143	3.1

Table 4.17. Results for Case M1C1S7 (Failure Probability $p_1 = 0.7$)

Tables 4.15, 4.16, and 4.17 give detailed results for failure probabilities 0.001, 0.07, and 0.7 and physical buffer sizes $C_{i,j} = 2$. (The results for the simulated buffer levels include the half-width of the 95% confidence intervals.)

We see in Table 4.15 that the simulated production rate of Machine M_2 is only 0.007 for $p_1 = 0.001$ and increases up to 0.6684 for $p_1 = 0.7$ (see Table 4.17). In this case (System M1C1S7), the isolated efficiency of Machine M_1 is only $\frac{0.1}{0.1+0.7} = 0.125$ which is close to the simulated production rate of 0.1214.

The accuracy of the production rate estimate is now no longer the same for all branches of the network. Tables 4.15 and 4.16 show that there may be high *relative* deviations for the *priority two input buffer* between Machines M_2 and M_3 if its production rate is very low in *absolute terms*. However, due to the low absolute values this appears to be only a minor problem in economic terms. It is interesting to compare some of the results for Structure M2 to results for two-machine lines. The production rate of a two-machine line consisting only of Machines M_3 and M_4 as given in Table 4.13 is 0.8409. It is an upper bound on the production rate of the system in Problem Class M1C1. Compare this upper bound with the result for Case M1C1S1 in Table 4.13: In this case, Machine M_3 is very rarely starved and the simulated production rate is very close to the upper bound. The approximated production rate also approaches this bound as Machine M_1 fails less frequently.

For the next two problem classes we varied p_2, the failure probability of the machine upstream of the priority two input buffer of Machine M_3. We see in Tables 4.18 and 4.19 that this does again lead to a slight reduction of

Class M1C3: $C_{i,j} = 2, \forall(i,j); \; p_i = 0.01, i = 1, 3, 4; r_i = 0.1, \forall i$						
Case	p_2	Iter.	BL %	PR_{Si}	PR_{Ap}	%
M1C3S1	0.001	18	20.565	.8331 ± .0030	.8409	0.9
M1C3S2	0.01	18	20.360	.8326 ± .0025	.8407	1.0
M1C3S3	0.04	19	19.470	.8277 ± .0021	.8398	1.5
M1C3S4	0.07	21	19.007	.8239 ± .0028	.8394	1.9
M1C3S5	0.1	20	19.059	.8190 ± .0016	.8399	2.6
M1C3S6	0.4	29	18.892	.8016 ± .0023	.8398	4.8
M1C3S7	0.7	32	18.484	.7975 ± .0019	.8394	5.3

Table 4.18. Results for Class M1C3 (Small Buffers)

Class M1C4: $C_{i,j} = 8, \forall(i,j); \; p_i = 0.01, i = 1, 3, 4; r_i = 0.1, \forall i$						
Case	p_2	Iter.	BL %	PR_{Si}	PR_{Ap}	%
M1C4S1	0.001	19	7.481	.8539 ± .0013	.8562	0.3
M1C4S2	0.01	19	8.002	.8503 ± .0022	.8562	0.7
M1C4S3	0.04	25	7.668	.8511 ± .0021	.8560	0.6
M1C4S4	0.07	28	7.262	.8502 ± .0016	.8557	0.6
M1C4S5	0.1	27	6.487	.8473 ± .0020	.8552	0.9
M1C4S6	0.4	32	9.011	.8425 ± .0026	.8560	1.6
M1C4S7	0.7	28	12.313	.8385 ± .0022	.8562	2.1

Table 4.19. Results for Class M1C4 (Larger Buffers)

the simulated production rate that is not very accurately predicted by the decomposition approach.

Class M1C5: $C_{i,j} = 2, \forall(i,j); \; p_i = 0.01, i = 3, 4; r_i = 0.1, \forall i$						
Case	$p_1 = p_2$	Iter.	BL %	PR_{Si}	PR_{Ap}	%
M1C5S1	0.001	21	24.845	.8352 ± .0028	.8409	0.7
M1C5S2	0.01	18	20.284	.8323 ± .0018	.8407	1.0
M1C5S3	0.04	18	12.561	.8079 ± .0024	.8283	2.5
M1C5S4	0.07	17	13.666	.7650 ± .0018	.8052	5.3
M1C5S5	0.1	18	7.272	.7204 ± .0023	.7567	5.0
M1C5S6	0.4	16	5.949	.3812 ± .0011	.3897	2.2
M1C5S7	0.7	15	4.135	.2468 ± .0007	.2488	0.8

Table 4.20. Results for Class M1C5 (Small Buffers)

In Classes M1C1 to M1C4, Machines M_1 and M_2 had a joint isolated production rate that was higher than the isolated production rates of Machines M_3 and M_4. We next studied the opposite situation and increased the failure probabilities p_1 and p_2 simultaneously in order to make these machines the bottleneck. Tables 4.20 and 4.21 show that there is now a strong decrease of the production rate. In Case M1C6S7 in Table 4.21, both bottleneck machines fail so often that they are almost never blocked and the simulated

Class M1C6: $C_{i,j} = 8, \forall(i,j); p_i = 0.01, i = 3,4; r_i = 0.1, \forall i$						
Case	$p_1 = p_2$	Iter.	BL %	PR_{Si}	PR_{Ap}	%
M1C6S1	0.001	22	9.868	.8530 ± .0021	.8562	0.4
M1C6S2	0.01	19	8.040	.8509 ± .0023	.8562	0.6
M1C6S3	0.04	24	5.285	.8437 ± .0016	.8527	1.1
M1C6S4	0.07	25	2.883	.8220 ± .0018	.8354	1.6
M1C6S5	0.1	26	4.960	.7873 ± .0020	.8128	3.2
M1C6S6	0.4	16	3.741	.3983 ± .0014	.3987	0.1
M1C6S7	0.7	14	1.816	.2499 ± .0007	.2500	0.0

Table 4.21. Results for Class M1C6 (Larger Buffers)

production rate of the network is very close to the sum of the two isolated production rates of Machines M_1 and M_2 of 0.25.

Class M1C7: $C_{i,j} = 2, \forall(i,j); p_i = 0.01, i = 1,2,3; r_i = 0.1, \forall i$						
Case	p_4	Iter.	BL %	PR_{Si}	PR_{Ap}	%
M1C7S1	0.001	18	23.363	.8985 ± .0021	.9021	0.4
M1C7S2	0.01	18	20.312	.8330 ± .0026	.8407	0.9
M1C7S3	0.04	20	27.667	.6687 ± .0029	.6758	1.1
M1C7S4	0.07	19	30.100	.5573 ± .0024	.5635	1.1
M1C7S5	0.1	19	30.865	.4801 ± .0026	.4830	0.6
M1C7S6	0.4	126	32.847	.1984 ± .0010	.1984	0.0
M1C7S7	0.7	****	****	.1245 ± .0006	****	****

Table 4.22. Results for Class M1C7 (Small Buffers)

Class M1C8: $C_{i,j} = 8, \forall(i,j); p_i = 0.01, i = 1,2,3; r_i = 0.1, \forall i$						
Case	p_4	Iter.	BL %	PR_{Si}	PR_{Ap}	%
M1C8S1	0.001	18	10.885	.9050 ± .0015	.9052	0.0
M1C8S2	0.01	19	8.052	.8518 ± .0022	.8562	0.5
M1C8S3	0.04	19	11.103	.6896 ± .0023	.6921	0.4
M1C8S4	0.07	20	12.156	.5731 ± .0026	.5761	0.5
M1C8S5	0.1	24	12.477	.4928 ± .0021	.4926	0.0
M1C8S6	0.4	****	****	.1993 ± .0008	****	****
M1C8S7	0.7	****	****	.1248 ± .0006	****	****

Table 4.23. Results for Class M1C8 (Larger Buffers)

In a last experiment for Structure M2, we varied the failure probability of Machine M_4, i.e. the bottleneck was now downstream of the merging machine. We found in Tables 4.22 and 4.23 that if the algorithm converged, the accuracy of the results was relatively high. However, for very low isolated production rates of Machine M_4 of $\frac{0.1}{0.1+0.4} = 0.2$ (System M1C8S6) or $\frac{0.1}{0.1+0.7} = 0.125$ (Systems M1C7S7 and M1C8S7), the algorithm failed to converge. In these situations, the probability to have *both* input buffers of the merging machine empty at the same time becomes very small and we conjecture that this causes problems in the flow rate-idle time equation as the probability that Machine M_3 is starved approaches 0.

4.4.3.2 Structure M2. We also examined the effect of having multiple merge operations and different priority assignments. The structure in Figure 4.19 contains two machines performing merge operations.

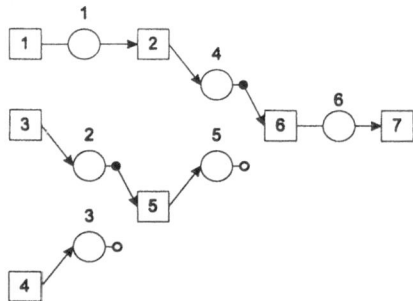

Fig. 4.19. Structure for Class M2C1

We first generated 100 problems with machines of roughly the same isolated efficiency. Due to the two merge machines in Structure M2, these systems were strongly unbalanced and both of the merge machines were very rarely starved. Out of these 100 random cases, the algorithm could solve only 25, i.e. it failed to converge for 75 cases out of 100. We conjectured that this was due to the fact that the systems tended to be very unbalanced. In order to test this hypothesis, we modified the failure probabilities of the machines upstream of merge machines in order to create more balanced systems. For the 100 random cases, we changed the failure probabilities of Machines M_1, M_2, and M_5 in a way that reduced their isolated efficiencies by exactly 50% and those of Machines M_3 and M_4 by exactly 75% unless this resulted in failure probabilities higher than 0.9. In these cases, we set the failure probabilities to 0.9. Due to the merge operations, the systems were now more evenly balanced.

All the repair probabilities and buffer sizes remained unchanged. Out of these 100 random problems, the algorithm could solve 85 and failed to converge for 15. The average absolute percentage error over the 85 solved cases was 2.27%. We conclude that structures with merge operations tend to be more difficult to solve with respect to convergence reliability. The production rates and percentage errors are displayed in Figures 4.20 and 4.21.

We next changed the priority assignments. The priority assignments for the input buffers in Figure 4.19 are as in Cases M2C1S1 and M2C1S2 in Table 4.24. The results for the production rate in the branch between Machines

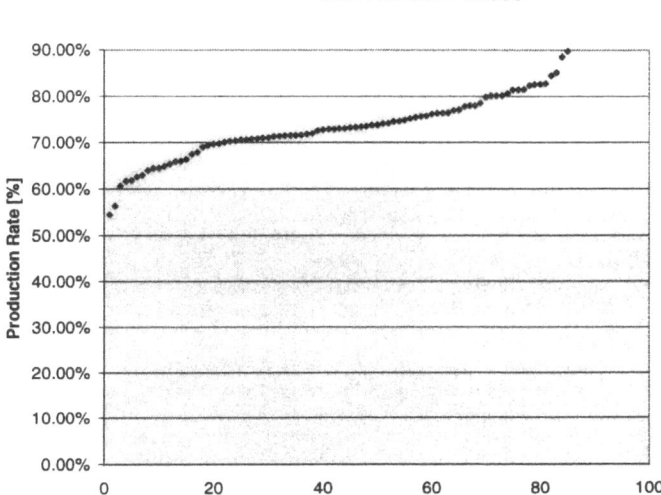

Fig. 4.20. Structure M2 - Simulated Production Rates for Random Problems

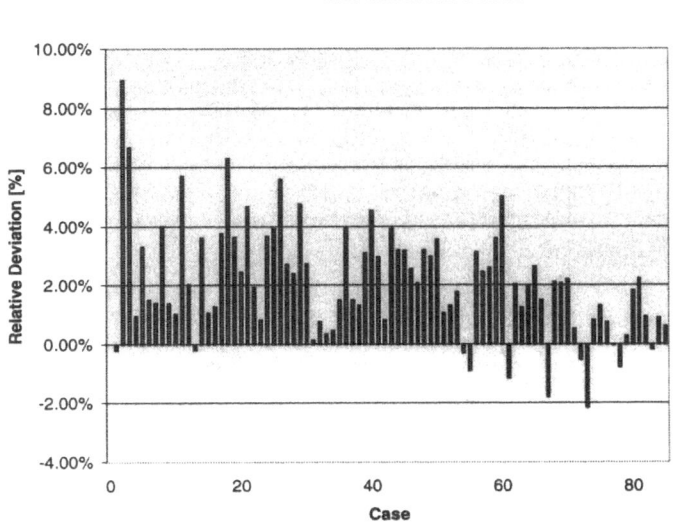

Fig. 4.21. Structure M2 - Percentage Errors for Random Problems

M_6 and M_7 in Table 4.25 suggest that the quality of the approximation deteriorates slightly as we add machines performing merge operations. However, for a physical buffer size $C_{i,j} = 8$, the maximum relative deviation between approximated and simulated production rate was only 4.3%. The priority assignment did not appear to have a major impact on the production rate.

$p_1 = p_2 = p_5 = 0.1, p_3 = p_4 = 0.2, p_6 = p_4 = 0.01$					
$r_i = 0.1, \forall i$					
System	$C_{i,j}, \forall(i,j)$	$B_{3,5}$	$B_{4,5}$	$B_{2,6}$	$B_{5,6}$
M2C1S1	2	Priority 1	Priority 2	Priority 1	Priority 2
M2C1S2	8	Priority 1	Priority 2	Priority 1	Priority 2
M2C2S1	2	Priority 1	Priority 2	Priority 2	Priority 1
M2C2S2	8	Priority 1	Priority 2	Priority 2	Priority 1
M2C3S1	2	Priority 2	Priority 1	Priority 1	Priority 2
M2C3S2	8	Priority 2	Priority 1	Priority 1	Priority 2
M2C4S1	2	Priority 2	Priority 1	Priority 2	Priority 1
M2C4S2	8	Priority 2	Priority 1	Priority 2	Priority 1

Table 4.24. Parameters for Cases M2C1S1 to M2C4S2

Case	Iter.	BL %	PR_{Si}	PR_{Ap}	%
M2C1S1	20	8.969	.6393 ± .0020	.6881	7.6
M2C1S2	18	8.024	.7470 ± .0018	.7786	4.2
M2C2S1	19	9.739	.6282 ± .0019	.6992	11.3
M2C2S2	27	7.862	.7407 ± .0017	.7619	2.9
M2C3S1	20	8.974	.6361 ± .0021	.6881	8.2
M2C3S2	18	8.009	.7463 ± .0019	.7786	4.3
M2C4S1	19	9.471	.6203 ± .0023	.6992	12.7
M2C4S2	27	7.435	.7323 ± .0019	.7619	4.0

Table 4.25. Results for Structure M2

We conclude that the algorithm is somewhat less accurate and reliable for pure merge networks than for pure split networks.

4.4.4 Structures with Loops

The last part of our numerical experiment was directed at more general structures that contain both split and merge operations and have loops. These loops in the flow of material occur if occasionally bad parts are produced, reworked, and sent back into the main line. Machines that perform split and merge operations are a building block of these more complex systems. Given that our decomposition works reasonably for pure split and pure merge structures, it remains to be shown that they can work together for a variety of different structures with random routing and loops.

4.4.4.1 Structure L1. We started with Structure L1, which is depicted in Figure 4.22. At Machine M_3, good parts are sent to Machine M_4 and finally leave the line. Bad parts are reworked at Machine M_5 and then sent to Machine M_2 where they have priority over the parts coming directly from Machine M_1. We asked for the production rate in the branch between Machines M_3 and M_4.

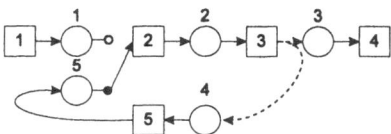

Fig. 4.22. Structure L1

We first tested the algorithm for 100 random cases with roughly similar efficiencies over all machines. In these random cases, we assumed that 90% of the parts are good and 10% are bad. Out of the 100 random cases, 99 could be solved analytically by our decomposition approach. These were sorted according to the simulated production rates and are displayed in Figures 4.23 and 4.24. The average absolute percentage error over the 99 cases was 0.99%.

Class L1C1			Class L1C3		
$p_i = 0.01, r_i = 0.1, \forall i, d_{3,4} = 0.9$			$p_i = 0.001, r_i = 0.01, \forall i, d_{3,4} = 0.9$		
System	$C_{i,j}, \forall(i,j)$	$N_{i,j}, \forall(i,j)$	System	$C_{i,j}, \forall(i,j)$	$N_{i,j}, \forall(i,j)$
L1C1S1	2	4	L1C3S1	2	4
$\vdots$	$\vdots$	$\vdots$	$\vdots$	$\vdots$	$\vdots$
L1C1S17	18	20	L1C3S17	18	20

Table 4.26. Parameters for Problem Classes L1C1 and L1C3

Class L1C2		Class L1C4	
$p_i = 0.01, r_i = 0.1, \forall i, C_{i,j} = 8, \forall(i,j)$		$p_i = 0.001, r_i = 0.01, \forall i, C_{i,j} = 8, \forall(i,j)$	
System	$d_{3,4}$	System	$d_{3,4}$
L1C2S1	0.9	L1C4S1	0.9
$\vdots$	$\vdots$	$\vdots$	$\vdots$
L1C2S9	0.1	L1C4S9	0.1

Table 4.27. Parameters for Problem Classes L1C2 and L1C4

We next studied systematically generated cases with different buffer sizes, frequencies of failures and repairs, and routing probabilities. Problem Classes

L1: Random Cases

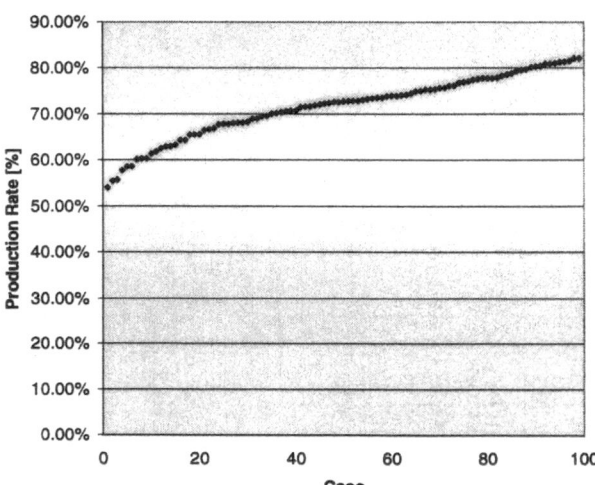

Fig. 4.23. Structure L1 - Simulated Production Rates for Random Problems

L1: Random Cases

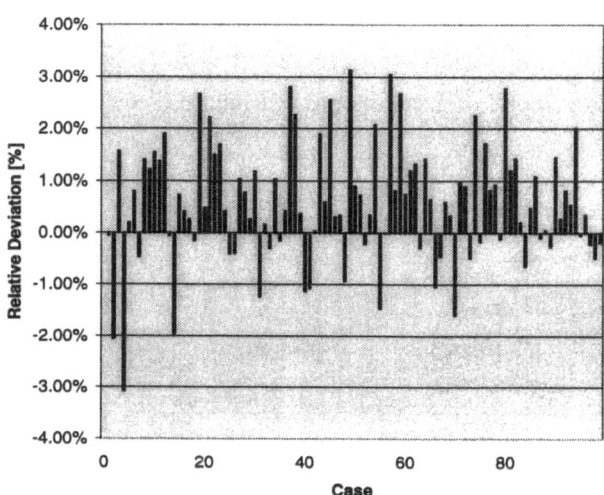

Fig. 4.24. Structure L1 - Percentage Errors for Random Problems

L1C1 and L1C3 in Table 4.26 differ with respect to the variability of failures and repairs. In both cases, all machines have an isolated production rate of about 0.9091. However, in Class L1C3 failures (and repairs) occur much less often than in Class L1C1. The expected duration $\frac{1}{r_i}$ of a repair is 10 time units in Class L1C1 and 100 time units in Class L1C3. Of the parts processed by Machine M_3, 90% were routed to Machine M_4.

The graph in Figure 4.25 for Class L1C1 shows that the production rate increases (from 0.666 to 0.767) as we add buffer spaces. The decomposition approach overestimates the production rate slightly and becomes very accurate for larger buffer sizes. Compare this to the graph for Class L1C3 in Figure 4.26 with the larger expected repair times: First, there is almost no increase of the (simulated) production rate as we add 16 spaces for each buffer. This is due to the very large repair times: If a failure occurs, buffers become full or empty long before the machine is repaired, even for 'larger' buffer sizes. This example shows how crucial the impact of infrequent failures with long repair times is. The graph also shows that the decomposition approach fails with respect to accuracy if buffer sizes are very small with respect to repair times.

We now study the impact of the routing probability and vary $d_{3,4}$ from 0.9 to 0.1 for a physical buffer sizes $C_{i,j} = 8$, i.e. $N(i,j) = 10$ in all two-machine models used in the decomposition. Figure 4.27 shows that there is an almost perfectly linear decrease of the production rate as $d_{3,4}$ decreases and $d_{3,5}$ increases. We also see that the approximation works reasonable over a wide range of values. Again, this picture changes completely with respect to accuracy if the variability increases: Figure 4.28 indicates that the decomposition should not be used due to large deviations if high variability and high feedback probabilities occur together.

Class L1C5	
$p_i = 0.01, r_i = 0.1, \forall i, C_{i,j} = 2, \forall (i,j)$	
System	$d_{3,4}$
L1C5S1	0.9
$\vdots$	$\vdots$
L1C5S9	0.1

Table 4.28. Parameters for Problem Class L1C5

In a last experiment for Structure L1 (see Table 4.28), we studied the case of small buffer sizes ($C_{i,j} = 2, \forall (i,j)$), low variability, and varying routing probabilities. The results in Figure 4.29 are satisfying for routing probabilities $d_{3,4}$ higher than 0.7. For the purpose of modeling rejection and rework, this may still be acceptable. However, given the results in Figure 4.28, it is not surprising that the approach failed to produce any useful results (which are

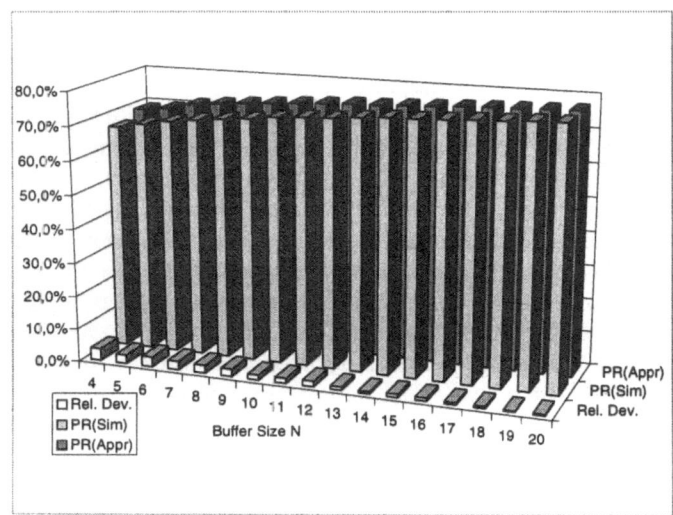

Fig. 4.25. Results for Class L1C1

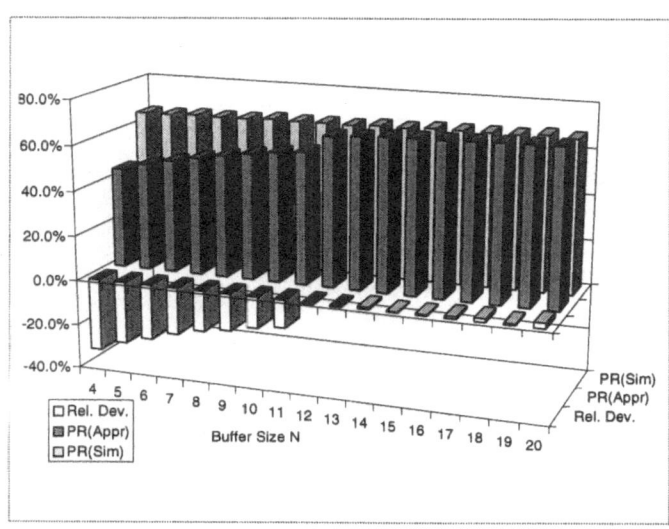

Fig. 4.26. Results for Class L1C3

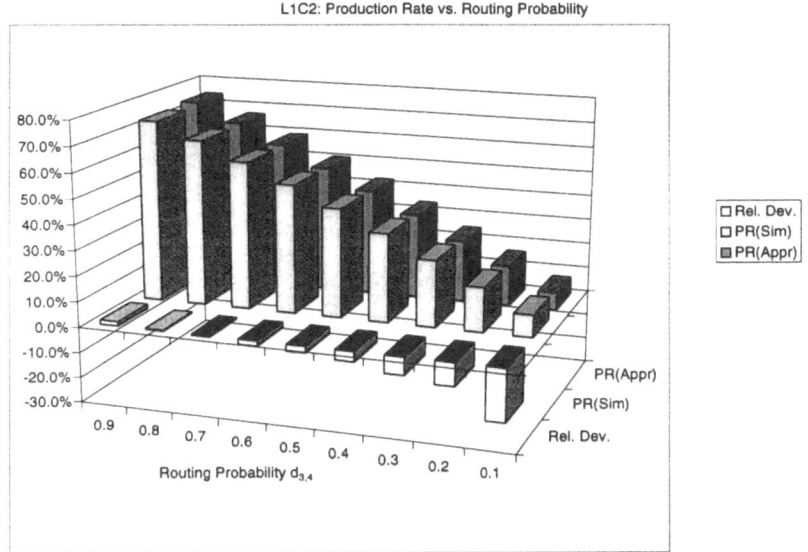

Fig. 4.27. Results for Class L1C2

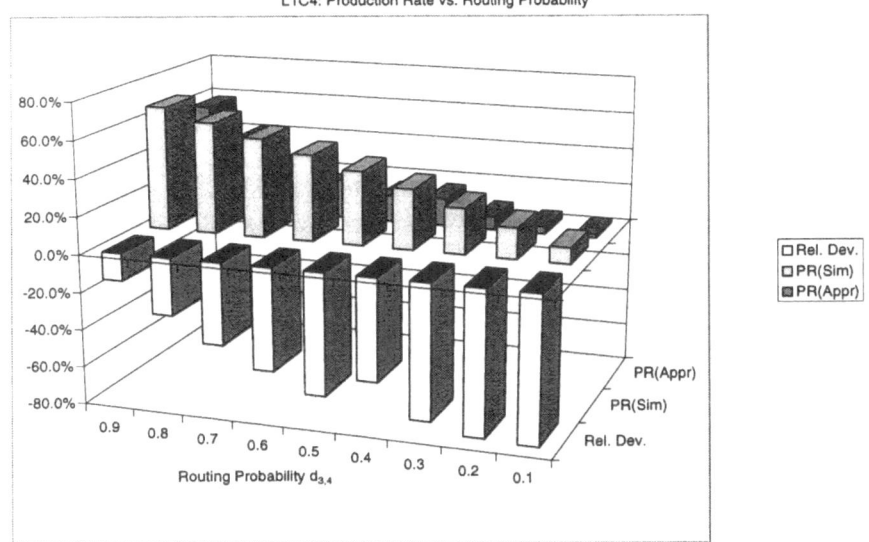

Fig. 4.28. Results for Class L1C4

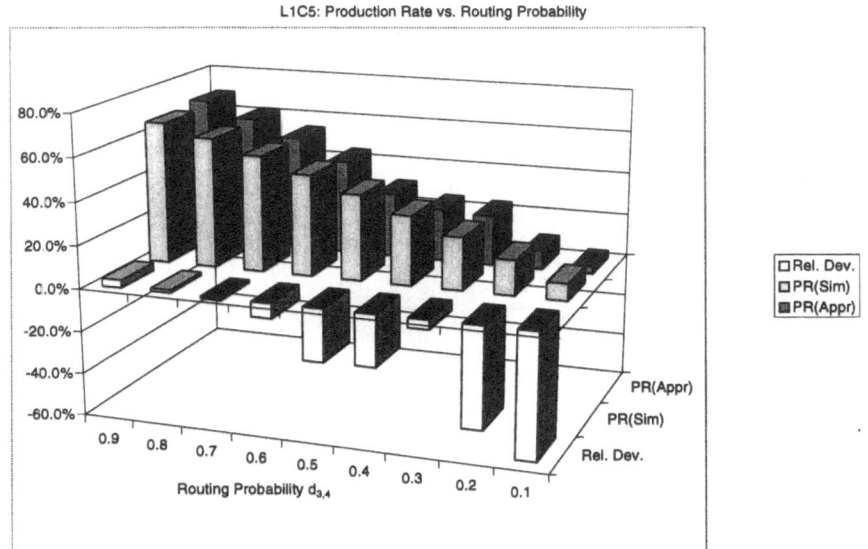

Fig. 4.29. Results for Class L1C5

not reported here) when we combined the small physical buffer size ($C_{i,j} = 2, \forall(i,j)$) with the high variability ($p_i = 0.001, r_i = 0.01, \forall i$).

4.4.4.2 Structure L2. To evaluate the effect of having a larger number of machines in the feedback loop, we next studied Structure L2 in Figure 4.30. Note that the loop in this structure consists of six machines as opposed to the three machines in Structure L1 in Figure 4.22. In the 100 random problems for this structure, we set the routing probabilities $d_{5,6} = 0.9$ and $d_{5,7} = 0.1$, i.e. we assumed that again 90% of the parts are good. The algorithm converged in 95 out of 100 cases and the average absolute value of the percentage error was 1.38% in these 95 cases. The results are depicted in Figures 4.31 and 4.32.

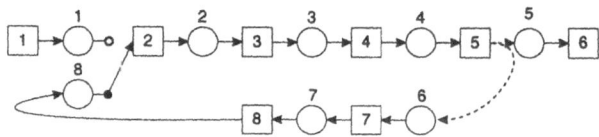

Fig. 4.30. Structure L2

In the systematic part of the study, we started with Problem Classes L2C1 and L2C3 (see Table 4.29) that are analogous to L1C1 and L1C3. Figure 4.33 suggests that the approximation is slightly less accurate than for

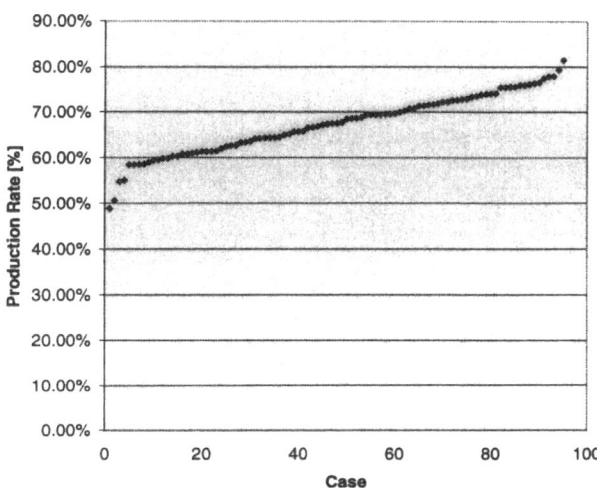

Fig. 4.31. Structure L2 - Simulated Production Rates for Random Problems

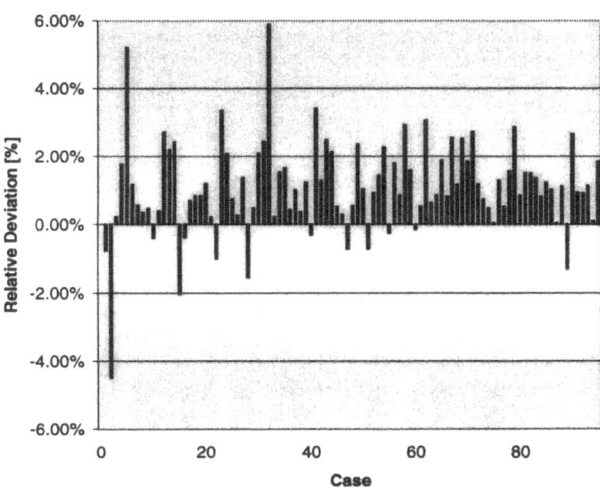

Fig. 4.32. Structure L2 - Percentage Errors for Random Problems

Structure L1 (compare with Figure 4.25). We observe a very similar increase of the production rate as we add buffer spaces. However, if we compare the results for the high-variability cases L2C3 in Figure 4.34 and L1C3 in Figure 4.26, we observe that for the larger network there are *fewer* cases where the decomposition resulted in very large deviations.

It appears to be rather difficult to describe the flow of material as seen by observers in a rework loop using virtual two-machine subsystems with geometrically distributed times to failure and to repair. In Section 5, a model with an additional parameter representing machine-specific processing rates will be analyzed. We will come back to Case L1C3 and show that this additional 'degree of freedom' helps to derive more accurate estimates even for 'small' loops.[7]

We again observe that in the high-variability cases of Class L2C3 adding only 16 spaces to each buffer has almost no impact on the production rate (see Figure 4.34) as the amount produced during an average failure is still more than five times higher than the maximum buffer content.

Class L2C1			Class L2C3		
$p_i = 0.01, r_i = 0.1, \forall i, d_{5,6} = 0.9$			$p_i = 0.001, r_i = 0.01, \forall i, d_{5,6} = 0.9$		
System	$C_{i,j}, \forall (i,j)$	$N_{i,j}, \forall (i,j)$	System	$C_{i,j}, \forall (i,j)$	$N_{i,j}, \forall (i,j)$
L2C1S1	2	4	L2C3S1	2	4
⋮	⋮	⋮	⋮	⋮	⋮
L2C1S17	18	20	L2C3S17	18	20

Table 4.29. Parameters for Problem Classes L2C1 and L2C3

Class L2C2		Class L2C4	
$p_i = 0.01, r_i = 0.1, \forall i, C_{i,j} = 8, \forall (i,j)$		$p_i = 0.001, r_i = 0.01, \forall i, C_{i,j} = 8, \forall (i,j)$	
System	$d_{5,6}$	System	$d_{5,6}$
L2C2S1	0.9	L2C4S1	0.9
⋮	⋮	⋮	⋮
L2C2S9	0.1	L2C4S9	0.1

Table 4.30. Parameters for Problem Classes L2C2 and L2C4

A similar picture emerged when we varied the routing probability $d_{5,6}$ (see Table 4.30). Compare Figures 4.35 and 4.27: The approximation appears to be more accurate for the larger loop. This does even hold for the high-variability case L2C4S1 with the highest routing probability $d_{5,6} = 0.9$ (see Figure 4.36).

---

[7] The comparison is given is Figure 5.2 on Page 186.

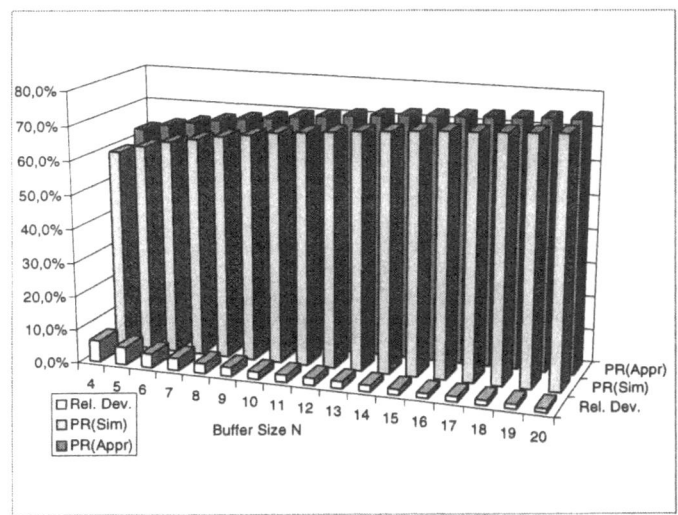

Fig. 4.33. Results for Class L2C1

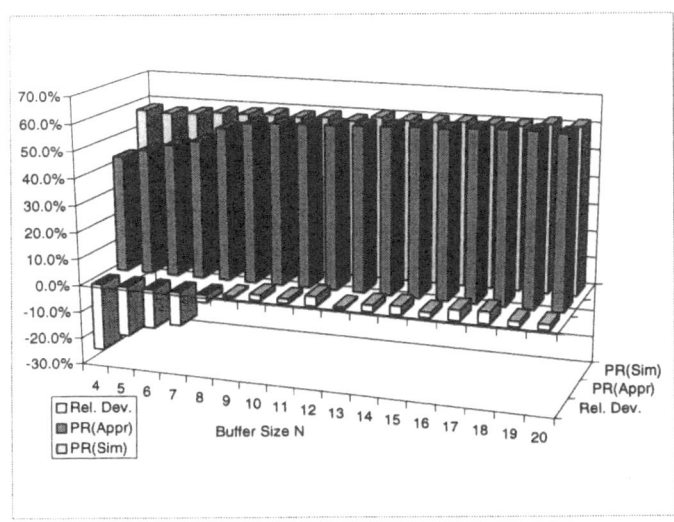

Fig. 4.34. Results for Class L2C3

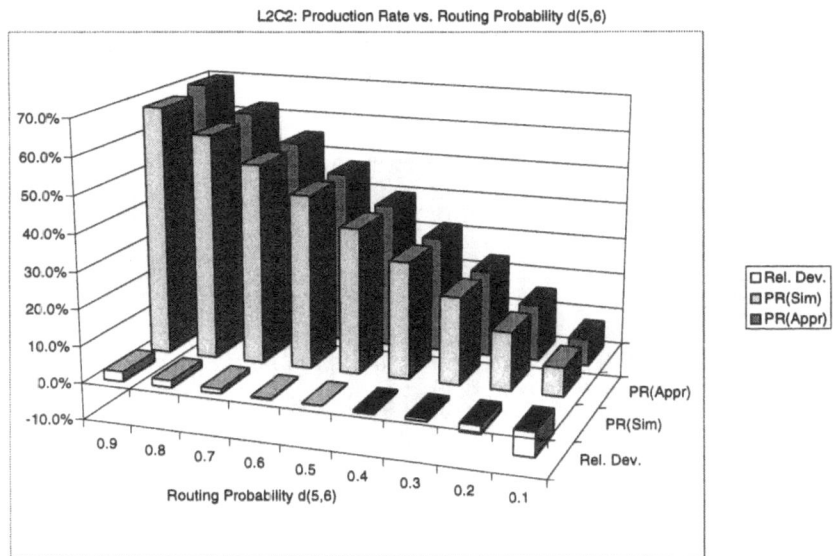

Fig. 4.35. Results for Class L2C2

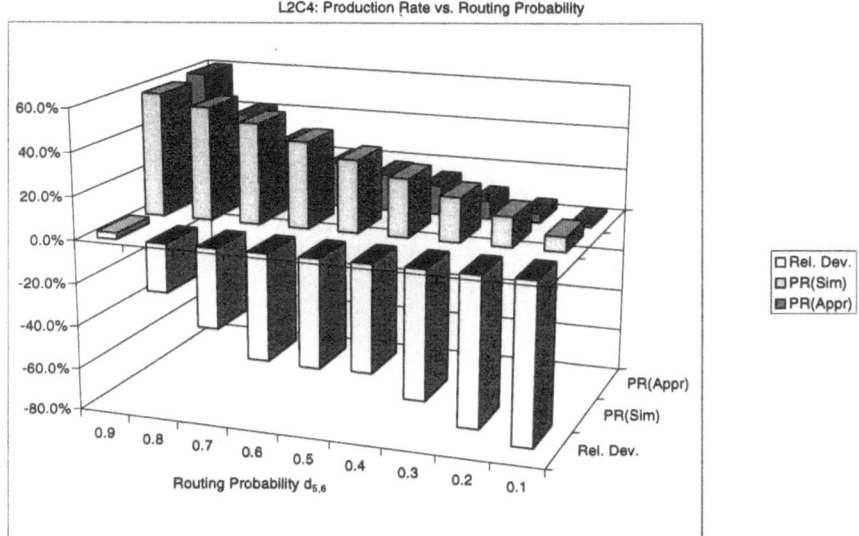

Fig. 4.36. Results for Class L2C4

Class L2C5	
$p_i = 0.01, r_i = 0.1, \forall i, C_{i,j} = 2, \forall (i,j)$	
System	$d_{3,4}$
L2C5S1	0.9
⋮	⋮
L2C5S9	0.1

Table 4.31. Parameters for Problem Class L2C5

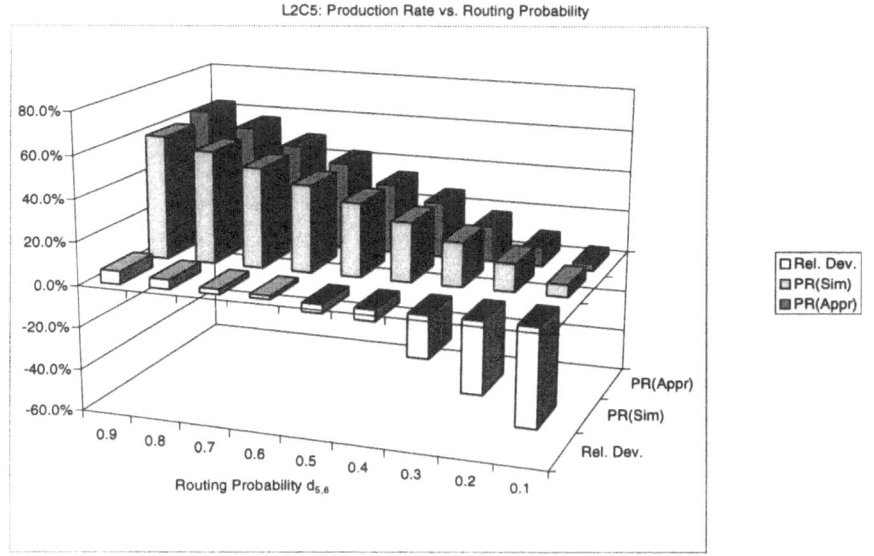

Fig. 4.37. Results for Class L2C5

In Class L2C5 we studied the impact of very small buffers (see Table 4.31). Figure 4.37 compared to Figure 4.29 shows again that the decomposition works better for the larger loop as there are fewer extreme deviations.

4.4.4.3 Structure L3. In Structures L1 and L2 we studied loops where a fraction of the material was randomly routed back to previous processing stages (feedback loop). We will now study two structures with feedforward loops using a similar experimental design. The first Structure L3 is depicted in Figure 4.38.

In the 100 random cases for this structure, we set the routing probabilities to $d_{2,3} = d_{2,4} = 0.5$. Out of the 100 random problems, the decomposition approach solved 78. The average error was 1.68%. This suggest that feedforward loops are more difficult to analyze than feedback loops. Other cases to be described below confirm this impression. The results are depicted in Figures 4.39 and 4.40.

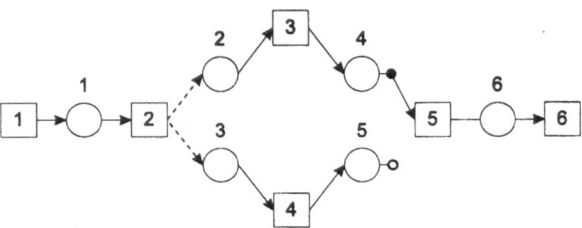

Fig. 4.38. Structure L3

In the next step, we again systematically varied buffer sizes and routing probabilities. Figures 4.41 and 4.42 give the results for the parameters in Table 4.32. They suggest that the decomposition results in less extreme deviations compared to what we saw for the feedback loops in Structures L1 and L3 (compare with Figures 4.25 and 4.26). Interestingly, the quality of the approximation does not become better as we add buffer spaces for this feedforward structure.

Class L3C1			Class L3C3		
$p_1 = p_2 = p_5 = p_6 = 0.01$			$p_1 = p_2 = p_5 = p_6 = 0.001$		
$p_3 = p_4 = 0.1$			$p_3 = p_4 = 0.01$		
$r_i = 0.1, \forall i, d_{2,3} = d_{2,4} = 0.5$			$r_i = 0.01, \forall i, d_{2,3} = d_{2,4} = 0.5$		
System	$C_{i,j}, \forall (i,j)$	$N_{i,j}, \forall (i,j)$	System	$C_{i,j}, \forall (i,j)$	$N_{i,j}, \forall (i,j)$
L3C1S1	2	4	L3C3S1	2	4
$\vdots$	$\vdots$	$\vdots$	$\vdots$	$\vdots$	$\vdots$
L3C1S17	18	20	L3C3S17	18	20

Table 4.32. Parameters for Problem Classes L3C1 and L3C3

Class L3C2		Class L3C4	
$p_1 = p_2 = p_5 = p_6 = 0.01$		$p_1 = p_2 = p_5 = p_6 = 0.001$	
$p_3 = p_4 = 0.1$		$p_3 = p_4 = 0.01$	
$r_i = 0.1, \forall i$		$r_i = 0.01, \forall i$	
$C_{i,j} = 8, \forall (i,j)$		$C_{i,j} = 8, \forall (i,j)$	
System	$d_{2,3}$	System	$d_{2,3}$
L3C2S1	0.1	L3C4S1	0.1
$\vdots$	$\vdots$	$\vdots$	$\vdots$
L3C2S9	0.9	L3C4S9	0.9

Table 4.33. Parameters for Problem Classes L3C2 and L3C4

We also varied the routing probability $d_{2,3}$, see Table 4.33. Figure 4.43 shows that the procedure worked reasonable for all values of $d_{2,3}$ but 0.9.

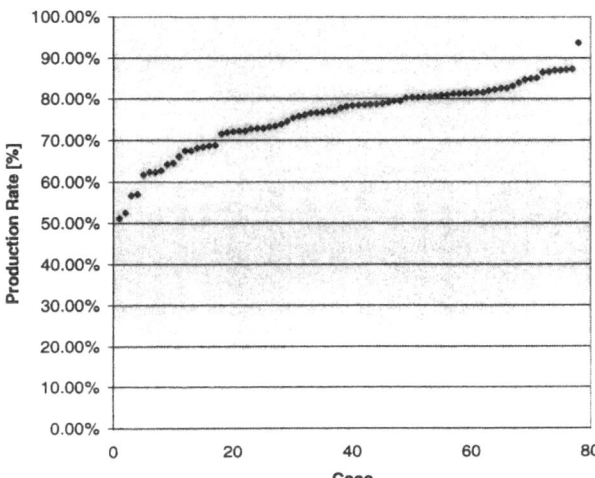

Fig. 4.39. Structure L3 - Simulated Production Rates for Random Problems

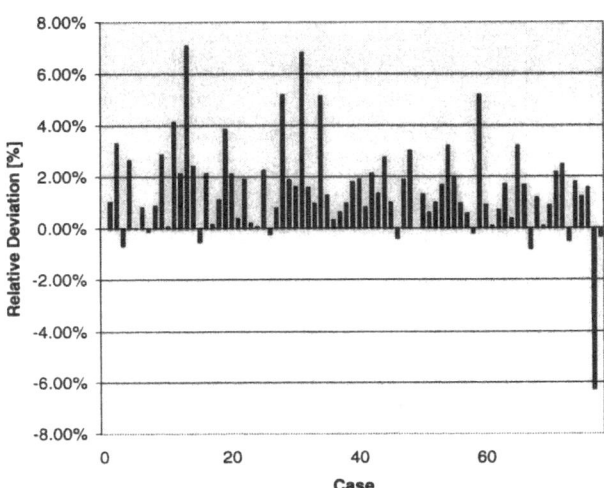

Fig. 4.40. Structure L3 - Percentage Errors for Random Problems

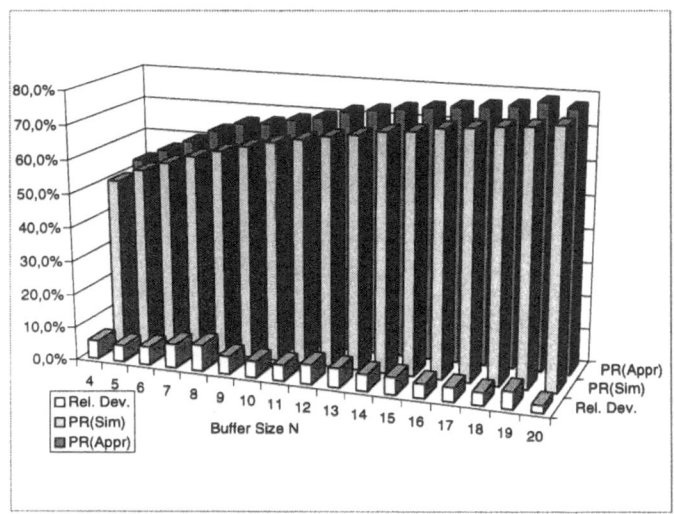

Fig. 4.41. Results for Class L3C1

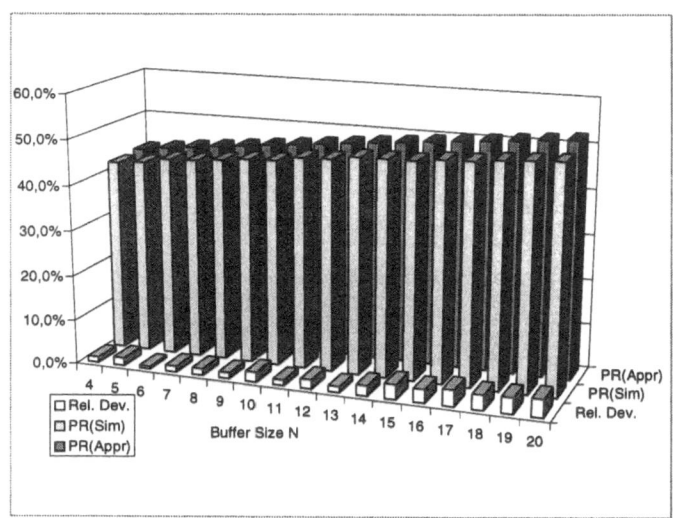

Fig. 4.42. Results for Class L3C3

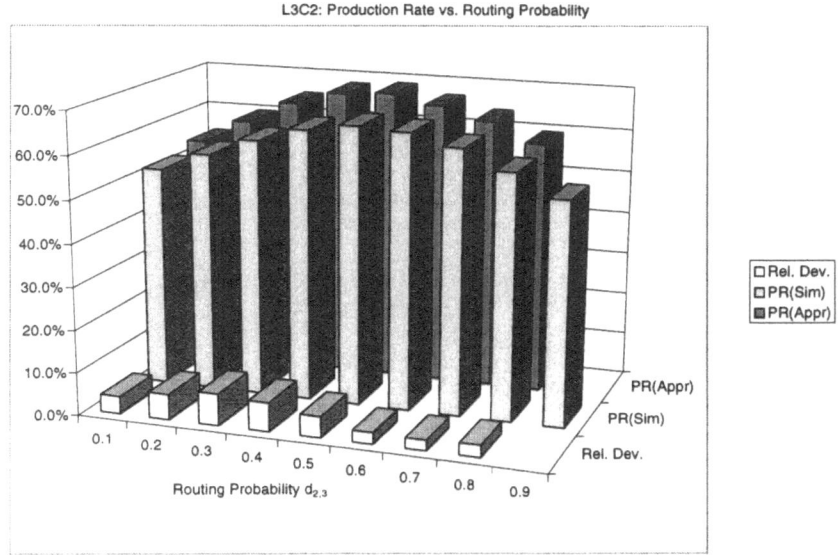

Fig. 4.43. Results for Class L3C2

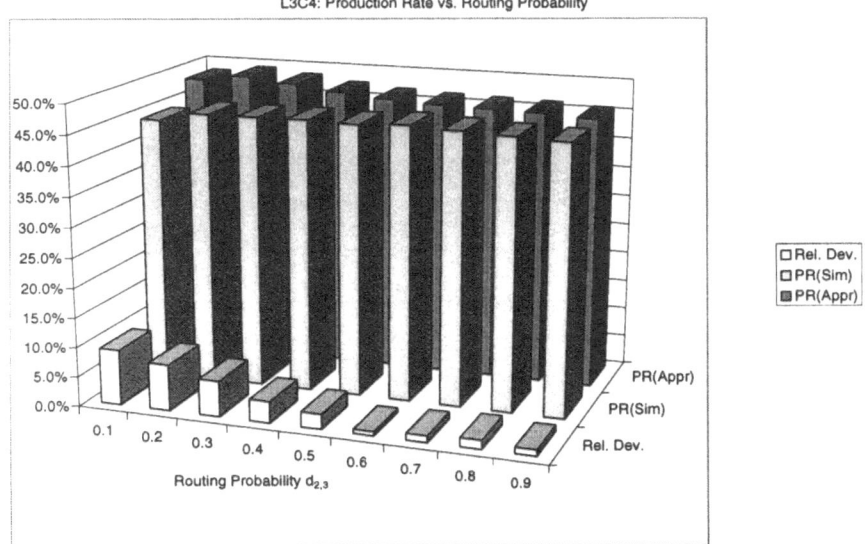

Fig. 4.44. Results for Class L3C4

In this case it failed to converge. Note that the simulated production rate is highest for $d_{2,3} = 0.5$. This is what we would expect for a system where the parallel branches have identical parameters. Unlike the results for structures with feedback loops, we have reasonable results even for the high-variability case in Figure 4.44.

4.4.4.4 Structure L4. We now study Structure L4 depicted in Figure 4.45 with four instead of two machines in the parallel branches. Remember that we observed a higher accuracy of the decomposition for larger feedback loops.

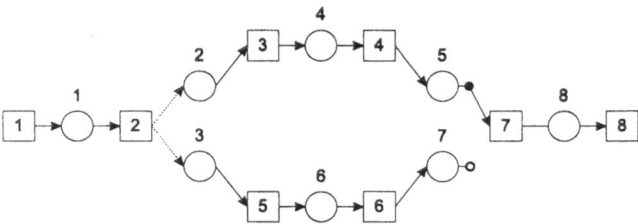

Fig. 4.45. Structure L4

Class L4C1			Class L4C3		
$p_1 = p_2 = p_7 = p_8 = 0.01$			$p_1 = p_2 = p_7 = p_8 = 0.001$		
$p_3 = p_4 = p_5 = p_6 = 0.1$			$p_3 = p_4 = p_5 = p_6 = 0.01$		
$r_i = 0.1, \forall i, d_{2,3} = d_{2,5} = 0.5$			$r_i = 0.01, \forall i, d_{2,3} = d_{2,5} = 0.5$		
System	$C_{i,j}, \forall(i,j)$	$N_{i,j}, \forall(i,j)$	System	$C_{i,j}, \forall(i,j)$	$N_{i,j}, \forall(i,j)$
L4C1S1	2	4	L4C3S1	2	4
$\vdots$	$\vdots$	$\vdots$	$\vdots$	$\vdots$	$\vdots$
L4C1S17	18	20	L4C3S17	18	20

Table 4.34. Parameters for Problem Classes L4C1 and L4C3

Class L4C2		Class L4C4	
$p_1 = p_2 = p_7 = p_8 = 0.01$		$p_1 = p_2 = p_7 = p_8 = 0.001$	
$p_3 = p_4 = p_5 = p_6 = 0.1$		$p_3 = p_4 = p_5 = p_6 = 0.01$	
$r_i = 0.1, \forall i$		$r_i = 0.01, \forall i$	
$C_{i,j} = 8, \forall(i,j)$		$C_{i,j} = 8, \forall(i,j)$	
System	$d_{2,3}$	System	$d_{2,3}$
L4C2S1	0.1	L4C4S1	0.1
$\vdots$	$\vdots$	$\vdots$	$\vdots$
L4C2S9	0.9	L4C4S9	0.9

Table 4.35. Parameters for Problem Classes L4C2 and L4C4

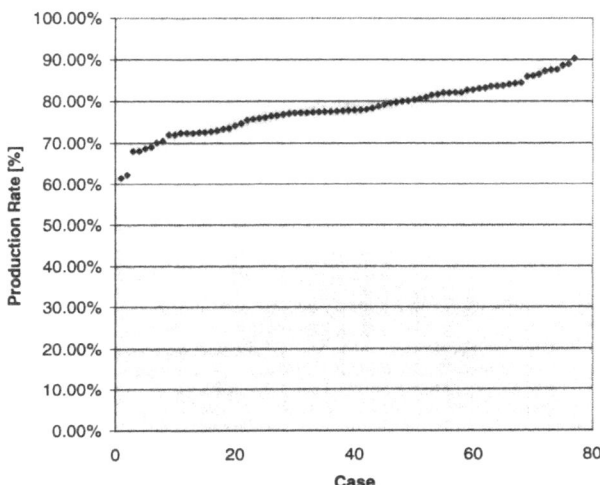

Fig. 4.46. Structure L4 - Simulated Production Rates for Random Problems

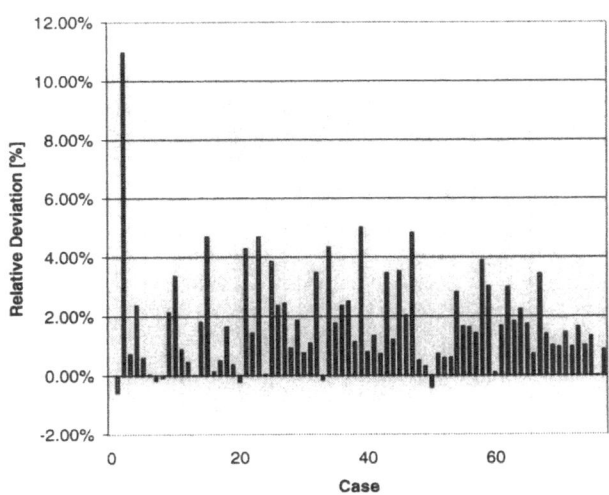

Fig. 4.47. Structure L4 - Percentage Errors for Random Problems

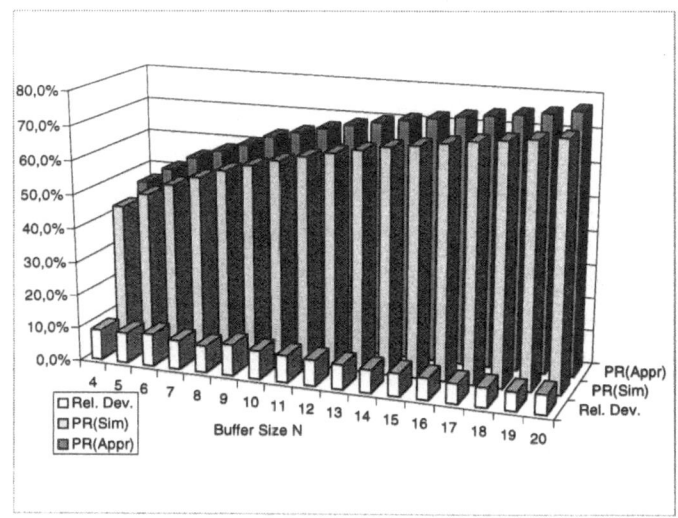

Fig. 4.48. Results for Class L4C1

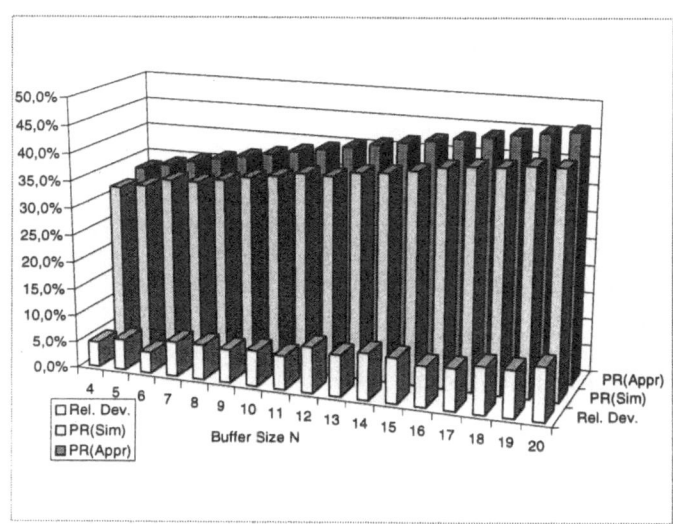

Fig. 4.49. Results for Class L4C3

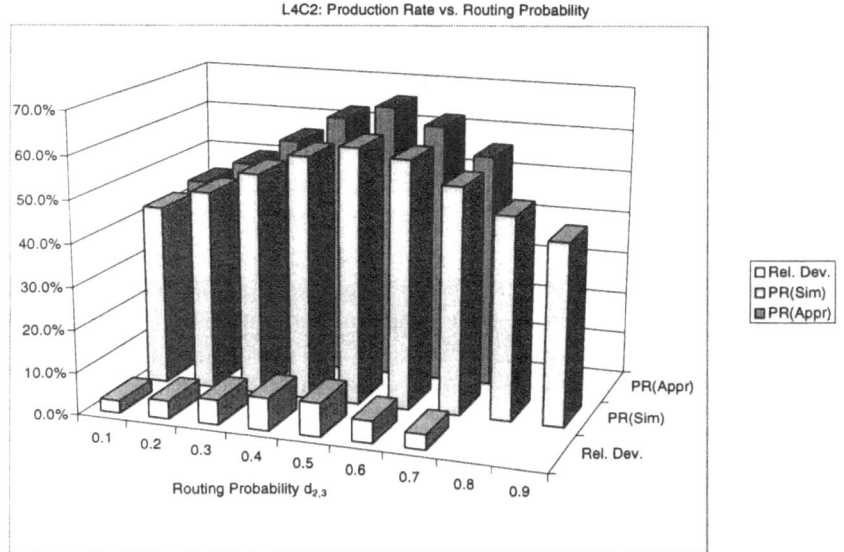

Fig. 4.50. Results for Class L4C2

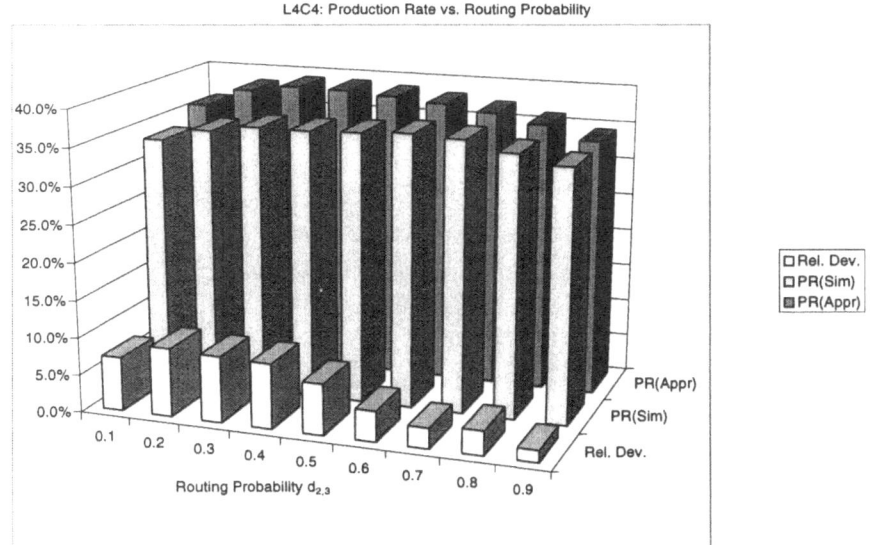

Fig. 4.51. Results for Class L4C4

The routing probabilities in the 100 random cases were $p_{2,3} = p_{2,5} = 0.5$. Out of these 100 random cases, 77 could be solved with an average absolute value of the percentage error of 1.77%. The results for the random cases are given in Figures 4.46 and 4.47. The results in Figure 4.48 and 4.49 for the parameters in Table 4.34 compared to Figures 4.41 and 4.42 indicate that the deviations get *larger* as we increase the number of machines in a feedforward loop.

Again, results for the cases of frequent failures and short repair times (L4C1 and L4C2) are more accurate than those for less frequent failures and long repair times (L4C3 and L4C4).

4.4.4.5 Structures L5 and L6. We next studied two different structures with multiple split and merge operations and a larger number of machines and buffers. Structure L5 is depicted in Figure 4.52. We asked for the production rate in the branch between Machines M_9 and M_{10}.

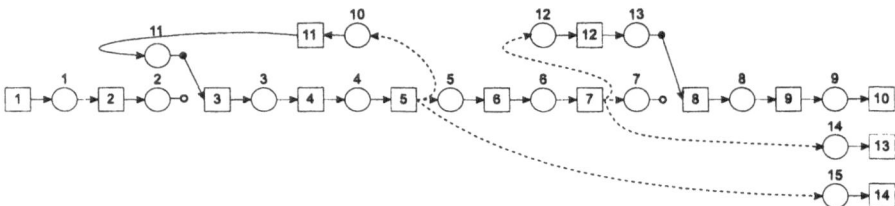

Fig. 4.52. Structure L5

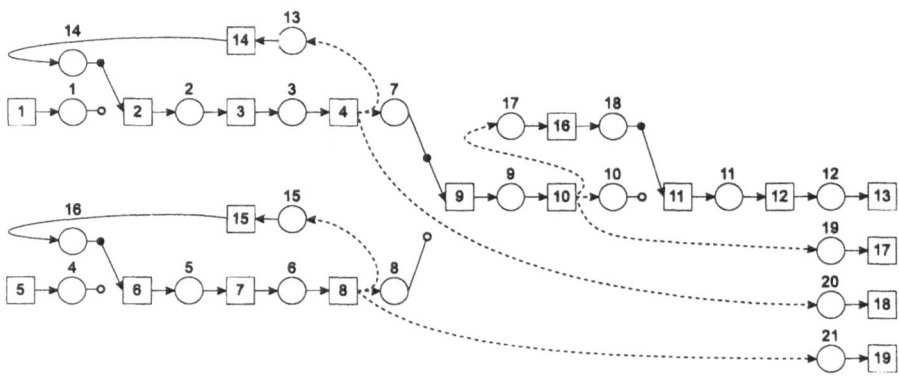

Fig. 4.53. Structure L6

The routing probabilities in the 100 random cases were identical to those of the systematic study of Class L5C1 as given in in the upper left part of Table 4.36. 99 out of 100 random cases for Structure L5 could be solved with an average absolute value of the percentage error of 1.55%. The results are given in Figures 4.54 and 4.55. For the systematically generated problems we used the parameters in the left column of Table 4.36.

Figure 4.56 shows that the decomposition works quite well even for this rather complex structure if the number of spaces per buffer is not too small.

Class L5C1			Class L6C1		
$p_i = 0.01, r_i = 0.1, \forall i$			$p_i = 0.02, r_i = 0.1, i = 1, ..., 8$		
			$p_i = 0.01, r_i = 0.1, i = 9, ..., 19$		
$d_{5,6} = 0.8, d_{5,11} = 0.15, d_{5,14} = 0.05$			$d_{4,9} = 0.7, d_{4,14} = 0.2, d_{4,18} = 0.1$		
$d_{7,8} = 0.7, d_{7,12} = 0.25, d_{7,13} = 0.05$			$d_{8,9} = 0.7, d_{8,15} = 0.2, d_{8,19} = 0.1$		
			$d_{10,11} = 0.65, d_{10,16} = 0.3, d_{10,17} = 0.05$		
System	$C_{i,j}, \forall(i,j)$	$N_{i,j}, \forall(i,j)$	System	$C_{i,j}, \forall(i,j)$	$N_{i,j}, \forall(i,j)$
L5C1S1	2	4	L6C1S1	2	4
$\vdots$	$\vdots$	$\vdots$	$\vdots$	$\vdots$	$\vdots$
L5C1S17	18	20	L6C1S17	18	20

Table 4.36. Parameters for Problem Classes L5C1 and L6C1

We finally studied Structure L6 depicted in Figure 4.53 where we asked for the production rate in the branch between machines M_{12} and M_{13}. Note that this is basically a large merge structure with some additional loops. We used the routing probabilities in the upper right part of Table 4.36.

Given our experience with Structure M2, we expected this Structure to be difficult to analyze if all machines have similar isolated efficiencies and the system is hence very unbalanced. In fact, the algorithm converged for 62 out of 100 cases. We conjectured that the convergence problems were due to the merge operation performed by Machine M_9. For this reason, we modified the failure probabilities of all machines upstream of Machine M_9 in such a way that their isolated efficiencies were reduced by exactly 50% wherever this was possible without exceeding an upper limit on the failure probabilities of 0.9. (Compare this to the generation of random cases for Structure M2). Out of these 100 modified random problems, the algorithm converged for 76 and failed for 24. The average absolute value of the percentage error was 1.65%. We conclude that the convergence reliability decreases somewhat as the complexity of the structure increases, but when the algorithm converges, it still tends to produce surprisingly accurate results.

Figure 4.57 shows that the production rate increases in a way very similar to Problem Class L5C1 for the parameters on the right hand side of Table 4.36.

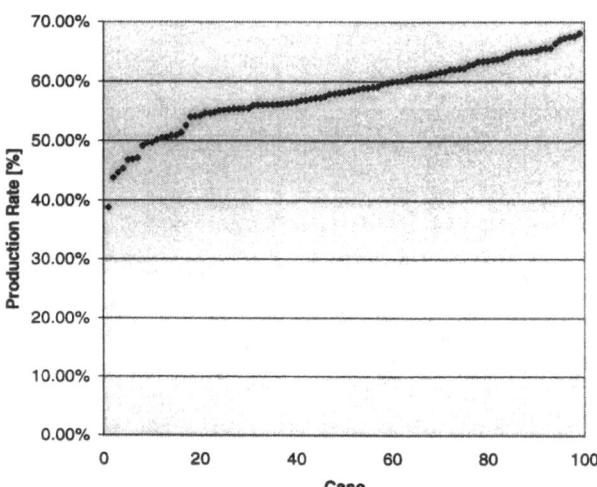

Fig. 4.54. Structure L5 - Simulated Production Rates for Random Problems

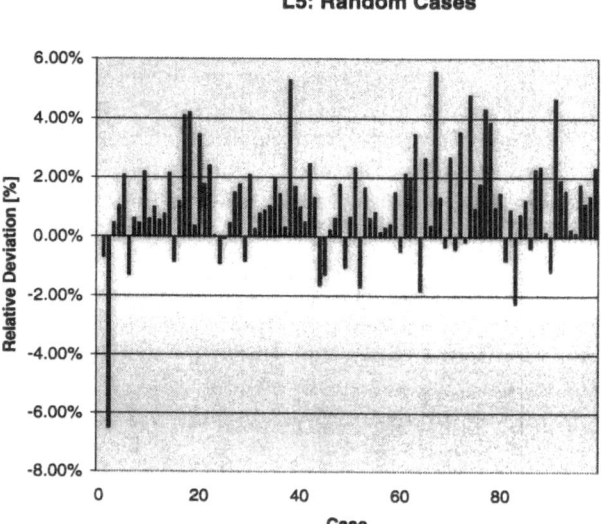

Fig. 4.55. Structure L5 - Percentage Errors for Random Problems

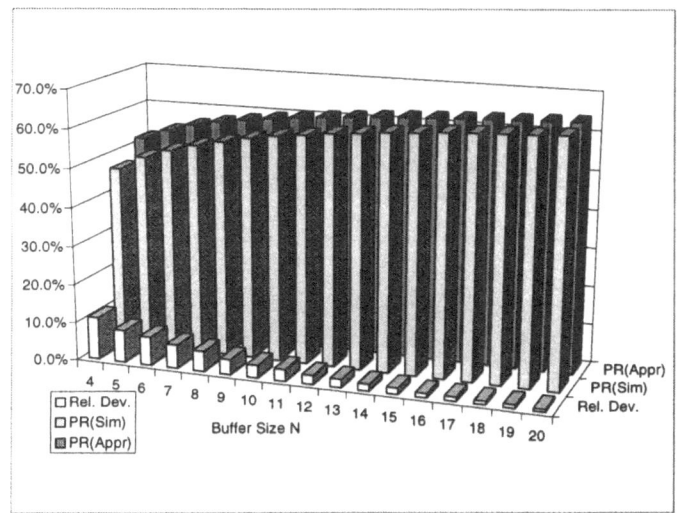

Fig. 4.56. Results for Class L5C1

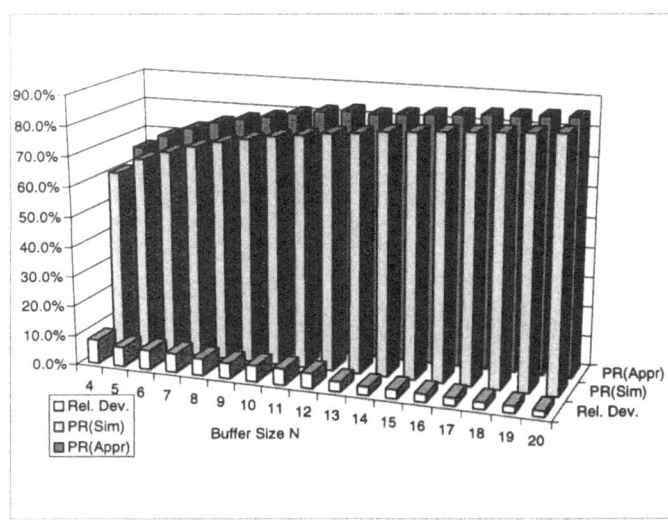

Fig. 4.57. Results for Class L6C1

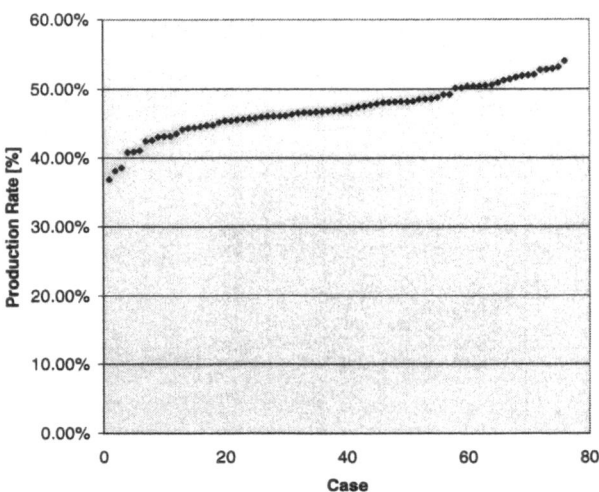

Fig. 4.58. Structure L6 - Simulated Production Rates for Random Problems

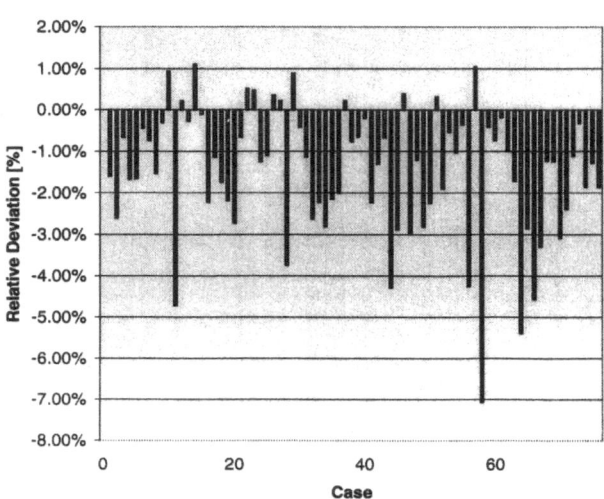

Fig. 4.59. Structure L6 - Percentage Errors for Random Problems

4.4.5 Summary of the Numerical Results

The numerical study shows that our decomposition approach provides good production rate estimates for a wide range of systems that perform split and/or merge operations. It appears to be very accurate and reliable for systems that perform split operations only, i.e. systems without rework of bad parts. For pure split structures, the accuracy increases as the buffer sizes increase and the algorithm converges very quickly. It is therefore a valuable tool for the evaluation of systems where bad parts are scrapped.

The decomposition for merge operations provides very precise production rate estimates for the machine that has two input buffers. However, if the throughput of the priority two buffer is very low in absolute terms, the production rate estimate for this priority two buffer tends to be very inaccurate in relative terms. Since we assume that reworked parts in systems with loops will always be routed to the priority one buffer, the priority two buffer will usually not have such a low production rate unless almost all parts need to be reworked.

The iterative algorithm may fail to converge if the merging machine is almost never starved, for example, because of an extreme bottleneck downstream. We conclude that the decomposition works well for a wide range of parameters that may be of practical interest.

Our study of more general systems with loops in the flow of material suggests that the method works well for a wide range of systems and parameters. For systems with feedback loops, the method tends to fail if buffer sizes are very small, compared to the number of parts processed during the expected repair time and a very large fraction of the parts has to be reworked. However, for reasonable buffer sizes and rejection probabilities, the method works well. Its accuracy appears to increase as buffer sizes increase.

For systems with feedforward loops, i.e. parallel branches, we observed fewer very large deviations, but also fewer very small deviations. Surprisingly, for this type of system the accuracy of the production rate estimate tended to decrease as the number of buffers increased. The convergence reliability appears to be lower than for feedback loops.

We eventually analyzed two larger systems with multiple loops and found that unless buffer sizes are very small, useful production rate estimates can be obtained even for systems with a much more complex flow of material than in a transfer line for which the first decomposition approaches have originally been developed.

4.5 Optimal Design of Systems with Loops and Identical Processing Times

In this section, some of the manufacturing system design problems are studied that arise only in systems with split and merge operations. As in previous sections, we ask for buffer allocations that lead to the maximum possible net present value (NPV) of the investment in machines, buffers, and material over the total lifetime of the system as defined in Section 2.3.

4.5.1 Impact of the Acceptance Probability at the Quality Inspection Station

One new question that arises in the context of flow lines with rework loops is the impact of the routing probability at the inspection station on the economic design and performance of the system. To illustrate the problem, consider the model of a transfer line with rework loop depicted in Figure 4.60. We want to know how the NPV of the investment, as well as the optimal buffer allocation, depends on the probability d_{56} that a part processed at Machine M_5 is 'good' and can leave the system after some final operations at Machine M_6.

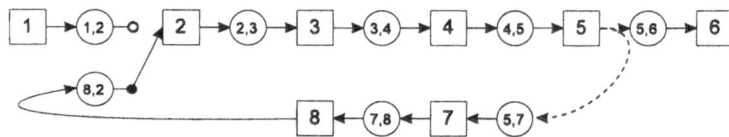

Fig. 4.60. Transfer Line with Rework Loop

Assume that all failure probabilities p_i are 0.01, all repair probabilities $r_i = 0.1$, and the initial buffer capacity is $C_{i,j} = 2$ for all buffers $B_{i,j}$. The isolated production rate of each machine is thus $10/11 \approx 0.909$ parts per period or production cycle. Each machine costs \$100,000 and each buffer \$2,000, the scrap value is 10% in each case. The raw material per product unit costs \$200, each unit is sold for \$230, and the annual fixed cost is \$220,000. The system is expected to operate for 4 years with 24,000 production cycles during each year and the interest rate on a perfect capital market is 10% per year.

To analyze the impact of the acceptance probability, we varied d_{56} from 98% to 80% in steps of 2%. For each case, we estimated the NPV for two buffer spaces between any two adjacent machines and for the (sub-)optimal buffer allocation yielding the maximum possible NPV.[8]

The results are summarized in Figure 4.61. The graphs indicate that the acceptance probability can have a high and almost linear impact on the net present value of the investment. This holds for the minimal buffer allocation (all $C_{i,j} = 2$), as well as for the optimized allocations. An observation that

---

[8] To determine this allocation, a gradient algorithm based on the work in Schor (1995) was used.

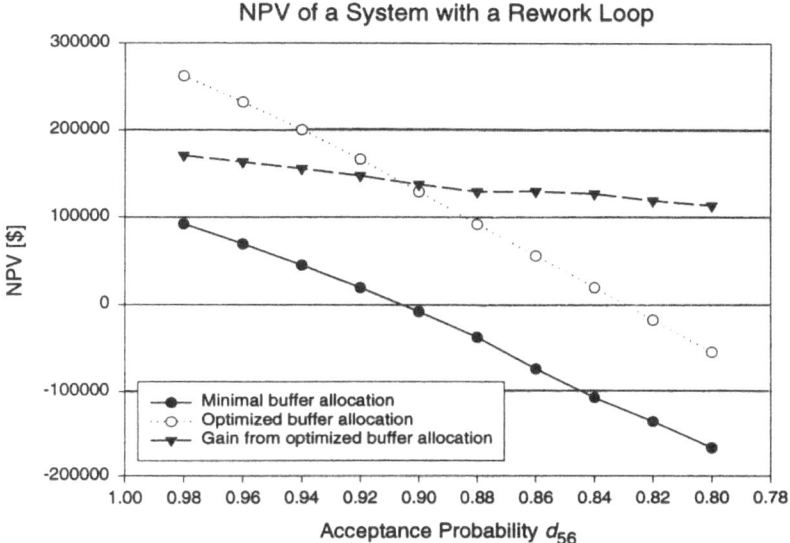

Fig. 4.61. NPV as a Function of the Acceptance Probability

does not appear to be obvious is that the *gain* from allocating the optimal number of buffers decreases as the acceptance probability decreases.[9]

To put it the other way around: As the acceptance probability d_{56} increases from 80% to 98%, the gain of the NPV that is due to adding the (close to) optimal number of buffers also increases, from about \$113,000 to about \$170,000. That is, as the product quality improves, it becomes *more worthwhile* to reduce starving and blocking by allocating buffers.

In Figures 4.62 and 4.63, the numbers between the machines represent the capacity of the corresponding buffers as proposed by a version of Schor's optimization algorithm. Note in Figure 4.62 that the system with the higher acceptance probability of 98% receives *more* buffers in the most profitable design than the system in Figure 4.63. In other words: As the product quality

---

[9] Note that from a mathematical point of view, we cannot state that any buffer allocation is optimal as the method used to evaluate any configuration other than a two-machine line is only an approximation and we have not *proved* that the NPV as a function of the buffer allocation exhibits the required properties for a gradient method to find the one and only local and global maximum. See Schor (1995) for an in-depth discussion of these problems. Whenever we use the terms 'optimal' or 'optimized', we do so with the understanding of these above-mentioned limitations.

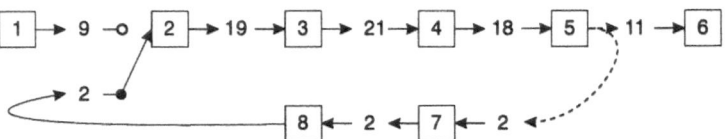

Fig. 4.62. Optimized Buffer Allocation for $d_{56} = 0.98$

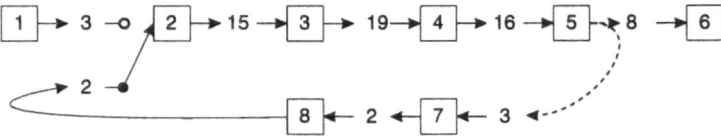

Fig. 4.63. Optimized Buffer Allocation for $d_{56} = 0.8$

gets better, i.e. d_{56} increases, the need for buffers also *increases* if one wants to maximize the NPV of the investment.

At first glance, this may seem to be highly implausible as improving product quality should make the manufacturing system more 'well-behaved' and buffers are required to reduce disruptions in the flow of material. One might think that as the acceptance probability, i.e. the product quality, improves, so does the production rate—which is right—and that this reduces the need for buffers—which can be wrong, if we are looking for the most profitable system design.

An intuitive explanation for the phenomenon that better product quality can call—other things being equal—for more instead of less buffers is as follows: Assume that for a *given* acceptance probability the optimal buffer allocation has been found and assume—to keep the argument simple—that it is possible to change each buffer size by an infinitely small amount. Since the buffer allocation is optimal, i.e. yields the maximum possible NPV, any such small change has no impact on the NPV. For each buffer, the NPV of the marginal revenue due to the additional output equals the NPV of the marginal investment in the buffer.

Now assume that the acceptance probability increases by a very small amount. In this situation, increasing any buffer size by an infinitely small amount leads to a slightly higher increase of the revenue due to the increase in the production rate, compared to the initial situation. However, the NPV of the marginal investment in the additional buffer space remains the same which implies that the net effect is positive. If the net effect is positive, the NPV still increases for increasing buffer sizes, which means that higher product quality makes additional buffers economically worthwhile.

It is not surprising that an error in estimating the fraction of bad parts, i.e. d_{56}, can lead to severe errors in the estimated NPV of the investment. However, the similarity of the buffer allocations in Figures 4.62 and 4.63 suggests that this does not have to have a major impact on the buffer allocation. The following two experiments confirm this impression.

Assume that the buffer allocation depicted in Figure 4.63 is chosen that is optimal if 80% of the parts are good quality ($d_{56} = 0.8$), but it turns out that the true value of the acceptance probability is much higher, say $d_{56} = 0.98$. In this case, this 'wrong' buffer allocation still leads to an NPV of about $253,000, not much less than the approximately $262,000 for the optimal buffer allocation depicted in Figure 4.62.

On the other hand, if one chooses the allocation in Figure 4.62, which is optimal for $d_{56} = 0.98$, but the acceptance probability turns out to be only 80%, the NPV of the project is about $-63,000 as opposed to $-54,000 for the then optimal allocation in Figure 4.62.

That is, the profitability of the investment depends heavily on the product quality, i.e. the acceptance probability, but the buffer allocation can be less dependent on the product quality, as the similarity of the allocations in Figures 4.62 and 4.63 suggests. This means that even if the wrong buffer allocation is chosen due to a wrong estimate of the acceptance probability, this may still lead to results close to those for the optimal buffer allocation corresponding to the true acceptance probability, as this single example shows.

4.5.2 Impact of the Placement of the Inspection Station

In the system depicted in Figure 4.60 of the previous section, we assumed that the quality inspection was performed at Machine M_5. In this section, we want to study what can be gained if bad parts are detected earlier. To keep the argument simple, assume that the yield is perfect at all machines but M_3. Assume furthermore, that the quality inspection could also be performed at Machines M_3 and M_4 instead of M_5 as in Figure 4.60. Thus, we want to evaluate the systems in Figures 4.64 and 4.65 compared to those in Figure 4.60 with respect to the maximum possible NPV of the investment.

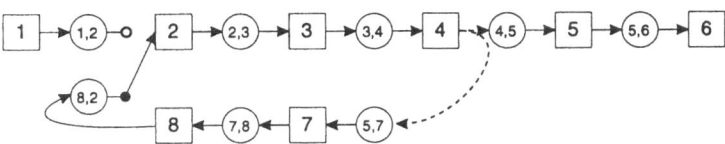

Fig. 4.64. Transfer Line with Rework Loop: Inspection at Machine M_4

The results are presented in Table 4.37. They indicate that it is indeed worthwhile to detect bad parts as early as possible since the NPV increases as the inspection is moved from Machine M_5 to Machine M_3. This happens for two reasons: The first reason is that the production rate increases as the number of machines in the rework loop decreases. This leads to a higher production rate for the good parts and to a higher NPV. Second, it also reduces the workload of Machine M_5 and—for the inspection at M_3—also

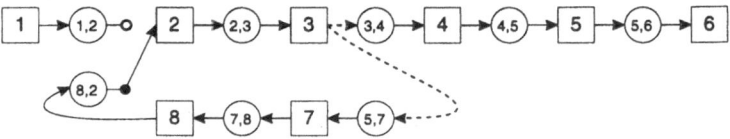

Fig. 4.65. Transfer Line with Rework Loop: Inspection at Machine M_3

of M_4 as these machines now only process good parts. Since failures are assumed to be operation dependent, this means that these machines fail less often. For this reason, they require fewer buffers. Removing these buffers makes the investment more profitable as well. The message for the design of manufacturing system is that—other things being equal—bad parts should not receive any further operations that have to be repeated anyway after performing the rework operations.

Inspection Station	Maximum NPV for Low Acceptance Probability (90%)	Maximum NPV for High Acceptance Probability (98%)
M_5	$129,000	$262,000
M_4	$159,000	$274,000
M_3	$196,000	$287,000

Table 4.37. Approximate Net Present Values for Different Placements of the Inspection Station

Note that in Table 4.37 the increase of the NPV is stronger if the fraction of good parts is low (90%) than when it is high (98%). The reason is that if the product quality is very high, detecting the few bad parts early in the manufacturing process does not lead to a strong workload reduction at the machines downstream of the inspection station. This means that the reduction in operation dependent failures and in the number of buffer spaces is not as strong as for the low product quality.

5. Flow Lines with Rework Loops and Machine-Specific Processing Times

5.1 Continuous-Material Flow Line Model with Machine-Specific Processing Times

In this section, the case of *machine-specific* deterministic processing times is studied. Machine-specific processing times are especially important in networks with split and/or merge operations. The reason is that these operations lead to an uneven distribution of the workload. The workload of a rework station, for example, is usually smaller than those of a station in the main part of a transfer line unless all parts require rework operations. Another possibility is that in a system two (slow) machines operate in parallel and that a merge in the flow of material is required downstream of the two machines. It is therefore important to have a tool that can take these machine-specific processing times into account.

In Burman (1995), an extremely powerful algorithm for the case of machine-specific processing times and a purely linear flow of material has been developed. This algorithm leads to very precise performance estimates for flow lines. It has been used in the process of designing a Hewlett-Packard printer production line yielding—according to Burman et al. (1998)—an increase in Hewlett-Packard's revenue from this operation of $280,000,000. This algorithm is also being used in a commercially available software program called *FlowEval*.[1]

The model and algorithm has been extended to deal with assembly and disassembly operations.[2] In what follows, Burman's model and algorithm are generalized to split and merge operations. The generalization for the case of split operations and continuous material alone, i.e. without merge operations, were first developed in Schicht (1996).

We assume that each machine M_i is completely characterized by three parameters:

- the processing rate μ_i,
- the failure rate p_i, and
- the repair rate r_i.

---

[1] Tempelmeier (1997)
[2] Gershwin and Burman (1995)

While processing times are deterministic such that each operation at Machine M_i takes $1/\mu_i$ time units, we assume that times to failure and times to repair are exponentially distributed with rates p_i and r_i, respectively.

The method used in this thesis to determine performance measures is based on the decomposition of flow lines or networks into systems of two-machine lines. The steady-state probabilities of these two-machine lines can often be evaluated analytically if the two-machine models are Markov process models. The standard approach to develop Markov process models of flow lines where processing times are machine-specific is to treat the material as a fluid.

If processing times are deterministic and machine-specific and the material is treated as a discrete quantity, the model loses the Markov property. The reason is that the knowledge about the history of the process, i.e. about how long a discrete part has already been processed, helps to predict when the operation will be completed. In this situation, the process is not memoryless, i.e. it does not possess the Markov property and cannot be analyzed as easily as, for example, the two-machine model in Section 3.2.

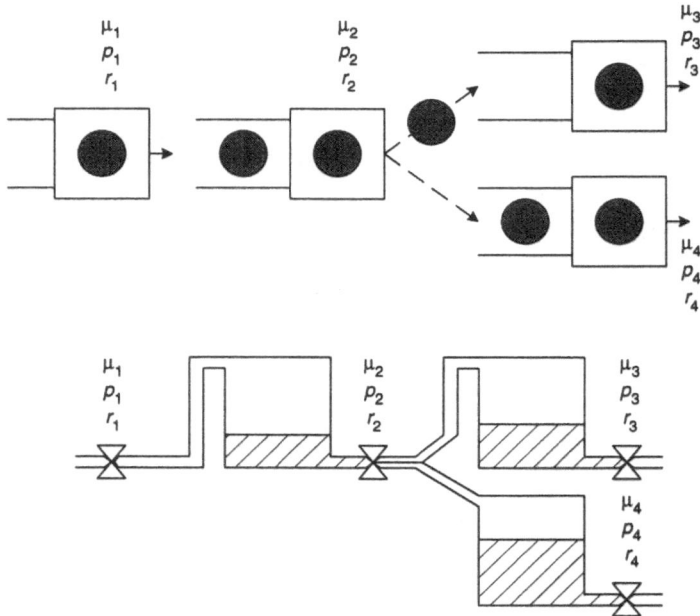

Fig. 5.1. Representation of Machines as Valves and Buffers as Tanks for a Split System

In order to develop a Markov process model where machines can operate with different processing rates, each machine that processes discrete material as depicted in the upper part of Figure 5.1 is modeled as a pump or valve through which a continuous fluid passes. This situation is depicted in the lower part of Figure 5.1. The buffers of the original system behave like tanks in a continuous material model and the discrete material of the original system can be considered to be melted, i.e. it is treated like a fluid. In this case, the buffer (or tank) levels are continuous quantities.

If we allow for machine-specific processing times in a *continuous* material model, the instantaneous rate $\mu_i(t)$ of Machine M_i at time t can be less than its maximum possible rate μ_i due to the possibility of *partial* starvation and blocking. A machine is partially starved if it is slowed down because it receives material at a rate that is lower than the rate at which it could process the material. This can only happen if the input buffer is empty. (Whenever a discrete part in the real system enters this input buffer, it is immediately transferred to Machine M_i that actually alternates between starvation and processing.) By the same token, it is partially blocked if it can only dispose material at a rate that is lower than its own processing rate. This requires a full downstream buffer $B_{i,q}$ upstream of a machine M_q that is working because otherwise M_i would be completely blocked.

If failures of Machine M_i are operation dependent, the effective failure rate of Machine M_i at time t is

$$p_i(t) = \frac{\mu_i(t)}{\mu_i}p_i, \tag{5.1}$$

i.e. the failure rate is reduced according to the reduction of the processing rate due to partial starvation or blocking.

Two-Machine models were developed in Zimmern (1956), Wijngaard (1979) and Gershwin and Schick (1980). We use the latter model that is reproduced in great detail in (Gershwin, 1994, p. 112-132) as it takes the reduction of failure rates due to partial starvation and blocking into account.

5.2 Decomposition Equations for Loops and Different Processing Times

The derivation of the decomposition equations for the continuous material model is more complicated, since three parameters have to be determined for each virtual machine in the virtual two-machine lines. The additional parameters are the processing rates $\mu_u(i,q)$ and $\mu_d(i,q)$ for each line $L(i,q)$. To determine the additional quantity, an additional decomposition equation is required. This is the *interruption of flow* equation, which describes the interruption in the flow of material, as seen by an observer between two adjacent machines.

Many of the arguments in the derivation of the equations are similar. For this reason we merely state those equations that are almost identical to those for the discrete-time discrete-material model in Section 4 and concentrate on those that are different.

5.2.1 Conservation of Flow Equation

Define $PR(j,i)$ and $PR(i,q)$ as the production rate in the decomposed two-machine lines $L(j,i)$ and $L(i,q)$, respectively. The decomposition must satisfy the following conservation of flow equation

$$PR(i) = \sum_{(i,q)\in D(i)} PR(i,q) = \sum_{(j,i)\in U(i)} PR(j,i), \ \forall i \qquad (5.2)$$

where $PR(i)$ is the total production rate related to Machine M_i in the two-machine lines $L(i,q)$ or $L(j,i)$.

In the real discrete material system, a part is randomly routed from Machine M_i to Machine M_m with probability $d_{i,m}$. We therefore have

$$d_{i,m} = \frac{PR(i,m)}{\sum_{(i,q)\in D(i)} PR(i,q)}, \ (i,m) \in D(i), \forall i \qquad (5.3)$$

for the fraction of continuous material that is flowing from Machine M_i into Buffer $B_{i,m}$ whenever material is leaving Machine M_i. The COF equation for the transfer line model in (Burman, 1995, p. 45) is a special case of (5.2) where each machine has no more than one input buffer and output buffer, respectively.

5.2.2 Flow Rate-Idle Time Equations

The flow rate-idle time (FRIT) relationship for a machine with a processing rate μ_i is approximately (omitting partial blockage and starvation)

$$PR_i \approx \mu_i e_i \ \text{prob} \left[\begin{array}{ll} \{n(l,i) > 0, & \text{some } (l,i) \in U(i)\} \text{ and} \\ \{n(i,q) < N(i,q), & \forall (i,q) \in D(i)\} \end{array} \right] \qquad (5.4)$$

where $n(l,i)$ denotes the buffer level in Buffer $B_{l,i}$. Assuming as second approximation that input buffer levels are independent, and using the two-machine flow rate-idle time equations for the continuous material two-machine line,[3] we find in a way similar to the previous section the first two decomposition equations

---

[3] Gershwin (1994, p. 126)

$$\frac{r_u(i,m) + p_u(i,m)}{r_u(i,m)\mu_u(i,m)} = K_1 \tag{5.5}$$

$$\frac{r_d(j,i) + p_d(j,i)}{r_d(j,i)\mu_d(j,i)} = K_2 \tag{5.6}$$

where we define

$$K_1 = \frac{\frac{PR(i)}{e_i\mu_i} + \prod_{(l,i)\in U(i)}\left(1 - \frac{PR(l,i)}{e_d(l,i)\mu_d(l,i)}\right) + \sum_{(i,q)\in D(i), q\neq m}\left(1 - \frac{PR(i,q)}{e_u(i,q)\mu_u(i,q)}\right)}{PR(i,m)}$$

$$\tag{5.7}$$

$$K_2 = \left[\frac{\frac{PR(i)}{e_i\mu_i} + \sum_{(i,q)\in D(i)}\left(1 - \frac{PR(i,q)}{e_u(i,q)\mu_u(i,q)}\right) - 1}{\prod_{(l,i)\in U(i), l\neq j}\left(1 - \frac{PR(l,i)}{e_d(l,i)\mu_d(l,i)}\right)} + 1\right]\frac{1}{PR(j,i)} \tag{5.8}$$

The FRIT equation for the continuous material transfer line model in (Burman, 1995, p. 70) is again a special case of (5.5) and (5.6).

5.2.3 Resumption of Flow Equations I: Split Operations

The reasoning to derive the resumption of flow equations for the continuous material appears—at first glance—to be very similar to the discrete material case treated in Section 4. In the discrete material-discrete time model we looked at transitions from period t to period $t + 1$, whereas here we analyze the behavior during infinitesimally short time intervals of length δt, i.e. from time t to time $t + \delta t$. While this may look like a minor difference, it leads to a completely different derivation of the decomposition equations as shown in the appendices in great detail.

5.2.3.1 Upstream Machine. To an observer in Buffer $B_{i,m}$ of the split system in Figure 4.4 on Page 96, the virtual upstream machine $M_u(i,m)$ is up at time t if material enters the buffer. For this to happen, Machine M_i must be up and it must not be *completely* starved or blocked.

Define $\{\alpha_u[(i,m),t] = 1\}$ as the event of the virtual upstream machine $M_u(i,m)$ being up at time t:

$$\{\alpha_u[(i,m),t]=1\} \quad \text{iff} \quad \{\alpha_i(t)=1\} \text{ and}$$

$$\Big\{\{n[(j,i),t]>0\} \text{ or}$$

$$\{n[(j,i),t]=0\} \text{ and } \{\alpha_u[(j,i),t]=1\}\Big\} \text{ and}$$

$$\Big\{\{n[(i,q),t]<N(i,q)\} \text{ or}$$

$$\{n[(i,q),t]=N(i,q)\} \text{ and } \{\alpha_d[(i,q),t]=1\},$$

$$\forall \text{(i,q)} \in D(i), q \neq m\Big\} \tag{5.9}$$

The first condition on the right hand side of (5.9) says that Machine M_i must be up and the second demands that the single input buffer is not empty while Machine $M_u(j,i)$ is down. The third condition says that Machine M_i must not be (completely) blocked.

This definition of virtual machine states for a continuous time-continuous material model differs in several ways from those for the discrete time-discrete material model in Equations (4.18) and (4.19) on Page 97.

First, buffer levels at time t are considered due to the continuous modeling of time, whereas in the previous model, we referred to buffer levels at the end of the discrete period $t-1$. Second, while in the discrete material model we have to take the routing of a particular part into account that is sent to one output buffer only, in the continuous material model, whenever material is leaving Machine M_i, a constant fraction $d_{i,m}$ of the total instantaneous flow goes to each buffer $B_{i,m}$ simultaneously. Third, in the continuous material model, a machine can be partially starved or blocked. That is, Machine M_i is not starved at all at time t if the input buffer level $n[(j,i),t]$ is positive. However, if it is zero while the virtual upstream machine $M_u(j,i)$ is up, continuous material is flowing into Buffer $B_{j,i}$ at a rate that is not higher than the rate at which Machine M_i can process material. Otherwise, the buffer level would become positive instantaneously. In this situation, Machine M_i is at most *partially* starved, i.e. slowed down by the instantaneous production rate of $M_u(j,i)$. A similar argument holds for blocking of Machine M_i.

Machine $M_u(i,m)$ is down if it is not up:

$$\{\alpha_u[(i,m),t]=0\} \quad \text{iff} \quad \{\alpha_i(t)=0\} \text{ or}$$

$$\{n[(j,i),t]=0\} \text{ and } \{\alpha_u[(j,i),t]=0\} \text{ or}$$

$$\Big\{\{n[(i,q),t]=N(i,q)\} \text{ and } \{\alpha_d[(i,q),t]=0\},$$

$$\text{for some } (i,q) \in D(i), q \neq m\Big\} \tag{5.10}$$

The different events that can force Machine $M_u(i, m)$ down are mutually disjoint: If Machine M_i is either starved or blocked it cannot fail. Due to the continuous material, a machine cannot be blocked and starved simultaneously and it cannot be blocked by more than one downstream machine at any moment in time.

The repair probability $r_u(i, m)\delta t$ of the virtual upstream machine $M_u(i, m)$ is the probability of seeing material flowing into Buffer $B_{i,m}$ at time $t + \delta t$ given that no continuous material was flowing into Buffer $B_{i,m}$ at time t, i.e. Machine $M_u(i, m)$ was down at time t. This can be written as

$$
r_u(i, m)\delta t = \text{prob}\Big[\{\alpha_u[(i, m), t + \delta t] = 1\} \mid \{\alpha_u[(i, m), t] = 0\}\Big] \quad (5.11)
$$

$$
= \text{prob}\Big[\{\alpha_u[(i, m), t + \delta t] = 1\} \Big|
$$

$$
\Big\{ \{\alpha_i(t) = 0\} \text{ or}
$$

$$
\{n[(j, i), t] = 0\} \text{ and } \{\alpha_u[(j, i), t] = 0\} \text{ or}
$$

$$
\Big\{ \{n[(i, q), t] = N(i, q)\} \text{ and } \{\alpha_d[(i, q), t] = 0\},
$$

$$
\text{for some } (i, q) \in D(i), q \neq m\}\Big\}\Big\}\Big] \quad (5.12)
$$

if we use the definition of an upstream machine being down (5.10). Since the different events that force Machine $M_u(i, m)$ down are mutually disjoint, we can break Equation (5.12) down in Appendix B.1.1 on Page 225 by decomposing the conditioning event to find

$$
r_u(i, m) = \left[r_i + K_3 \frac{r_u(i, m)\mu_u(i, m)}{p_u(i, m)} \right] \quad (5.13)
$$

where we define

$$
K_3 = \Big[(r_u(j, i) - r_i)\mathbf{p}[(j, i); 001]
$$

$$
+ \sum_{(i,q) \in D(i), q \neq m} (r_d(i, q) - r_i)\mathbf{p}[(i, q); N(i, q)10] \Big] \frac{1}{PR(i, m)} \quad (5.14)
$$

This is again a generalization of the corresponding equation for the continuous material transfer line model in Burman (1995). Compare this result to Equation (4.22) on Page 98 for the discrete material network. The expression for the continuous material split model is less complicated since the random routing in the real system is here modeled as a continuous split in the flow of material. In the discrete material model split model, and additional term is required that takes care of the random routing of discrete material.

5.2.3.2 Downstream Machine. To an observer in Buffer $B_{j,i}$, the virtual downstream machine $M_d(j,i)$ is up at time t if the continuous material flows out of the buffer at time t. Define $\{\alpha_d[(j,i),t] = 1\}$ as the event of the virtual downstream machine $M_d(j,i)$ being up at time t. Machine $M_d(j,i)$ is up if Machine M_i is up and not (completely) blocked, i.e.

$$\{\alpha_d[(j,i),t] = 1\} \quad \text{iff} \quad \{\alpha_i(t) = 1\} \text{ and}$$
$$\left\{ \begin{array}{l} \{n[(i,q),t] < N(i,q)\} \text{ or} \\ \{n[(i,q),t] = N(i,q)\} \text{ and } \{\alpha_d[(i,q),t] = 1\}, \\ \forall (i,q) \in D(i)\} \end{array} \right\} \tag{5.15}$$

as it is completely blocked only if $\{n[(i,q),t] = N(i,q)\}$ and $\{\alpha_d[(i,q),t] = 0\}$ hold simultaneously. Machine $M_d(j,i)$ is down if it is not up, i.e:

$$\{\alpha_d[(j,i),t] = 0\} \quad \text{iff} \quad \{\alpha_i(t) = 0\} \text{ or}$$
$$\left\{ \begin{array}{l} \{n[(i,q),t] = N(i,q)\} \text{ and } \{\alpha_d[(i,q),t] = 0\}, \\ \text{for some } (i,q) \in D(i)\} \end{array} \right\} \tag{5.16}$$

The resumption of flow probability $r_d(j,i)\delta t$ defined as

$$r_d(j,i)\delta t = \text{prob}\Big[\{\alpha_d[(j,i),t+\delta t] = 1\} \mid \{\alpha_d[(j,i),t] = 0\}\Big] \tag{5.17}$$

is evaluated in Appendix B.1.2 on Page 228 to find

$$r_d(j,i) = r_i + K_4 \frac{r_d(j,i)\mu_d(j,i)}{p_d(j,i)} \tag{5.18}$$

with

$$K_4 = \sum_{(i,q)\in D(i)} (r_d(i,q) - r_i)\mathbf{p}[(i,q); N(i,q)10]\frac{1}{PR(j,i)} \tag{5.19}$$

which is almost identical to the corresponding equation (4.29) for the discrete material model on Page 100.

5.2.4 Resumption of Flow Equations II: Merge Operations

5.2.4.1 Upstream Machine. To an observer in the buffer corresponding to Line $L(i, q)$ in Figure 4.5 on Page 101, the virtual machine $M_u(i, q)$ is up if Machine M_i is up and not starved. It is starved if both input buffers are empty and the respective virtual upstream machines are down simultaneously, i.e.

$$\{\alpha_u[(i, q), t] = 1\} \quad \text{iff} \quad \{\alpha_i(t) = 1\} \text{ and}$$
$$\left\{ \begin{array}{l} \{n[(j, i), t] > 0\} \text{ or} \\[2mm] \{n[(j, i), t] = 0\} \text{ and } \{\alpha_u[(j, i), t] = 1\} \\[2mm] \text{for some } j \in \{j_1, j_2\} \end{array} \right\} \tag{5.20}$$

Machine $M_u(i, q)$ is down if it is not up, i.e

$$\{\alpha_u[(i, q), t] = 0\} \quad \text{iff} \quad \{\alpha_i(t) = 0\} \text{ or}$$
$$\left\{ \begin{array}{l} \{n[(j, i), t] = 0\} \text{ and } \{\alpha_u[(j, i), t] = 0\}, \\[2mm] \forall j \in \{j_1, j_2\} \end{array} \right\} \tag{5.21}$$

The different events that can force Machine $M_u(i, q)$ down are mutually disjoint since Machine M_i cannot fail if it is (completely) starved.

To derive the resumption of flow equation

$$r_u(i, q)\delta t \;=\; \text{prob}\Big[\{\alpha_u[(i, q), t + \delta t] = 1\}\big|\{\alpha_u[(i, q), t] = 0\}\Big] \tag{5.22}$$
$$= \; \text{prob}\Big[\{\alpha_u[(i, q), t + \delta t] = 1\}\big|$$
$$\Big\{ \{\alpha_i(t) = 0\} \text{ or}$$
$$\Big\{ \{n[(j, i), t] = 0\} \text{ and } \{\alpha_u[(j, i), t] = 0\},$$
$$\forall j \in \{j_1, j_2\}\Big\}\Big\}\Big] \tag{5.23}$$

we break Equation (5.23) down in Appendix B.2.1 on Page 230 to find

$$r_u(i, q) = r_i + K_3 \frac{r_u(i, q)\mu_u(i, q)}{p_u(i, q)} \tag{5.24}$$

with

$$K_3 = \left(\sum_{j \in \{j_1, j_2\}} r_u(j,i) - r_i \right) \prod_{j \in \{j_1, j_2\}} \mathbf{p}[(j,i); 001] \frac{1}{PR(i,q)} \quad (5.25)$$

If there is just one line $L(j,i)$ with $j \in \{j_1, j_2\}$, equation (5.24) reduces to the corresponding equation of the continuous material model for the linear transfer line (Burman, 1995, p. 69). Note that this term differs from the corresponding one for the discrete material model in Equation (4.36) on Page 102). The reason is that in the continuous material model, second order terms including $(\delta t)^2$ can be omitted for infinitesimally short time intervals of length δt, whereas the corresponding terms cannot be omitted in the discrete material model.

5.2.4.2 Downstream Machine. *Priority One Line.* The repair behavior of Machine $M_d(j_1, i)$ is exactly as in a the continuous material transfer line model for which the resumption of flow equation

$$r_d(j_1, i)) = r_i + K_4 \frac{r_d(j_1, i)\mu_d(j_1, i)}{p_d(j_1, i)} \quad (5.26)$$

with

$$K_4 = \left(r_d(i,q) - r_i \right) \mathbf{p}[(i,q); N(i,q), 10] \frac{1}{PR(j_1, i)} \quad (5.27)$$

is given in (Burman, 1995, p. 69).

Priority Two Line. Three conditions must hold for Machine $M_d(j_2, i)$ to be up: Machine M_i must be up, its priority one upstream buffer $B_{j_1 i}$ must be empty, and it must not be (completely) blocked due to a failure of its downstream machine $M_d(i, q)$. Formally

$$\{\alpha_d[(j_2, i), t] = 1\} \quad \text{iff} \quad \{\alpha_i(t) = 1\} \text{ and} \quad (5.28)$$
$$\{n[(j_1, i), t] = 0\} \text{ and}$$
$$\Big\{ \{n[(i,q), t] < N(i,q)\} \text{ or}$$
$$\{n[(i,q), t] = N(i,q)\} \text{ and } \{\alpha_d[(i,q), t] = 1\} \Big\}$$

Machine $M_d(j_2, i)$ is down if it is not up:

$$\{\alpha_d[(j_2,i),t] = 0\} \quad \text{iff} \quad \{\alpha_i(t) = 0\} \text{ or} \tag{5.29}$$
$$\{n[(j_1,i),t] > 0\} \text{ or}$$
$$\{n[(i,q),t] = N(i,q)\} \text{ and } \{\alpha_d[(i,q),t] = 1\}$$

The different reasons for Machine $M_d(j_2,i)$ to be down as given in (5.29) are not disjoint: Machine M_i can fail while processing material from the non-empty priority one buffer. In this case, Machine $M_d(j_2,i)$ is down for two reasons at the same time. If this happens at time t, then Machine $M_d(j_2,i)$ cannot possibly be up at time $t + \delta t$. Even if Machine M_i were repaired at time $t + \delta t$, the priority one buffer would still be non-empty as nothing was removed during the interval $[t, t + \delta t]$.

The resumption of flow probability $r_d(j_2,i)$ defined as

$$r_d(j_2,i) = \text{prob}\Big[\{\alpha_d[(j_2,i),t+\delta t] = 1\} \mid \{\alpha_d[(j_2,i),t] = 0\}\Big] \tag{5.30}$$

and is evaluated in Section B.2.2 on Page 231 to find

$$r_d(j_2,i) = \left[r_i + K_4 \frac{r_d(j_2,i)\mu_d(j_2,i)}{p_d(j_2,i)}\right] \tag{5.31}$$

where we define

$$K_4 = \left[\begin{array}{l}\left[\begin{array}{ll} \mathbf{p}[(j_1,i);011]p_d(j_1,i)\frac{\mu_u(j_1,i)}{\mu_d(j_1,i)} & \text{if } \mu_d(j_2,i) > \mu_u(j_2,i) \\ \mathbf{p}[(j_1,i);001]r_u(j_1,i) & \text{if } \mu_d(j_2,i) \leq \mu_u(j_2,i) \end{array}\right]\cdot \\[4mm] \left[e_u(i,q)\Big(1 - \mathbf{p}[(i,q);N(i,q),10] - \mathbf{p}[(i,q);N(i,q),11]\Big) + \right. \\[2mm] \left. \Big(e_u(i,q) - 1\Big)\frac{\mu_d(i,q)}{\mu_u(i,q)}\mathbf{p}[(i,q);N(i,q),11] + \mathbf{p}[(i,q);N(i,q),11]\right] + \\[3mm] \Big[r_d(i,q)\mathbf{p}[(j_1,i);001]\mathbf{p}[(i,q);N(i,q)10]\Big] - \\[3mm] r_i\Big[\Big(1 - \mathbf{p}[(j_1,i);001] - \mathbf{p}[(j_1,i);011]\Big)\Big(1 - \mathbf{p}[(i,q);N(i,q)10]\Big) \\[3mm] \left. + \mathbf{p}[(i,q);N(i,q)10]\Big]\right]\frac{1}{PR(j_2,i)} \end{array}\right] \tag{5.32}$$

Note that the first term in K_4 in (5.32) depends on the relative speed of the virtual machines up- and downstream of the priority one buffer. Now all resumption of flow equations have been determined.

5.2.5 Interruption of Flow Equations I: Split Operations

The interruption of flow equations describe the failure process of virtual machines as seen by an observer in the buffer between two adjacent machines. This is an additional set of equations that we did not have to determine for the discrete material model since we had to determine 'only' two parameters for each virtual machine. In the continuous material model, however, we need an additional parameter to take the machine specific processing rate of the real system into account. We first analyze the failure processes related to split operations.

5.2.5.1 Upstream Machine. If Machine M_i performs a split operation as depicted in Figure 4.4 on Page 96, there are three reasons why an observer in Buffer $B_{i,m}$ can see a failure of the virtual upstream machine $M_u(i, m)$. First, Machine M_i can fail. Second, Machine M_i can be starved due to a failure of a machine upstream of M_i. Third, it can be blocked if Buffer $B_{i,q}, q \neq m$ is full and $M_d(i, q)$ is down. Using the definition of the virtual machine states in (5.9) and (5.10), the interruption of flow equation can be written as:

$$p_u(i, m)\delta t = \text{prob} \left[\{\alpha_u[(i, m), t + \delta t] = 0\} \Big| \right. \tag{5.33}$$
$$\left. \{\alpha_u[(i, m), t] = 1\} \text{ and } \{n[(i, m), t] < N(i, m)\} \right]$$

In the conditioning event of this interruption of flow equation (5.33), Buffer $B_{i,m}$ must be non-full in order to determine the failure rate $p_u(i, m)$ of Machine $M_u(i, m)$ if it operates in isolation. If Buffer $B_{i,m}$ is full while Machine $M_d(i, m)$ is up, Machine $M_u(i, m)$ is partially blocked and failures occur less frequently. In order to exclude these effects, we demand $\{n[(i, m), t] < N(i, m)\}$ in the conditioning event of (5.33).

In Appendix B.3 on Page 246, Equation (5.33) is approximated as

$$p_u(i, m) = p_i + K_5 \mu_u(i, m) \tag{5.34}$$

with

$$K_5 = \left[p_i \left(\frac{\mu_u(j,i)}{\mu_d(j,i)} - 1 \right) \mathbf{p}[(j,i);011] + r_u(j,i)\mathbf{p}[(j,i);001] + \right.$$

$$\sum_{\substack{(i,q)\in D(i) \\ q\neq m}} \left(p_i \left(\frac{\mu_d(i,q)}{\mu_u(i,q)} - 1 \right) \mathbf{p}[(i,q); N(i,q), 11] + \right.$$

$$\left. \left. r_d(i,q)\mathbf{p}[(i,q); N(i,q), 10] \right) \right] \frac{1}{PR(i,m)} \qquad (5.35)$$

which is a generalization of the respective equation in (Burman, 1995, p. 79).

5.2.5.2 Downstream Machine. There are two reasons why an observer in Buffer $B_{j,i}$ upstream of a split machine M_i (depicted in Figure 4.4 on Page 96) can observe a failure of the virtual downstream machine $M_d(j,i)$: Machine M_i can be down or it can be blocked if one Buffer $B_{i,q}$ is full while Machine $M_d(i,q)$ is down. Using the definition of virtual machine states in (5.15) and (5.16), the interruption of flow equation for the virtual machine $M_d(j,i)$ can be written as

$$p_d(j,i)\delta t = \text{prob} \left[\{\alpha_d[(j,i), t + \delta t] = 0\} \right| \qquad (5.36)$$
$$\{\alpha_d[(j,i), t] = 1\} \text{ and } \{n[(j,i), t] > 0\} \right]$$

A derivation that is very similar to those for $p_i(i,m)$ in the previous section leads to the following approximation:

$$p_d(j,i) = p_i + K_6 \mu_d(j,i) \qquad (5.37)$$

with

$$K_6 = \left[\sum_{(i,q)\in D(i)} \left(p_i \left(\frac{\mu_d(i,q)}{\mu_u(i,q)} - 1 \right) \mathbf{p}[(i,q); N(i,q), 11] + \right. \right.$$

$$\left. \left. r_d(i,q)\mathbf{p}[(i,q); N(i,q), 10] \right) \right] \frac{1}{PR(j,i)} \qquad (5.38)$$

5.2.6 Interruption of Flow Equations II: Merge Operations

5.2.6.1 Upstream Machine. From the perspective of an observer in Buffer $B_{i,q}$ downstream of a merge machine M_i depicted in Figure 4.5 on Page 101, Machine $M_u(i,q)$ is down if either Machine M_i is down or both input buffers $B_{j_1,i}$ and $B_{j_2,i}$ are empty and both upstream virtual machines $M_u(j_1,i)$ and

$M_u(j_2, i)$ are down. Using the definition of virtual machine states (5.20) and (5.21), the interruption of flow equation for Machine $M_u(i, q)$ with two input buffers can be expressed as

$$p_u(i, q)\delta t = \text{prob} \left[\{\alpha_u[(i, q), t + \delta t] = 0\} \mid \right.$$
$$\left. \{\alpha_u[(i, q), t] = 1\} \text{ and } \{n[(i, q), t] < N(i, q)\} \right] \tag{5.39}$$

which is approximated in Appendix B.4.1 on Page 254 as

$$p_u(i, q) = p_i + K_5 \, \mu_u(i, q) \tag{5.40}$$

with

$$K_5 = \left[p_i \left(\frac{\mu_u(j_1, i) + \mu_u(j_2, i)}{\mu_d(j_1, i) + \mu_d(j_2, i)} - 1 \right) \mathbf{p}[(j_1, i); 011] \mathbf{p}[(j_1, i); 011] + \right.$$
$$p_i \left(\frac{\mu_u(j_1, i)}{\mu_d(j_1, i)} - 1 \right) \mathbf{p}[(j_1, i); 011] \mathbf{p}[(j_2, i); 001] + \tag{5.41}$$
$$p_i \left(\frac{\mu_u(j_2, i)}{\mu_d(j_2, i)} - 1 \right) \mathbf{p}[(j_1, i); 001] \mathbf{p}[(j_2, i); 011] +$$
$$\left. (r_u(j_1, i) + r_u(j_2, i)) \, \mathbf{p}[(j_1, i); 001] \mathbf{p}[(j_2, i); 001] \right] \frac{1}{PR(i, q)}$$

and which is *not* a generalization of Burman's equations.

5.2.6.2 Downstream Machine. *Priority One Line.* To an observer in Buffer $B_{j_1, i}$, there are two possible reasons to see Machine $M_d(j_1, i)$ down: First, Machine M_i can be down. Second, Machine M_i can be blocked because Buffer $B_{i, q}$ is full and Machine $M_d(i, q)$ is down. This is similar to the situation of an observer in a linear flow line without split or merge operations. The interruption of flow equation for the linear flow line is

$$p_d(j_1, i) = p_i + K_6 \, \mu_d(j_1, i) \tag{5.42}$$

with

$$K_6 = \left[p_i \left(\frac{\mu_d(i, q)}{\mu_u(i, q)} - 1 \right) \mathbf{p}[(i, q); N(i, q); 11] + \right.$$
$$\left. r_d(i, q) \, \mathbf{p}[(i, q); N(i, q); 10] \right] \frac{1}{PR(j_1, i)}$$

This equation is based on the assumption that Machine M_i cannot experience an operation dependent failure if Buffer $B_{j_1,i}$ is empty and Machine $M_u(j_1, i)$ is down. However, if Machine M_i has two input buffers, it may fail while processing material from the priority two buffer $B_{j_2,i}$. If this happens, the observer in the priority one buffer sees a failure of the downstream machine even though his virtual downstream machine $M_d(j_1, i)$ was starved. From the perspective of the observer in the priority one buffer, this is not an operation dependent failure. As an approximation, we ignore this type of failure in (5.42). In order to include this type of failure in the two-machine decomposition approach, we would have to develop a new two-machine model that allows for multiple failure modes and we would have to derive a complete additional set of decomposition equations dealing with this second failure mode. Ignoring these failures can be justified if the numerical results are satisfactory.

Priority Two Line. From the perspective of an observer in the priority two buffer $B_{j_2,i}$, the virtual downstream machine $M_d(j_2, i)$ is down if at least one of following three conditions holds: Machine M_i is down, the priority one buffer $B_{j_1,i}$ is non-empty, or the buffer $B_{i,q}$ downstream of Machine M_i is full and Machine $M_d(i, q)$ is down. If the priority one buffer $B_{j_1,i}$ is non-empty, all the capacity of Machine M_i is directed at the material in this buffer and therefore $M_d(j_2, i)$ is down.

The reasons for Machine $M_d(j_2, i)$ to be down at time t as given above and in (5.29) are not disjoint. For example, Machine M_i can fail while processing material from a non-empty priority one buffer.

In Appendix B.4.2 on Page 264 the interruption of flow equation

$$p_d(j_2, i)\delta t = \text{prob}\ \Big[\{\alpha_d[(j_2, i), t + \delta t] = 0\} \Big| \tag{5.43}$$
$$\{\alpha_d[(j_2, i), t] = 1\} \text{ and } \{n[(j_2, i), t] > 0\} \Big]$$

is approximated as

$$p_d(j_2, i) = p_i + K_6\, \mu_d(j_2, i) \tag{5.44}$$

with

K_6

$$
= \left[p_i \left(\frac{\mu_d(i,q)}{\mu_u(i,q)} - 1 \right) \mathbf{p}[(i,q); N(i,q); 11] \Big(\mathbf{p}[(j_1,i); 011] + \mathbf{p}[(j_1,i); 001] \Big) + \right.
$$

$$
\left[\begin{array}{ll} r_u(j_1,i) & \text{if } \mu_u(j_1,i) > \mu_d(j_1,i) \\ 0 & \text{otherwise} \end{array} \right] \mathbf{p}[(j_1,i); 001] +
$$

$$
\left. r_d(i,q) \, \mathbf{p}[(i,q); N(i,q); 10] \Big(\mathbf{p}[(j_1,i); 011] + \mathbf{p}[(j_1,i); 001] \Big) \right] \frac{1}{PR(j_2,i)}
$$

Note that in K_6 failures of $M_d(j_2, i)$ due to a priority one buffer becoming non-empty can only occur if Machine $M_u(j_1, i)$ is faster than $M_d(j_2, i)$.

5.2.7 Simultaneous Solution of the Decomposition Equations

The decomposition equations for upstream machines have the following general structure

$$
\mu_u(i,m) = \frac{1}{K_1} \frac{r_u(i,m) + p_u(i,m)}{r_u(i,m)\mu_u(i,m)}
$$

$$
r_u(i,m) = r_i + K_3 \frac{\mu_u(i,m) r_u(i,m)}{p_u(i,m)}
$$

$$
p_u(i,m) = p_i + K_5 \mu_u(i,m)
$$

and differ only with respect to the value of K_1, K_3, and K_5. This set of equations can be solved simultaneously[4] to find:

$$
\mu_u(i,m) = \frac{p_i + r_i}{K_1 r_i + K_3 - K_5} \tag{5.45}
$$

$$
r_u(i,m) = \frac{K_1 p_i r_i + K_3 p_i + K_5 r_i}{K_1 p_i - K_3 + K_5} \tag{5.46}
$$

$$
p_u(i,m) = \frac{K_1 p_i r_i + K_3 p_i + K_5 r_i}{K_1 r_i + K_3 - K_5} \tag{5.47}
$$

A similar set of decomposition equations for downstream machines has the following general structure

---

[4] (Burman, 1995, p. 85-86)

$$\mu_d(j,i) = \frac{1}{K_2}\frac{r_d(j,i)+p_d(j,i)}{r_d(j,i)\mu_d(j,i)}$$

$$r_d(j,i) = r_i + K_4\frac{\mu_d(j,i)r_d(j,i)}{p_d(j,i)}$$

$$p_d(j,i) = p_i + K_6\mu_d(j,i)$$

and can be solved simultaneously to find:

$$\mu_d(j,i) = \frac{p_i + r_i}{K_2 r_i + K_4 - K_6} \tag{5.48}$$

$$r_d(j,i) = \frac{K_2 p_i r_i + K_4 p_i + K_6 r_i}{K_2 p_i - K_4 + K_6} \tag{5.49}$$

$$p_d(j,i) = \frac{K_2 p_i r_i + K_4 p_i + K_6 r_i}{K_2 r_i + K_4 - K_6} \tag{5.50}$$

Using this closed-form solution proposed by Burman leads to an improved convergence behavior of the algorithm.

5.3 The Algorithm to Determine Performance Measures

The algorithm to solve the decomposition equations for the continuous material model in order to determine average production rate and inventory level estimates is very similar to the one described in great detail in Section 4.3. For this reason, we only describe the following differences between the two algorithms:

1. We have to determine three instead of two parameters for each virtual machine in the decomposition. To do so, we first initialize the virtual two-machine lines according to

$$\mu_u(i,m) = \mu_i d_{i,m}, \tag{5.51}$$

$$p_u(i,m) = p_i, \tag{5.52}$$

$$r_u(i,m) = r_i, \tag{5.53}$$

$$\mu_d(i,m) = \mu_m, \tag{5.54}$$

$$p_d(i,m) = p_m, \tag{5.55}$$

$$r_d(i,m) = r_m \tag{5.56}$$

such that the speed of $M_u(i,m)$ takes the fraction $d_{i,m}$ of material into account that is routed from M_i to M_m in the real system.

2. In order to improve the convergence of the algorithm with respect to merge structures, we found it useful to reduce the initial guess $\mu_d(i,m)$

whenever Machine M_m in the real system performs a merge operation. For these cases, we set

$$\mu_d(i, m) \ = \ 0.1\mu_m \tag{5.57}$$

for both priority one and priority two input buffers $B_{i,m}$.

3. Since in the continuous material model nothing is assumed to be stored at the workspace of any machine, we set the buffer size $N(i, m)$ for each two-machine line

$$N(i, m) = C_{i,m} \tag{5.58}$$

equal to the number $C_{i,m}$ of physical buffer spaces between the machines.

4. To compute steady state probabilities and production rate estimates $PR(i, m)$ for any two-machine line $L(i, m)$, we use the two-machine model developed in Gershwin and Schick (1980), which is presented in great detail in (Gershwin, 1994, p. 112-133).

5. In equations that propagate production rate estimates like (4.60), any expression of the type $E(.)$ is replaced by $PR(.)$.[5]

6. During the upstream phase of each iteration, we compute coefficients K_1 according to (5.5). To determine K_3, we use (5.14) or (5.25) for split or merge operations at the upstream machines, respectively. In a similar way, K_5 is computed from either (5.35) or (5.41).

7. Preliminary values for upstream parameters $\mu_u^*(i, m)$, $p_u^*(i, m)$, as well as $r_u^*(i, m)$ can then be derived from the simultaneous solution of the decomposition equations as given in (5.45), (5.46), and (5.47). They may be exponentially smoothed as in (4.62) and (4.63) if convergence problems occur.

8. Occasionally, even smoothed parameter updates $\mu_u^{**}(i, m)$, $p_u^{**}(i, m)$, and $r_u^{**}(i, m)$ may occasionally be negative. Since negative processing, failure and repair rates are meaningless, we impose a set of hard constraints according to

---

[5] In the discrete time-discrete material model of Section 4, the average production rate $PR_u(i, m)$ of a virtual machine $M_u(i, m)$ is equal to the fraction of time $E_u(i, m)$ it is up and operating since the processing time is the time unit. In the continuous time-continuous material model, however, these quantities are not identical unless the processing rate is one part per time unit. We therefore have to distinguish these quantities and use the production rate $PR(.)$ as determined by the Schick and Gershwin procedure.

$$\mu_u^k(i,m) = \begin{cases} \mu_u^{**}(i,m) & \text{if } \mu_u^{**}(i,m) > 0 \\ 0.9\mu_u^{k-1}(i,m) & \text{if } \mu_u^{**}(i,m) \le 0 \end{cases} \quad (5.59)$$

$$p_u^k(i,m) = \begin{cases} p_u^{**}(i,m) & \text{if } p_u^{**}(i,m) > 0 \\ 0.9p_u^{k-1}(i,m) & \text{if } p_u^{**}(i,m) \le 0 \end{cases} \quad (5.60)$$

$$r_u^k(i,m) = \begin{cases} r_u^{**}(i,m) & \text{if } r_u^{**}(i,m) > 0 \\ 0.9r_u^{k-1}(i,m) & \text{if } r_u^{**}(i,m) \le 0 \end{cases} \quad (5.61)$$

to determine values for $\mu_u^k(i,m)$, $p_u^k(i,m)$, and $r_u^k(i,m)$ that are used in iteration k to update performance measures for Line $L(i,m)$. The procedure for downstream parameters $\mu_d^k(i,m)$, $p_d^k(i,m)$, and $r_d^k(i,m)$ is analogous.

5.4 Numerical Results: Algorithm and Flow Line Behavior

In order to evaluate the decomposition method for the continuous material model with split and merge operations, we performed two large-scale numerical studies based on artificial problems with random parameters. Given that the impact of split and merge operations had already been studied in Section 4 in great detail, the focus of the study in this section was on the accuracy and reliability of the algorithm for the continuous material model.

In the first part of the study, we used the continuous material model to re-evaluate again all random problems already analyzed in Section 4. The continuous material model had been developed in order to analyze systems with different, but deterministic, processing times at the machines. In the systems studied in Section 4, however, all processing times were assumed to be deterministic and *identical*, which is a special case of the more general problem treated in the continuous material model.

In order to use the continuous material model to solve the random problems in Section 4, we set all the processing rates $\mu_i = 1.0$ and interpreted the failure and repair parameters p_i and r_i as *rates* corresponding to the exponential distribution in continuous time instead of probabilities in discrete time. When solving the two-machine lines in the decomposition, we set the two-machine-line buffer size $N(i,j)$ of the virtual two-machine lines equal to the capacity $C_{i,j}$ of the buffer between Machines M_i and M_j in the real system.[6]

---

[6] This is because in the continuous material two-machine model, no material is assumed to be stored at the workspace of a machine. In the discrete material models, however, we assume that a workpiece can actually be stored at the workspace of a machine while it is being processed, while the machine is under repair, or while the downstream buffer is full and the machine is blocked. In the latter case, we therefore set $N(i,j) = C_{i,j} + 2$ for the extended storage of the two-machine line that takes into account the possibility that a processed part

Structure	# of problems solved		Rel. Dev.		Superior accuracy	
	DMM	CMM	DMM [%]	CMM [%]	DMM	CMM
S1	100	100	0.81	0.38	13	87
S2	99	100	0.60	0.34	19	80
S3	83	100	1.26	0.65	10	73
M1	91	45	0.29	0.19	19	24
M2	85	71	2.27	2.34	28	33
L1	99	90	0.99	0.74	36	53
L2	95	87	1.38	0.93	18	64
L3	78	71	1.68	1.52	18	46
L4	77	74	1.77	1.37	12	49
L5	99	79	1.55	1.18	27	51
L6	76	49	1.65	4.09	37	1

Table 5.1. Comparison of the Discrete and the Continuous Material Model: Identical Processing Times

The results of this first study of random problems for the continuous material model are presented in Table 5.1. For each structure S1–L6 introduced in Section 4, 100 random systems were analyzed using both a simulation model and approximation techniques for the discrete material model (DMM) developed in Section 4 and the continuous material model (CMM) developed in this section. The second and third column in Table 5.1 give the number of systems for which the DMM and the CMM algorithm converged. The fourth and fifth column is the respective average over the absolute values of the relative deviation of the production rate estimate provided by the analytical and the simulation method. For most structures, at least one of the two analytical approaches occasionally failed to converge. However, most problems could be solved using both the DMM and the CMM approach. In all these cases where both approaches converged, we asked for the method that led to the more accurate production rate estimate. The results are reported in the two last columns of Table 5.1.

This first part of the study suggests that for systems with identical processing times and pure split structures (S1–S3), the CMM leads to results that are both more accurate and that are achieved more reliably. The decomposition for the CMM was able to solve all 300 cases, whereas the DMM failed for some of the more complicated split systems based on Structures S2 and S3.

For pure merge systems, i.e. Structures M1 and M2, convergence reliability is the major problem. Notice that for system M1, the DMM converged for 91 out of 100 cases, whereas the CMM converged for 45 cases only. The results for the DMM indicate that pure merge structures are more difficult

---

cannot leave the upstream machine because the downstream buffer is full and the workspace at the downstream machine is also occupied.

to analyze than pure split structures, and these structures appear to be especially difficult for the CMM. However, the results for the accuracy indicate that if both approaches led to a result, the CMM tended to be more accurate.

The situation for systems with loops is similar, but less dramatic. For Structures L1 to L5, the CMM was slightly less reliable than the DMM with respect to convergence, but whenever both algorithms converged, the CMM tended to produce the more accurate result and the average accuracy is rather high. Structure L6, however, is basically a very large merge network that is very difficult for both methods to analyze, and here both methods reach their limits.

The conclusion from this first part of the numerical study is that even if processing times are identical at all machines and we could hence use the discrete material model developed in Section 4, one can expect to get more accurate results from the continuous material model. The exception from this general rule is that for (especially pure) merge systems, the discrete material model should be used as it appears to suffer less from convergence problems.

Besides random problems, we also studied in Section 4 systems where one or several parameters were varied in a systematic way in order to explore systematic effects of both algorithm and manufacturing system behavior. For some of these systematic cases, the algorithm for the DMM performed rather poorly. Most of them were related to very infrequent failures and very long average repair times. Consider the results for the problem class L1C3 in Figure 4.26 on Page 138. For small buffer sizes, the DMM failed to produce useful results. Now look at Figure 5.2 that compares the DMM and the CMM production rate estimate as a function of the (extended) buffer size N. Note that the CMM produces more accurate results than the DMM even for small buffer sizes.

Another problem class where the DMM failed dramatically is L1C4, see Figure 4.28 on Page 139. Especially for very high rejection probabilities, the performance of the DMM was very poor. Figure 5.3 shows that the CMM provides less inaccurate production rate estimates than the DMM for the whole range of acceptance probabilities $d_{3,4}$.

Problem Class L4C4[7] is a feedforward loop with infrequent failures and repairs for which the DMM produced rather inaccurate results as shown in Figure 4.51 on Page 153. The CMM, however, leads to very accurate results for routing probabilities $d_{3,4}$ between 0.9 and 0.1, as Figure 5.4 indicates.

For the large network of Structure L5, Figure 5.5 on Page 187 shows that even for small buffer sizes the CMM leads to much more accurate results than the DMM.

The conclusion is that for identical processing times at all machines, the CMM not only provides better results than the DMM on the average, but also for those cases that appear to be particularly difficult for the DMM.

---

[7] See Figure 4.45 on Page 150.

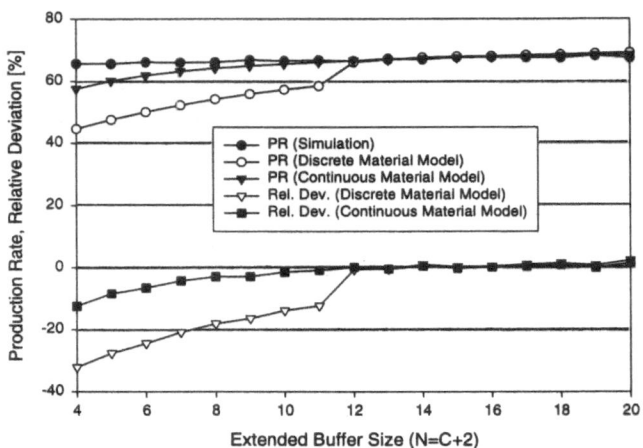

Fig. 5.2. Comparison for Problem Class L1C3

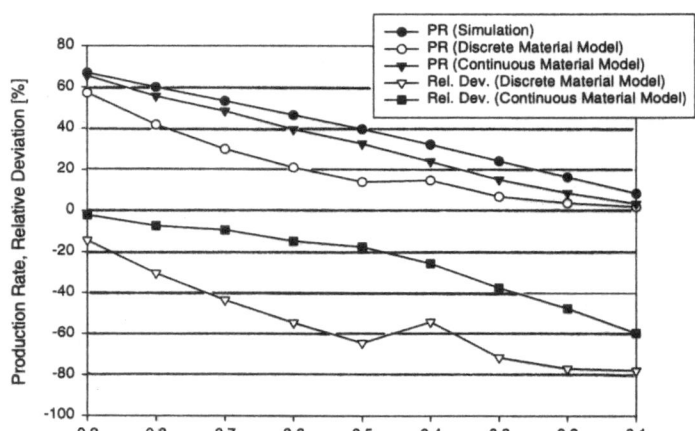

Fig. 5.3. Comparison for Problem Class L1C4

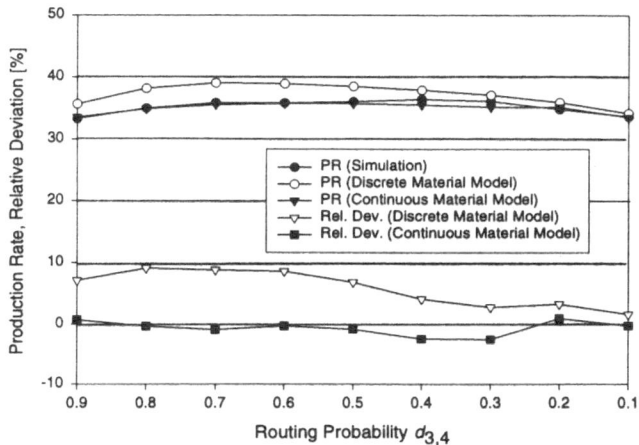

Fig. 5.4. Comparison for Problem Class L4C4

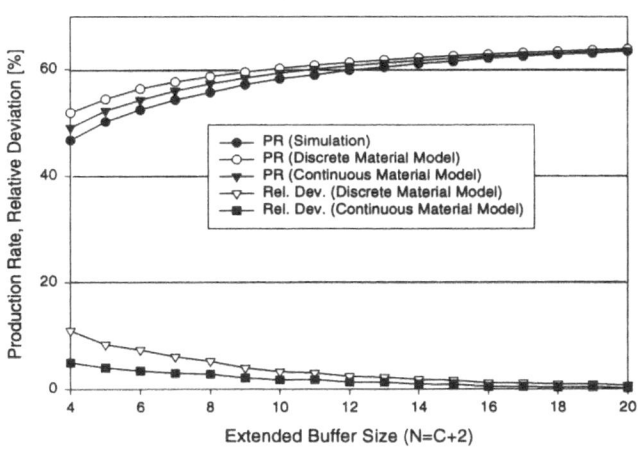

Fig. 5.5. Comparison for Problem Class L5C1

Most of the systems studied above are fairly unbalanced due to the assumption of identical processing times.[8] In a system with a rework loop, for example, a rework station tends to be less busy than a station in the main part of the line. If we assume identical processing times at all stations or machines, this rework station is idle most of the time. In order to study random problems that are more realistic in the sense that the system is more balanced, we created an additional set of 100 random problems for each structure S1–L6 using the approach in (Burman, 1995, p. 92–94). In this second set of 1,100 random problems, the processing rates of all the machines are random as well, in addition to random failure and repair rates and buffer sizes.

To create problems with machine-specific random processing times, we first generated for each problem a (pseudo-)random variable

$$PROD = 0.1 + RAN_a \tag{5.62}$$

where RAN_a denotes a call to a random-number generator yielding a number uniformly distributed between 0 and 1. In a second step, $PROD$ was used to generate a set of preliminary processing rates μ_i^p according to

$$\mu_i^p = PROD(3.6 + 0.8RAN_{b,i}) \tag{5.63}$$

for each machine M_i. Note that for each machine there is a separate call $RAN_{b,i}$ of the random number generator, i.e. these preliminary processing rates are different for each machine of the system, but within the same order of magnitude.

Due to the split and merge operations, we had to modify the preliminary production rates μ_i^p using a factor f_i for each machine in order to arrive at roughly balanced systems. We computed processing rates μ_i using

$$\mu_i = f_i\mu_i^p \tag{5.64}$$

where factors f_i for each machine and each structure are given in Table 5.2.

This table is to be read as follows: For Structure S1 depicted in Figure 4.7 on Page 117, the processing rate of the third machine (which receives 90% of the material) is $\mu_3 = 0.9\mu_3^p$ and the processing rate of Machine M_4 is computed as $\mu_4 = 0.1\mu_4^p$ as 10% of the parts are sent from Machine M_3 to M_4. The other factors in Table 5.2 have been chosen in a similar way to create roughly balanced systems with random processing times.

---

[8] Remember that for Structures M2 and L6 we had to adjust the failure and repair parameters to create more balanced systems in order to get the algorithm for the DMM to converge, see Pages 132 and 155, respectively.

Failure and repair rates were computed as described in Section 4.4.1.2 and the number of physical buffer spaces between adjacent machines M_i and M_j was determined according to

$$C_{i,j} = 1 + MAX[\frac{\mu_j}{r_i}, \frac{\mu_i}{r_j}]3RAN_{6,j,i} \qquad (5.65)$$

as a function of the material processed during an average repair.

	S1	S2	S3	M1	M2	L1	L2	L3	L4	L5	L6
M_1	1	1	1	0.5	0.5	0.9	0.9	1	1	0.85	0.8
M_2	1	1	1	0.5	0.5	1	1	1	1	0.85	1
M_3	0.9	0.9	0.9	1	0.25	1	1	0.5	0.5	1	1
M_4	0.1	0.05	0.9	1	0.25	0.9	1	0.5	0.5	1	1
M_5		0.05	0.1		0.5	0.1	1	1	0.5	1	0.8
M_6			0.05		1		0.9	1	0.5	0.8	1
M_7			0.05		1		0.1		1	0.8	1
M_8							0.1		1	0.76	1
M_9										0.76	1.4
M_{10}										0.76	1.4
M_{11}										0.15	1.33
M_{12}										0.2	1.33
M_{13}										0.04	1.33
M_{14}										0.05	0.2
M_{15}											0.2
M_{16}											0.42
M_{17}											0.07
M_{18}											0.1
M_{19}											0.1

Table 5.2. Processing Rate Adjustment Factors f_i for Each Machine and Structure

Using the decomposition equations and the algorithm for the continuous material model, we tried to evaluate 1,100 random problems, 100 for each of the 11 structures S1–L6.

These results in Table 5.3 compared to those in Table 5.1 on Page 183 show again that convergence is especially difficult for large networks with merge structures such as M2 and L6. The overall degree of convergence reliability appears to be comparable. For both algorithms there are many cases where the method fails and where further work appears to be worthwhile. The accuracy of the production rate estimate appears to be higher if all processing times are identical.

There were a few cases with an extreme deviation of more than 20% between the production rate estimates obtained using simulation and the analytical approach. This happened twice for Structures S3 and L6 and once for Structure L5. See the fourth column in Table 5.3. The two last columns in Table 5.3 show that these were relatively rare events. Consider Structure

	# Conv.	Rel. dev. [%]	Max. dev. [%]	Dev. < 2%		Dev. < 5%	
				abs.	rel. [%]	abs.	rel. [%]
S1	100	0.49	3.88	99	99.0	100	100.0
S2	100	2.65	18.58	65	65.0	83	83.0
S3	100	5.40	22.81	31	31.0	60	60.0
M1	90	1.22	7.53	71	78.9	88	97.8
M2	58	1.55	6.10	41	70.7	55	94.8
L1	77	1.66	12.87	61	79.2	70	90.9
L2	69	2.26	14.15	47	68.1	64	92.8
L3	85	2.11	10.55	51	60.0	79	92.9
L4	90	2.71	9.36	37	41.1	80	88.9
L5	66	4.22	23.39	20	30.3	46	69.7
L6	39	8.00	98.31	11	28.2	25	64.1

Table 5.3. Random Problems: Results for Machine-Specific Processing Times

M2 where the algorithm converged only for 58 out of 100 random problems. In 55 out of these 58 cases—or 94.8%—the relative deviation was less than 5%.

We conclude that it is apparently more difficult to analyze random problems with machine-specific random processing time than those with identical processing times studied in Table 5.1. However, even for these rather difficult problems the algorithm often converged and produced reasonably accurate results. It depends on the structure and parameters of the system whether the algorithm can be used in practice or whether an improved method is required. If problems occur, they are usually related to merge structures. Thus, the way merge structures are analyzed ist the weak spot of the method and the place where improvement appears to be most urgent.

5.5 Optimal Design of Systems with Loops and Different Processing Times

The optimal design of a flow line depends on the processing, failure and repair rate of each machine in the system. A model that allows for machine-specific processing rates can be used to evaluate the economic impact of different processing rates at a given machine. Consider again the eight-machine line with a rework loop depicted in Figure 5.6 where the numbers above the buffers represent buffer indices. Assume that all machines but M_1 have already been selected and that the parameters are as shown in the example on Page 160 with the exceptions that p_i and r_i are now interpreted as failure and repair rates instead of probabilities and that in the initial minimal buffer allocation each buffer can hold one part between adjacent machines. The processing rates for all machines but M_1 were set to $\mu_i = 1$, M_1's failure and repair rates to 0.01 and 0.1, respectively, and we assumed that 90% of the parts are good, i.e. $d_{5,6} = 0.9$.

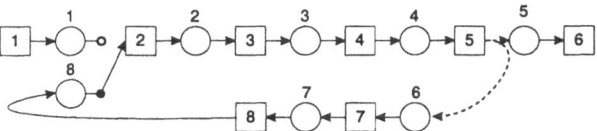

Fig. 5.6. Structure L2

One might ask how the speed of the first machine, which has not yet been chosen, affects the profitability of the investment for the rest of the system, especially with respect to the optimal buffer allocation. This question arises if several different machines are available to perform a given task.

In order to answer this question, we varied the Machine M_1 processing rate μ_1 between 0.2 and 2.0 parts per time unit. For each value of μ_1, we calculated estimates of the production rate and the net present value (NPV) as defined in Equation (2.5) on Page 14. The NPV estimates included Machines M_2-M_8, but not Machine M_1, which has yet to be chosen. The first set of estimates was for the initial minimal buffer allocation, i.e. the $C_{i,j} = 1$ case. The second set of estimates was for the buffer allocation that yields the maximum NPV.

The NPV of the complete investment including all machines but M_1 can be interpreted as the maximum justifiable NPV of the isolated investment for Machine M_1. The latter can be calculated as $-A_1 + L_1 e^{-iT}$ as we have to take both the initial payment A_1 and the discounted value of any possible scrap value L_1 into account. If both investments taken together have an NPV of zero, a risk-neutral investor would find the project just as worthwhile as investing his or her money on the perfect capital market.

The results with respect to the NPV as a function of μ_1 are depicted in Figure 5.7. The graphs indicate that for a very low value of μ_1 the NPV of the whole investment except M_1 is already negative, which means that there is no possibility to buy a machine M_1 such that the NPV of the investment including M_1 becomes positive. In this situation, M_1 is the bottleneck of the system and the other machines are idle most of the time, which is why so little revenue is produced. Figure 5.8 shows that the production rate is basically determined by μ_1 and allocating the economically optimal number of buffer spaces does not have a major impact.

However, as μ_1 approaches values around 0.8, the system becomes balanced and economically sound. For the given parameters, this leads to a positive NPV, even for the case of the minimal buffer allocation. Increasing μ_1 above values of 1.0 parts per time unit does not lead to any significant increase of the NPV anymore since now the rest of the system is too slow to keep up with Machine M_1. Note that in Figure 5.7, the gain from optimizing the buffer allocation has a maximum at $\mu_1 = 0.8$, which

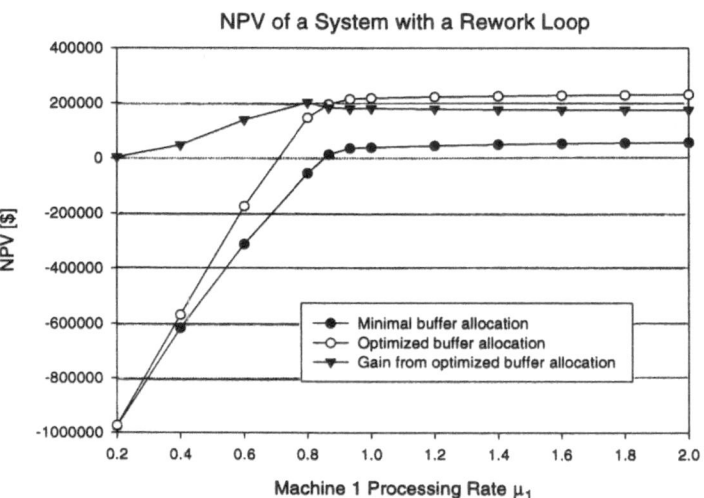

Fig. 5.7. NPV as a Function of μ_1

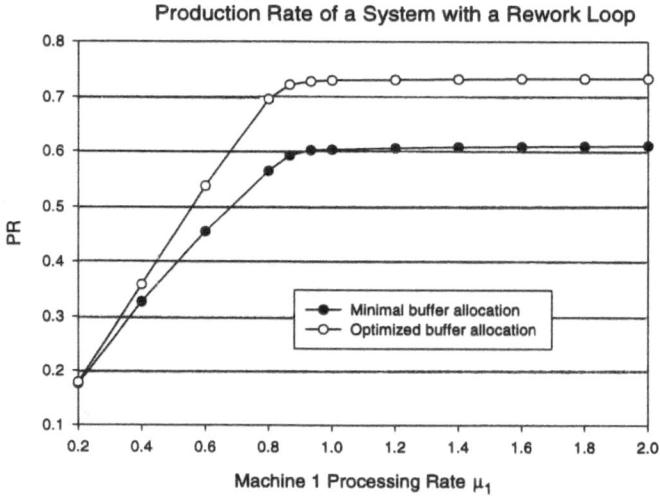

Fig. 5.8. PR as a Function of μ_1

is when the system is most evenly balanced. The economically optimal production rate of the system is about 0.733 parts per time unit. Note that Machines M_2, M_3, M_4, and M_5 actually process $10/9 \cdot 0.733 = 0.814$ parts per time unit, including the reworked parts. Since their isolated production rate $\mu_i r_i/(r_i + p_i)$ is approximately $1.0 \cdot 10/11 = 0.909$, these machines are idle $(0.909 - 0.814)/0.909 = 10.5\%$ of the time they are operational. It is possible to increase the production rate by adding more buffers, but this leads to a decrease of the NPV and is, therefore, not economically worthwhile with respect to the decision problem as formulated on **Page 18** of this thesis.

Optimized Buffer Allocation as a Function of Machine 1 Speed

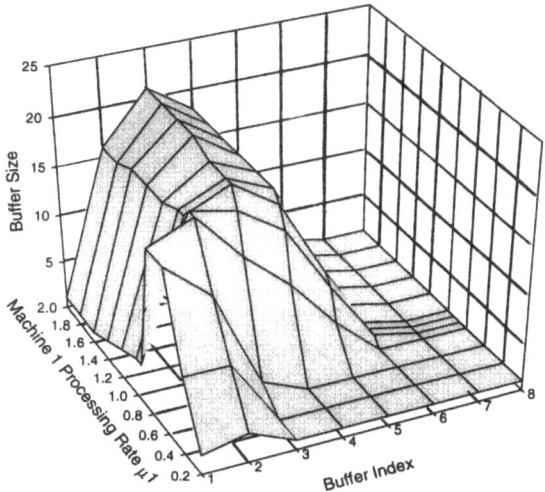

Fig. 5.9. Optimal Buffer Allocation as a Function of μ_1

The optimal buffer allocation for different processing rates μ_1 is depicted in Figure 5.9. The graph shows that for low values of μ_1 up to 0.6, only the first two or three buffers receive more than one buffer space. The reason is that for low values of μ_1, Machine M_1 is the bottleneck and attracts the few buffers that can be economically justified. Only a few additional buffer spaces lead to an increase of the NPV compared to the initial minimal buffer allocation, since the main part of the system is so much faster than M_1 that disruptions in the flow of material in the main part of the system rarely affect M_1.

However, as μ_1 approaches 1.0, M_1 is becoming too fast for the rest of the system and now the required number of buffer spaces between M_1 and M_2 decreases again as μ_1 increases. In this situation, most of the buffers are required in the main or inner part of the system between Machines M_2 and M_5 as these are now the bottleneck machines. For values of μ_1 between 1.4 and 2.0, the optimal buffer allocations are almost identical.

The analysis of this example illustrates the economic importance of designing balanced systems and to allocate the right number of buffer spaces. This can have a major impact on the profitability of the investment in a flow line, much higher than the relative difference of the (technical) production rates might suggest. In the presence of rework loops, the isolated production rates of the machines have to be different. We see again that in an optimal design even bottleneck machines are not always busy, unless buffers are free.

To evaluate and design such systems, models of the continuous material type are required. Therefore, the method developed in this section provides some potentially valuable insight into the behavior and design of systems with rework loops and machine-specific processing times. However, as the analysis of random cases suggests, the convergence reliability of the method is not yet really satisfactory. More work is required to model merge systems in a way that is as reliable as the method for split systems already is.

6. Conclusions and Suggestions for Further Research

The design of flow or transfer lines should take the randomness of manufacturing processes into account. Random processing times, machine failures, or quality problems lead to disruptions in the flow of material that can have a significant impact on production rates, inventory levels, and eventually on the associated net present value of the project.

The focus in this thesis was on flow lines with limited buffer capacity and non-linear flow of material. Due to the limited buffers, disruptions in the flow of material propagate through the system, causing idleness of machines and waiting times of parts. While there is a lot of literature on the analysis of purely serial arrangements of machines, there is relatively little on systems that perform assembly and/or disassembly operations or on systems with rework and scrapping of bad parts at dedicated machines. In order to study the effect of these patterns in the flow of material, we presented several analytical models from which we derived decomposition equations to calculate performance measures.

Many of these equations are based on approximations. To solve these equations, we used algorithms for which a proof of convergence or accuracy is not available. For this reason, we had to perform some large-scale numerical studies to explore the possibilities and limitations of the analytical approaches.

From an economic point of view, the allocation of buffer spaces is a major economic decision variable. We introduced a decision model to determine the optimal buffer allocation that yields the highest possible value of the expected net present value of the investment for a given set of technical and economical parameters. Several numerical examples showed that an apparently 'minor' change in the production rate due to the optimal buffer allocation can actually lead to a major change of the net present value. It is, therefore, important to allocate buffer spaces efficiently. To optimize the buffer allocation, one needs fast and accurate tools to evaluate a large number of similar system designs. While this is hardly possible with currently available computers if one uses simulation, this is relatively easy and painless with analytical methods once the appropriate models and algorithms are available. This leads to several suggestions for further research:

Combination of assembly/disassembly and split/merge operations. Assembly/disassembly operations may occur in systems with loops due to scrapping or rework. It should be straightforward to combine the split/merge model in Section 4 with the assembly/disassembly model in Gershwin (1991). One might also analyze how the continuous material model in Section 5 behaves if it is combined with a closed-loop model like the one developed in Frein et al. (1996). In this model, a fixed number of pallets circulates in a transfer line and the total number of pallets, as well as the size of the buffers between adjacent work stations, are the decision variables. This model might also be combined with a continuous material model of an assembly/disassembly system to analyze configurations like the Hewlett-Packard printer production line studied and improved in Burman et al. (1998).

Convergence problems for structures with merge operations. The accurate and reliable algorithm in Burman (1995) for machine-specific processing times has been extended in this thesis to cover both split and merge operations. While this works very well for split systems and to some extent for systems with rework loops, the decomposition for merge systems is clearly the weak spot of the method. The numerical experiments show that the decomposition tends to have convergence problems, especially if the merge station is almost never starved. If the algorithm converges, however, the production rate estimates tend to be rather accurate.

Three different approaches to solve this problem are

1. to develop a different set of decomposition equations for merge operations based on the two-machine decomposition approach,
2. to develop a different algorithm to solve the equations given in this thesis numerically, and
3. to develop a new building block of the decomposition for the merge operation based on the explicit analysis of a three-machine merge system. In this case, it would also be necessary to develop a new set of decomposition equations for merge operations, but these might be easier to derive.

In the last of these three options, one has to develop an exact model of a three-machine model with one or two buffers as depicted in Figure 6.1.

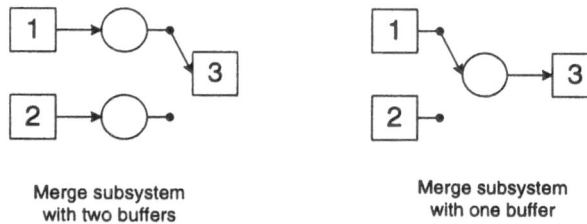

Merge subsystem
with two buffers

Merge subsystem
with one buffer

Fig. 6.1. Models of a Merge Process

Note that the subsystem depicted on the left side of Figure 6.1 is identical to the type of merge systems studied in this thesis while in the right system, there is only one buffer that receives material from both upstream machines. The dynamics of these systems will be more difficult to analyze than a two-machine model, but once such a building block is available, the convergence problems might be reduced. In a different context, Bonvik (1996, p. 61) develops an exact model of a three-machine system with two buffers where one of the machines works as a perfectly reliable and infinitesimally fast synchronization stage to match demand and material. While this exact model of a three-machine system has nothing to do with a merge process, it might be possible to analyze a three-machine merge system in a similar way.

Random processing times in reliable systems. Another possibly worthwhile path of research addresses non-linear flow of material in reliable systems with random processing times. In some systems, the disruptions in the flow of material may be due to random changes in the productivity of workers. In other systems, the workload per product unit and production stage may not be identical if several variants of a product are produced simultaneously. From the perspective of a work station, the workload of arriving parts can appear random, but there might be no randomness due to failures or repairs.

To analyze these systems for a linear flow of material, several methods are available. For systems with small coefficients of variation of the processing times, the approaches by Buzacott and Shanthikumar (1993, p. 204), Buzacott et al. (1995), and Bürger (1997, p. 69) can be used. They are based on a decomposition in $GI/G/1/N$ stopped-arrival queuing models. For coefficients of variation larger than one, the accuracy of these approaches appears to decrease. In this situation, one might consider a two-machine decomposition where the processing times are Cox-2 distributed. Two methods to solve the respective two-machine models are given in Buzacott and Kostelski (1987). One might develop decomposition equations for systems with assembly/disassembly, as well as split and merge operations for these systems with reliable workstations and random processing times.

Multiple product types. A transfer line may be used to produce a set of similar products simultaneously. It would be useful to be able to distinguish the different product types at the different production stages. The split systems studied in this thesis can be interpreted as producing different product types, but they do so randomly in fixed proportions. This may or may not be a reasonable assumption in a multiple-product transfer line. An alternative is that there are priorities for different product types at different production stages. Priorities have also been used in the model of merge systems derived in this thesis. It might be possible to modify these methods in order to analyze multiple-product transfer lines.

Variance of output. Another field of research that is highly important, but apparently very difficult to study is related to the variance of the output of flow lines. Many methods like these developed in this thesis provide only average values of production rates and inventory levels. However, production rates can fluctuate strongly in the short run and it would be useful to determine not only the average but also the variance of these quantities. In the context of two-machine decompositions, there is limited literature on the variance of the output of unreliable transfer line, see Gershwin (1993) and Carrascosa (1995). The variance of the output is especially important in the short run. The buffer sizes in the previously discussed models represented physical limitations of the production systems. However, they can also represent logical limitations of the work in process between adjacent work stations. In this case, they are a control parameter that can be changed in the short run in order to respond to changes in the environment, such as lower or higher demand than expected.

Uncertainty and the design of transfer lines. It is not only important to link technical performance measures like production rates to economic quantities like the expected net present value of the investment in a transfer line. It is also important to assess the uncertainty of any such estimate. Given that analytical methods allow to evaluate many different systems quickly, it is possible to deal with the uncertainty issue in a systematic way. In a *sensitivity analysis* one may ask how the value of a proposed system design changes if some of the system parameters like mean times to failure change. In a *risk analysis* one may assume that some of the parameters are subject to uncertainty and can be interpreted as random variables. Assume that for a single machine there are three estimates for mean times to failure, representing a pessimistic, a medium, and an optimistic estimate with assigned (subjective) probabilities. In this situation, any given system design leads to a distribution of expected net present values. A quick and reasonably accurate method for the performance evaluation allows to explore the shape of this distribution in order to assess the uncertainty associated with a given transfer line design.

A. Derivation for the Discrete Material Flow Line

This appendix presents the complete derivation of the resumption of flow equations for the discrete material model of a flow line with split and merge operations and identical deterministic processing times presented in Section 4.

A.1 Resumption of Flow Equations: Split System

A.1.1 Upstream Machine

Since the different events that force Machine $M_u(i, m)$ down are approximately mutually disjoint, we can break equation (4.21) down by decomposing the conditioning event to find:

$$
\begin{aligned}
r_u(i, m) \quad \approx \quad & A(i, m)W(i, m) + B(i, m)X(i, m) \\
& + \sum_{(i,q) \in D(i), q \neq m} \Big[C_{i,q}(i, m)Y_{i,q}(i, m) + D_{i,q}(i, m)Z_{i,q}(i, m) \Big]
\end{aligned}
\tag{A.1}
$$

where we define

$$
\begin{aligned}
A(i, m) \quad = \quad & \mathrm{prob}\Big[\{\alpha_u[(i, m), t + 1] = 1\} \mid \\
& \{\alpha_i(t) = 0\} \text{ and } \{n[(i, m), t - 1] < N(i, m)\}\Big]
\end{aligned}
\tag{A.2}
$$

$$
\begin{aligned}
W(i, m) \quad = \quad & \mathrm{prob}\Big[\{\alpha_i(t) = 0\} \text{ and } \{n[(i, m), t - 1] < N(i, m)\} \mid \\
& \{\alpha_u[(i, m), t] = 0\} \text{ and } \{n[(i, m), t - 1] < N(i, m)\}\Big]
\end{aligned}
\tag{A.3}
$$

$$
\begin{aligned}
B(i,m) \;=\;\; &\mathrm{prob}\Big[\{\alpha_u[(i,m),t+1]=1\}\mid \\
&\{n[(j,i),t-1]=0\}\text{ and }\{n[(i,m),t-1]<N(i,m)\}\Big] \quad\text{(A.4)}
\end{aligned}
$$

$$
\begin{aligned}
X(i,m) \;=\;\; &\mathrm{prob}\Big[\{n[(j,i),t-1]=0\}\text{ and }\{n[(i,m),t-1]<N(i,m)\}\mid \\
&\{\alpha_u[(i,m),t]=0\}\text{ and }\{n[(i,m),t-1]<N(i,m)\}\Big]
\end{aligned}
$$

$$
\begin{aligned}
C_{i,q}(i,m) \;=\;\; &\mathrm{prob}\Big[\{\alpha_u[(i,m),t+1]=1\}\mid \qquad\qquad\qquad\text{(A.5)} \\
&\{n[(i,q),t-1]=N(i,q)\}\text{ and }\{n[(i,m),t-1]<N(i,m)\}\Big]
\end{aligned}
$$

$$
\begin{aligned}
Y_{i,q}(i,m) \;=\;\; &\mathrm{prob}\Big[\{n[(i,q),t-1]=N(i,q)\}\text{ and }\qquad\qquad\text{(A.6)} \\
&\{n[(i,m),t-1]<N(i,m)\}\mid \\
&\{\alpha_u[(i,m),t]=0\}\text{ and }\{n[(i,m),t-1]<N(i,m)\}\Big] \quad\text{(A.7)}
\end{aligned}
$$

$$
\begin{aligned}
D_{i,q}(i,m) \;=\;\; &\mathrm{prob}\Big[\{\alpha_u[(i,m),t+1]=1\}\mid \\
&\{\beta_i(t)=q\}\text{ and }\{n[(i,m),t-1]<N(i,m)\}\Big] \qquad\text{(A.8)}
\end{aligned}
$$

$$
\begin{aligned}
Z_{i,q}(i,m) \;=\;\; &\mathrm{prob}\Big[\{\beta_i(t)=q\}\text{ and }\{n[(i,m),t-1]<N(i,m)\}\mid \\
&\{\alpha_u[(i,m),t]=0\}\text{ and }\{n[(i,m),t-1]<N(i,m)\}\Big] \quad\text{(A.9)}
\end{aligned}
$$

We now determine the conditional probabilities. Probabilities $A(i,m)$ and $W(i,m)$ deal with a possible failure of Machine M_i itself. In (A.2), $A(i,m)$ is the probability that flow resumes into Buffer $B_{i,m}$ at time $t+1$ given that Machine M_i was down at time t. For this to happen, Machine M_i must be repaired (with probability r_i) and the part that is then produced must be sent into Buffer $B_{i,m}$ (with probability $d_{i,m}$), i.e.

$$
A(i,m) = r_i d_{i,m} \qquad\qquad\qquad\text{(A.10)}
$$

In (A.3), $W(i,m)$ is the probability that a failure of the virtual machine $M_u(i,m)$ is due to a failure of Machine M_i. It can be expressed in terms of the conditional probabilities of all the other events that may lead to a failure of Machine $M_u(i,m)$

$$
W(i,m) = 1 - X(i,m) - \sum_{(i,q)\in D(i),q\neq m}(Y_{i,q}(i,m) + Z_{i,q}(i,m)) \qquad\text{(A.11)}
$$

since all these conditional probabilities add up to 1.

Probabilities $B(i, m)$ and $X(i, m)$ account for starvation of Machine M_i. In (A.4), $B(i, m)$ is the probability that flow resumes into Buffer $B_{i,m}$ at time $t + 1$, given that Machine M_i was starved at time t. The reason for an empty upstream buffer is an upstream machine failure. Therefore, the virtual upstream machine $M_u(j, i)$ must be repaired and Buffer $B_{i,m}$ must be selected, so

$$B(i, m) = r_u(j, i)d_{i,m} \tag{A.12}$$

In (A.5), $X(i, m)$ is the probability that Machine M_i is starved given that Machine $M_u(i, m)$ is down. Since event $\{n[(j, i), t - 1] = 0\}$ implies event $\{\alpha_u[(i, m), t] = 0\}$, we can write

$$
\begin{aligned}
X(i, m) \quad = \quad & \text{prob}\Big[\{n[(j, i), t - 1] = 0\} \text{ and } \{n[(i, m), t - 1] < N(i, m)\} \\
& \text{and } \{\alpha_u[(i, m), t] = 0\} \mid \\
& \{\alpha_u[(i, m), t] = 0\} \text{ and } \{n[(i, m), t - 1] < N(i, m)\}\Big] \quad \text{(A.13)}
\end{aligned}
$$

Using the definition of conditional probability, this can be written as a quotient, i.e.

$$
\begin{aligned}
X(i, m) \quad = \quad & \text{prob}\left[\begin{array}{l}\{n[(j, i), t - 1] = 0\} \text{ and } \{n[(i, m), t - 1] < N(i, m)\} \\ \text{and } \{\alpha_u[(i, m), t] = 0\}\end{array}\right] : \\
& \text{prob}\Big[\{\alpha_u[(i, m), t] = 0\} \text{ and } \{n[(i, m), t - 1] < N(i, m)\}\Big]
\end{aligned}
$$

$$\tag{A.14}$$

Since we assume that the probability of Machine M_i being starved and blocked simultaneously is negligible, the event $\{n[(j, i), t - 1] = 0\}$ implies $\{n[(i, m), t - 1] < N(i, m)\}$. It furthermore implies that Machine $M_u(j, i)$ is down as the only reason for Machine M_i to be starved is a failure of $M_u(j, i)$. For this reason, the numerator in (A.14) is approximately $\mathbf{p}[(j, i); 001]$ where $\mathbf{p}[(j, i); n\alpha_u\alpha_d]$ denotes the probability of finding the virtual upstream machine $M_u(j, i)$ in Line $L(j, i)$ in state α_u, the downstream machine $M_d(j, i)$ in state α_d and the buffer at level n.

As there must be exactly one repair for each failure, the following equation (Gershwin, 1994, p. 81-82) for the two-machine model by Gershwin and Schick holds exactly as in the transfer line decomposition in Gershwin (1987)

$$
\begin{aligned}
r_u(i, m) \, & \text{prob}\Big[\{\alpha_u[(i, m), t] = 0\} \text{ and } \{n[(i, m), t - 1] < N(i, m)\}\Big] \\
= \quad & p_u(i, m) \, \text{prob}\Big[\{\alpha_u[(i, m), t] = 1\} \text{ and } \{n[(i, m), t - 1] < N(i, m)\}\Big] \\
= \quad & p_u(i, m)E(i, m) \tag{A.15}
\end{aligned}
$$

where $E(i, m)$ is the production rate of Line $L(i, m)$ in the decomposition. The denominator in (A.14) can hence be written as:

$$\text{prob}\Big[\{\alpha_u[(i, m), t] = 0\} \text{ and } \{n[(i, m), t - 1] < N(i, m)\}\Big]$$
$$= \frac{p_u(i, m)E(i, m)}{r_u(i, m)} \qquad \text{(A.16)}$$

Using these expressions for the numerator and denominator of (A.13), we find:

$$X(i, m) = \frac{\mathbf{p}[(j, i); 001]r_u(i, m)}{p_u(i, m)E(i, m)} \qquad \text{(A.17)}$$

In (A.6), $C_{i,q}(i, m)$ is the probability that flow resumes into Buffer $B_{i,m}$ after a blocking of Machine M_i due to a failure of Machine $M_d(i, q)$. It is the repair probability of $M_d(i, q)$ times the probability of sending the part then processed at Machine M_i to M_m:

$$C_{i,q}(i, m) = r_d(i, q)d_{i,m} \qquad \text{(A.18)}$$

In (A.7), $Y_{i,q}(i, m)$ is the probability that Machine M_i is blocked due to a full buffer $B_{i,q}$, given that $M_u(i, m)$ is down. The expression for $Y_{i,q}(i, m)$ is derived like the one for $X(i, m)$ to find:

$$Y_{i,q}(i, m) = \frac{\mathbf{p}[(i, q); N(i, q)10]r_u(i, m)}{p_u(i, m)E(i, m)} \qquad \text{(A.19)}$$

We now derive an expression for $Z_{i,q}(i, m)$, the probability that a failure of Machine $M_u(i, m)$ is due to sending a part from Machine M_i to $M_q, q \neq m$. Since event $\{\beta_i(t) = q\}$ implies events $\{\alpha_u[(i, m), t] = 0\}$ and $\{n[(i, m), t - 1] < N(i, m)\}$, we can write

$$Z_{i,q}(i, m) = \text{prob}\left[\begin{array}{c} \{\beta_i(t) = q\} \text{ and } \{n[(i, m), t - 1] < N(i, m)\} \\ \text{and } \{\alpha_u[(i, m), t] = 0\}| \\ \{\alpha_u[(i, m), t] = 0\} \text{ and } \{n[(i, m), t - 1] < N(i, m)\} \end{array}\right].$$
$$\text{(A.20)}$$

Using the definition of conditional probability and equation (A.15), this can be written as

$$Z_{i,q}(i,m) = \text{prob}\left[\begin{array}{l} \{\beta_i(t) = q\} \text{ and } \{n[(i,m),t-1] < N(i,m)\} \\ \text{and } \{\alpha_u[(i,m),t] = 0\} \end{array}\right] :$$

$$\text{prob}\left[\{\alpha_u[(i,m),t] = 0\} \text{ and } \{n[(i,m),t-1] < N(i,m)\}\right]$$

$$= \text{prob}[\{\beta_i(t) = q\}]\frac{r_u(i,m)}{p_u(i,m)E(i,m)}. \tag{A.21}$$

To determine $\text{prob}[\{\beta_i(t) = q\}]$, we first analyze in Table A.1 the impli-cations of sending a part from Machine M_i to M_q at time t, i.e. the event $\{\beta_i(t) = q\}$: If at time t a part is processed by Machine M_i and sent to M_q, the virtual upstream machine $M_u(i,q)$ seen by an observer in the correspond-ing buffer $B_{i,q}$ is up. Since Machine M_i is not blocked at time t, this implies that the buffer level $n[(i,q),t-1]$ is in the interval $[0, N(i,q) - 1]$. However, there is no implication concerning the state of the virtual downstream ma-chine $M_d(i,q)$. These implications are summarized in the first row of Table A.1.

Line	Buffer Level	Machine State
$L(i,q)$	$0 \le n[(i,q),t-1] \le N(i,q) - 1$	$\alpha_u[(i,q),t] = 1,$ $\alpha_d[(i,q),t] \in \{0,1\}$
$L(i,k),$ $k \ne q,$ $k \ne m$	$0 \le n[(i,k),t-1] \le N(i,k) - 1$	$\alpha_u[(i,k),t] = 0,$ $\alpha_d[(i,k),t] \in \{0,1\}$
$L(j,i)$	$1 \le n[(j,i),t-1] \le N(j,i)$	$\alpha_u[(j,i),t] \in \{0,1\},$ $\alpha_d[(j,i),t] = 1$

Table A.1. Implications of Event $\{\beta_i(t) = q\}$

If the part processed at Machine M_i at time t is sent to Machine M_q with $q \ne m$, then no part is sent into any other buffer $B_{i,k}$ downstream of M_i. For this reason, the virtual upstream machine $M_u(i,k)$ corresponding to any line $L(i,k)$ is down. The buffer corresponding to Line $L(i,k)$ cannot be full since Machine M_i is not blocked. However, nothing is implied with respect to Machine $M_d(i,k)$. It can be up or down. See the second row of Table A.1.

The last set of implications we have to study are those on the single upstream line $L(j,i)$. If Machine M_i is processing a part, then the virtual downstream machine $M_d(j,i)$ corresponding to Line $L(j,i)$ is up and not starved, i.e. the buffer level $n[(j,i),t-1]$ is in the interval $[1, N(j,i)]$. However, nothing is implied with respect to the state of Machine $M_u(j,i)$.

We use the implications listed in Table A.1 to write:

$$\text{prob}[\{\beta_i(t) = q\}] = \text{prob} \begin{bmatrix} \{0 \leq n[(i,q), t-1] \leq N(i,q) - 1\} \text{ and} \\ \{\alpha_u[(i,q), t] = 1\} \text{ and } \{\alpha_d[(i,q), t] \in \{0,1\}\} \\ \text{and} \\ \left\{ \begin{array}{l} \{0 \leq n[(i,k), t-1] \leq N(i,k) - 1\} \text{ and} \\ \{\alpha_u[(i,k), t] = 0\} \text{ and } \{\alpha_d[(i,k), t] \in \{0,1\}\}, \\ \forall (i,k) \in D(i), k \neq m, k \neq q \end{array} \right\} \\ \text{and} \\ \{1 \leq n[(j,i), t-1] \leq N(j,i)\} \text{ and} \\ \{\alpha_u[(j,i), t] \in \{0,1\}\} \text{ and } \{\alpha_d[(j,i), t] = 1\} \end{bmatrix}$$

We approximate the probability on the right hand side of this equation by treating the events for Line $L(i,q)$, Lines $L(i,k), k \neq m, k \neq q$, and Line $L(j,i)$ as if they were independent:

$$\text{prob}[\{\beta_i(t) = q\}]$$
$$\approx \quad \text{prob} \begin{bmatrix} \{0 \leq n[(i,q), t-1] \leq N(i,q) - 1\} \text{ and} \\ \{\alpha_u[(i,q), t] = 1\} \text{ and } \{\alpha_d[(i,q), t] \in \{0,1\}\} \end{bmatrix} \cdot$$
$$\text{prob} \begin{bmatrix} \left\{ \begin{array}{l} \{0 \leq n[(i,k), t-1] \leq N(i,k) - 1\} \text{ and} \\ \{\alpha_u[(i,k), t] = 0\} \text{ and } \{\alpha_d[(i,k), t] \in \{0,1\}\}, \\ \forall (i,k) \in D(i), k \neq m, k \neq q \end{array} \right\} \end{bmatrix} \cdot$$
$$\text{prob} \begin{bmatrix} \{1 \leq n[(j,i), t-1] \leq N(j,i)\} \text{ and} \\ \{\alpha_u[(j,i), t] \in \{0,1\}\} \text{ and } \{\alpha_d[(j,i), t] = 1\} \end{bmatrix} \quad \text{(A.22)}$$

These three probabilities can be obtained from the solutions of the two-machine lines $L(i,q)$, $L(i,k)$, and $L(j,i)$. We have to add over the probabilities of the respective states in each of the two-machine lines. In case of Line $L(i,q)$ that receives the part, these are all states where the upstream machine is up and not blocked. We find

$$\text{prob} \begin{bmatrix} \{0 \leq n[(i,q), t-1] \leq N(i,q) - 1\} \text{ and} \\ \{\alpha_u[(i,q), t] = 1\} \text{ and } \{\alpha_d[(i,q), t] \in \{0,1\}\} \end{bmatrix}$$
$$= \begin{bmatrix} \sum_{n=0}^{N(i,q)-1} \mathbf{p}[(i,q); n11] + \mathbf{p}[(i,q); n10] \end{bmatrix} \quad \text{(A.23)}$$

for Line $L(i,q)$. As an approximation, we treat Lines $L(i,k), k \neq m, k \neq q$, as if they were independent and add over all states where the upstream machine is down and not blocked:

$$\text{prob} \left[\begin{array}{l} \{\{0 \leq n[(i,k), t-1] \leq N(i,k) - 1\} \text{ and} \\ \{\alpha_u[(i,k), t] = 0\} \text{ and } \{\alpha_d[(i,k), t] \in \{0,1\}\}, \\ \forall (i,k) \in D(i), k \neq m, k \neq q\} \end{array} \right]$$

$$\approx \prod_{\substack{(i,k) \in D(i) \\ k \neq m \\ k \neq q}} \text{prob} \left[\begin{array}{l} \{0 \leq n[(i,k), t-1] \leq N(i,k) - 1\} \text{ and} \\ \{\alpha_u[(i,k), t] = 0\} \text{ and } \{\alpha_d[(i,k), t] \in \{0,1\}\} \end{array} \right]$$

$$= \prod_{\substack{(i,k) \in D(i) \\ k \neq m \\ k \neq q}} \sum_{n=0}^{N(i,k)-1} [\mathbf{p}[(i,k); n00] + \mathbf{p}[(i,k); n01]] \tag{A.24}$$

For the single Line $L(j,i)$ immediately upstream of Machine M_i, we add over all states where the downstream machine is up and the buffer not empty, i.e.

$$\text{prob} \left[\begin{array}{l} \{1 \leq n[(j,i), t-1] \leq N(j,i)\} \text{ and} \\ \{\alpha_u[(j,i), t] \in \{0,1\}\} \text{ and } \{\alpha_d[(j,i), t] = 1\} \end{array} \right]$$

$$= \left[\sum_{n=1}^{N(j,i)} \mathbf{p}[(j,i); n11] + \mathbf{p}[(j,i); n01] \right] \tag{A.25}$$

and therefore

$$\text{prob}[\{\beta_i(t) = q\}]$$

$$\approx \left[\sum_{n=0}^{N(i,q)-1} \mathbf{p}[(i,q); n11] + \mathbf{p}[(i,q); n10] \right] \cdot$$

$$\left[\prod_{\substack{(i,k) \in D(i) \\ k \neq m \\ k \neq q}} \sum_{n=0}^{N(i,k)-1} \mathbf{p}[(i,k); n00] + \mathbf{p}[(i,k); n01] \right] \cdot$$

$$\left[\sum_{n=1}^{N(j,i)} \mathbf{p}[(j,i); n11] + \mathbf{p}[(j,i); n01] \right] \tag{A.26}$$

which eventually leads to

$$Z_{i,q}(i,m)$$

$$\approx \left[\sum_{n=0}^{N(i,q)-1} \mathbf{p}[(i,q); n11] + \mathbf{p}[(i,q); n10] \right] \cdot$$

$$\left[\prod_{\substack{(i,k)\in D(i) \\ k\neq m \\ k\neq q}} \sum_{n=0}^{N(i,k)-1} \mathbf{p}[(i,k); n00] + \mathbf{p}[(i,k); n01] \right] \cdot$$

$$\left[\sum_{n=1}^{N(j,i)} \mathbf{p}[(j,i); n11] + \mathbf{p}[(j,i); n01] \right] \cdot$$

$$\frac{r_u(i,m)}{p_u(i,m)E(i,m)} \tag{A.27}$$

$Z_{i,q}(i,m)$ is now expressed in terms of quantities of the two-machine models in the decomposition that are related to Machine M_i and that are available in the course of the decomposition.

We now have to find an expression for the conditional probability $D_{i,q}(i,m)$

$$D_{i,q}(i,m) \quad = \quad \text{prob}\Big[\{\alpha_u[(i,m),t+1]=1\} \mid$$
$$\{\beta_i(t)=q\} \text{ and } \{n[(i,m),t-1] < N(i,m)\}\Big] \quad \text{(A.28)}$$

of processing a part at Machine M_i at time $t+1$ and sending it to M_m, given that at time t a part was sent to Machine $M_q, q \neq m$. The implications of the conditioning event $\{\beta_i(t) = q\}$ are given in Table A.1 on Page 203. To reduce the notational effort when decomposing the conditioning event, auxiliary events $a_1, a_2, a_3, b, c_1, c_2$, and c_3 are defined in Table A.2.

The mutually exclusive and collectively exhaustive events a_1, a_2, and a_3 describe the possible states of Line $L(i,q)$ that are implied by event $\{\beta_i(t) = q\}$. In a similar manner, the auxiliary event b describes the possible states of Lines $L(i,k)$ and c_1, c_2, and c_3 those of Line $L(j,i)$. Given the definition of the auxiliary variables, we find

$$\text{prob}[\{\beta_i(t) = q\}] = \text{prob}[\{a_1 \text{ or } a_2 \text{ or } a_3\} \text{ and } b \text{ and } \{c_1 \text{ or } c_2 \text{ or } c_3\}]$$

and therefore

$$a_1 = \{0 \le n[(i,q), t-1] \le N(i,q) - 2\} \text{ and }$$
$$\{\alpha_u[(i,q), t] = 1\} \text{ and } \{\alpha_d[(i,q), t] \in \{0,1\}\}$$
$$a_2 = \{n[(i,q), t-1] = N(i,q) - 1\} \text{ and }$$
$$\{\alpha_u[(i,q), t] = 1\} \text{ and } \{\alpha_d[(i,q), t] = 1\}$$
$$a_3 = \{n[(i,q), t-1] = N(i,q) - 1\} \text{ and }$$
$$\{\alpha_u[(i,q), t] = 1\} \text{ and } \{\alpha_d[(i,q), t] = 0\}$$

$$b = \{0 \le n[(i,k), t-1] = N(i,k) - 1\} \text{ and }$$
$$\{\alpha_u[(i,k), t] = 0\} \text{ and } \{\alpha_d[(i,k), t] \in \{0,1\}\},$$
$$(i,k) \in D(i), k \ne m, k \ne q$$

$$c_1 = \{2 \le n[(j,i), t-1] \le N(j,i)\} \text{ and }$$
$$\{\alpha_u[(j,i), t] \in \{0,1\}\} \text{ and } \{\alpha_d[(j,i), t] = 1\}$$
$$c_2 = \{n[(j,i), t-1] = 1\} \text{ and }$$
$$\{\alpha_u[(j,i), t] = 1\} \text{ and } \{\alpha_d[(j,i), t] = 1\}$$
$$c_3 = \{n[(j,i), t-1] = 1\} \text{ and }$$
$$\{\alpha_u[(j,i), t] = 0\} \text{ and } \{\alpha_d[(j,i), t] = 1\}$$

$$e = \{\alpha_u[(i,m), t+1] = 1\}$$
$$f = \{n[(i,m), t-1] < N(i,m)\}$$

Table A.2. Definition of Auxiliary Events

$$D_{i,q}(i,m) = \text{prob}\Big[e \,\Big|\, \{a_1 \text{ or } a_2 \text{ or } a_3\} \text{ and } b \text{ and } \{c_1 \text{ or } c_2 \text{ or } c_3\} \text{ and } f \Big]$$
$$= \text{prob}\Big[e \,\Big|\, a_1\, b\, c_1\, f \text{ or } a_1\, b\, c_2\, f \text{ or } a_1\, b\, c_3\, f \text{ or }$$
$$a_2\, b\, c_1\, f \text{ or } a_2\, b\, c_2\, f \text{ or } a_2\, b\, c_3\, f \text{ or }$$
$$a_3\, b\, c_1\, f \text{ or } a_3\, b\, c_2\, f \text{ or } a_3\, b\, c_3\, f \Big]$$

Each of the events $a_k\, b\, c_l\, f$ with $k, l \in \{1, 2, 3\}$ implies event $\{\beta_i(t) = q\}$ and each of the events a_k, $k \in \{1, 2, 3\}$ implies event f, i.e. $\{n[(i,m), t-1] < N(i,m)\}$. As an approximation, we assume that the events related to different lines are independent. In this case, $D_{i,q}(i,m)$ can be decomposed to find

$$
\begin{aligned}
D_{i,q}(i,m) \;=\; \Big[&\text{prob}[\,e\mid a_1\,b\,c_1\,]\;\text{prob}[\,a_1\,]\;\text{prob}[\,b\,]\;\text{prob}[\,c_1\,]\;+\\
&\text{prob}[\,e\mid a_1\,b\,c_2\,]\;\text{prob}[\,a_1\,]\;\text{prob}[\,b\,]\;\text{prob}[\,c_2\,]\;+\\
&\text{prob}[\,e\mid a_1\,b\,c_3\,]\;\text{prob}[\,a_1\,]\;\text{prob}[\,b\,]\;\text{prob}[\,c_3\,]\;+\\
&\text{prob}[\,e\mid a_2\,b\,c_1\,]\;\text{prob}[\,a_2\,]\;\text{prob}[\,b\,]\;\text{prob}[\,c_1\,]\;+\\
&\text{prob}[\,e\mid a_2\,b\,c_2\,]\;\text{prob}[\,a_2\,]\;\text{prob}[\,b\,]\;\text{prob}[\,c_2\,]\;+\\
&\text{prob}[\,e\mid a_2\,b\,c_3\,]\;\text{prob}[\,a_2\,]\;\text{prob}[\,b\,]\;\text{prob}[\,c_3\,]\;+\\
&\text{prob}[\,e\mid a_3\,b\,c_1\,]\;\text{prob}[\,a_3\,]\;\text{prob}[\,b\,]\;\text{prob}[\,c_1\,]\;+\\
&\text{prob}[\,e\mid a_3\,b\,c_2\,]\;\text{prob}[\,a_3\,]\;\text{prob}[\,b\,]\;\text{prob}[\,c_2\,]\;+\\
&\text{prob}[\,e\mid a_3\,b\,c_3\,]\;\text{prob}[\,a_3\,]\;\text{prob}[\,b\,]\;\text{prob}[\,c_3\,]\,\Big]\cdot\\
&\frac{1}{\text{prob}[\{\beta_i(t)=q\}]}.
\end{aligned}
\tag{A.29}
$$

The conditional probabilities in (A.29) can be determined using the information in Table A.2 on Page 207. The first conditional probability, $\text{prob}[\,e\mid a_1\,b\,c_1\,]$, is the probability that Machine $M_u(i,m)$ is up at time $t+1$, given that the buffer level in Line $L(i,q)$ that receives the part at time t, is below $N(i,q)-1$ at time $t-1$ (event a_1), that none of the buffers in any of the lines $L(i,k)$ is full (event b), and that $n[(j,i),t-1]>1$ holds for the input buffer level (event c_1).

The following has to happen in order for Machine $M_u(i,m)$ to be up at time $t+1$:

- Machine M_i in the real system must not fail, with probability $(1-p_i)$.
- The part must be sent to Machine M_m, with probability $d_{i,m}$.

Given the buffer levels in Lines $L(i,q)$, $L(i,k)$, and $L(j,i)$, Machine M_i cannot be blocked or starved at time $t+1$. Since routing decisions and machine failures are assumed to be independent, we find

$$
\text{prob}[\,e\mid a_1\,b\,c_1\,] = (1-p_i)\,d_{i,m}.
\tag{A.30}
$$

In the second conditional probability in (A.29), $\text{prob}[\,e\mid a_1\,b\,c_2\,]$, event c_2 says that Buffer $B_{j,i}$ that looses a part at time t is almost empty at time $t-1$, i.e. $n[(j,i),t-1]=1$, and that Machine $M_u(j,i)$ is up at time t. In this situation, an additional condition must be met in order for Machine $M_u(i,m)$ to be up at time $t+1$:

- Machine $M_u(j,i)$ must not fail (because otherwise Machine M_i would be starved), with probability $(1-p_u(j,i))$.

The second conditional probability is therefore

$$\text{prob}[\, e \mid a_1 \, b \, c_2 \,] = (1 - p_i) \, d_{i,m} \, (1 - p_u(j, i)). \tag{A.31}$$

In the third conditional probability in (A.29), $\text{prob}[\, e \mid a_1 \, b \, c_3 \,]$, event c_3 says that Buffer $B_{j,i}$ is almost empty at time $t-1$, and that Machine $M_u(j, i)$ is down at time t. If $M_u(j, i)$ is down at time t, it has to be repaired in order for $M_u(i, m)$ to be up at time $t + 1$, with probability $r_u(j, i)$, so

$$\text{prob}[\, e \mid a_1 \, b \, c_1 \,] = (1 - p_i) \, d_{i,m} \, r_u(j, i). \tag{A.32}$$

The derivation for the six remaining conditional probabilities is completely analogous and we find:

$$
\begin{aligned}
\text{prob}[\, e \mid a_2 \, b \, c_1 \,] &= (1 - p_i) \, d_{i,m} \, (1 - p_d(i, q)) \\
\text{prob}[\, e \mid a_2 \, b \, c_2 \,] &= (1 - p_i) \, d_{i,m} \, (1 - p_d(i, q)) \, (1 - p_u(j, i)) \\
\text{prob}[\, e \mid a_2 \, b \, c_3 \,] &= (1 - p_i) \, d_{i,m} \, (1 - p_d(i, q)) \, r_u(j, i) \\
\text{prob}[\, e \mid a_3 \, b \, c_1 \,] &= (1 - p_i) \, d_{i,m} \, r_d(i, q) \\
\text{prob}[\, e \mid a_3 \, b \, c_2 \,] &= (1 - p_i) \, d_{i,m} \, r_d(i, q) \, (1 - p_u(j, i)) \\
\text{prob}[\, e \mid a_3 \, b \, c_3 \,] &= (1 - p_i) \, d_{i,m} \, r_d(i, q) \, r_u(j, i)
\end{aligned}
$$

The special structure of these nine conditional probabilities allows to factor (A.29). This yields

$$
\begin{aligned}
D_{i,q}&(i, m) \\
&= (1 - p_i) d_{i,m} \\
&\quad \Big[\, \text{prob}[\, a_1 \,] + (1 - p_d(i, q)) \, \text{prob}[\, a_2 \,] + r_d(i, q)) \, \text{prob}[\, a_3 \,] \Big] \cdot \\
&\quad \text{prob}[\, b \,] \cdot \\
&\quad \Big[\, \text{prob}[\, c_1 \,] + (1 - p_u(j, i)) \, \text{prob}[\, c_2 \,] + r_u(j, i)) \, \text{prob}[\, c_3 \,] \Big] \cdot \\
&\quad \frac{1}{\text{prob}[\{\beta_i(t) = q\}]}.
\end{aligned}
\tag{A.33}
$$

Using the information in Table A.2 on Page 207, the unconditional probabilities can be expressed as

$$\text{prob}[\,a_1\,] \;=\; \sum_{n=0}^{N(i,q)-2} \Big(\mathbf{p}[(i,q);n11] + \mathbf{p}[(i,q);n10]\Big)$$

$$\text{prob}[\,a_2\,] \;=\; \mathbf{p}[(i,q);N(i,q)-1,11]$$

$$\text{prob}[\,a_3\,] \;=\; \mathbf{p}[(i,q);N(i,q)-1,10]$$

$$\text{prob}[\,b\,] \;=\; \left[\prod_{\substack{(i,k)\in D(i) \\ k\neq m \\ k\neq q}} \sum_{n=0}^{N(i,k)-1} \Big(\mathbf{p}[(i,k);n00] + \mathbf{p}[(i,k);n01]\Big) \right]$$

$$\text{prob}[\,c_1\,] \;=\; \sum_{n=2}^{N(j,i)} \Big(\mathbf{p}[(j,i);n11] + \mathbf{p}[(j,i);n01]\Big)$$

$$\text{prob}[\,c_2\,] \;=\; \mathbf{p}[(j,i);111]$$

$$\text{prob}[\,c_3\,] \;=\; \mathbf{p}[(j,i);101]$$

This leads to the following result

$$D_{i,q}(i,m) \approx (1-p_i)d_{i,m}\frac{F(i,q)}{\text{prob}[\{\beta_i(t)=q\}]} \tag{A.34}$$

where we define

$$
\begin{aligned}
F(i,q) \;=\; &\left[\sum_{n=0}^{N(i,q)-2} \Big(\mathbf{p}[(i,q);n11] + \mathbf{p}[(i,q);n10]\Big) \right.\\
&\left. + (1-p_d(i,q))\mathbf{p}[(i,q);N(i,q)-1,11] \right.\\
&\left. + r_d(i,q)\mathbf{p}[(i,q);N(i,q)-1,10] \vphantom{\sum} \right] \cdot \\[4pt]
&\left[\prod_{\substack{(i,k)\in D(i) \\ k\neq m \\ k\neq q}} \sum_{n=0}^{N(i,k)-1} \Big(\mathbf{p}[(i,k);n00] + \mathbf{p}[(i,k);n01]\Big) \right] \cdot \\[4pt]
&\left[\sum_{n=2}^{N(j,i)} \Big(\mathbf{p}[(j,i);n11] + \mathbf{p}[(j,i);n01]\Big) \right.\\
&\left. + (1-p_u(j,i))\mathbf{p}[(j,i);111] \right.\\
&\left. + r_u(j,i)\mathbf{p}[(j,i);101] \vphantom{\sum} \right]
\end{aligned} \tag{A.35}
$$

The fraction $F(i,q)/\text{prob}[\{\beta_i(t)=q\}]$ in (A.34) has a value between 0 and 1, as the comparison of (A.35) and (A.26) reveals. The probability $D_{i,q}(i,m)$ that Machine M_i receives a part from Machine M_i at time $t+1$, given that at time t a part was routed to Machine M_q, is therefore smaller than $(1-p_i)d_{i,m}$, as we have to take the possibility of blocking and starvation into account.

Now all components of the resumption of flow equation for the virtual upstream machine $M_u(i,m)$ are expressed in terms of parameters and performance measures of the real system or the two-machine lines in the decomposition, i.e. of quantities that are available in the course of the iterative solution of the decomposition equations. The equation can be written as

$$r_u(i,m) \;=\; \left[r_i + K_3 \frac{r_u(i,m)}{p_u(i,m)} \right] d_{i,m} \tag{A.36}$$

where we define

$$
\begin{aligned}
K_3 \;=\; & \left[(r_u(j,i) - r_i)\mathbf{p}[(j,i);001] \right. \\
+\;& \sum_{(i,q)\in D(i),q\neq m} (r_d(i,q) - r_i)\mathbf{p}[(i,q);N(i,q)10] \\
+\;& \left. \sum_{(i,q)\in D(i),q\neq m} (1 - p_i)F(i,q) - r_i\,\mathrm{prob}[\{\beta_i(t) = q\}] \right] \frac{1}{E(i,m)}
\end{aligned}
\tag{A.37}
$$

and $d_{i,m}$ is the fraction of parts sent from Machine M_i to M_m. The term $F(i,q)$ is given in (A.35) and $\mathrm{prob}[\{\beta_i(t) = q\}]$ in (A.26).

A.1.2 Downstream Machine

The resumption of flow probability $r_d(j,i)$ is defined as

$$
\begin{aligned}
r_d(j,i) = \mathrm{prob}\, \Big[& \{\alpha_d[(j,i),t+1] = 1\}\,| \\
& \{\alpha_d[(j,i),t] = 0\} \text{ and } \{n[(j,i),t-1] > 0\} \Big]
\end{aligned}
\tag{A.38}
$$

and is evaluated by decomposing the conditioning event

$$r_d(j,i) = A(j,i)X(j,i) + \sum_{(i,q)\in D(i)} B_{i,q}(j,i)Y_{i,q}(j,i) \tag{A.39}$$

where we define

$$A(j,i) = \text{prob} \left[\{\alpha_d[(j,i), t+1] = 1\} \,| \right. \tag{A.40}$$
$$\left. \{\alpha_i(t) = 0\} \text{ and } \{n[(j,i), t-1] > 0\} \right]$$

$$X(j,i) = \text{prob} \left[\{\alpha_i(t) = 0\} \text{ and } \{n[(j,i), t-1] > 0\} \,| \right. \tag{A.41}$$
$$\left. \{\alpha_d[(j,i), t] = 0\} \text{ and } \{n[(j,i), t-1] > 0\} \right]$$

$$B_{i,q}(j,i) = \text{prob} \left[\{\alpha_d[(j,i), t+1] = 1\} \,| \right. \tag{A.42}$$
$$\left. \{n[(i,q), t-1] = N(i,q)\} \text{ and } \{n[(j,i), t-1] > 0\} \right]$$

$$Y_{i,q}(j,i) = \text{prob} \left[\{n[(i,q), t-1] = N(i,q)\} \text{ and } \{n[(j,i), t-1] > 0\} \,| \right.$$
$$\left. \{\alpha_d[(j,i), t] = 0\} \text{ and } \{n[(j,i), t-1] > 0\} \right] \tag{A.43}$$

We find that $A(j,i)$ in (A.40) is the repair probability of Machine M_i, i.e.

$$A(j,i) = r_i \tag{A.44}$$

and $B_{i,q}(j,i)$ in (A.42) is the repair probability of the blocking machine $M_d(i,q)$, i.e.

$$B_{i,q}(j,i) = r_d(i,q) \tag{A.45}$$

After a derivation similar to that leading to (A.19), we find

$$Y_{i,q}(j,i) = \frac{\mathbf{p}[(i,q); N(i,q)10]r_d(j,i)}{p_d(j,i)E(j,i)} \tag{A.46}$$

and

$$X(j,i) = 1 - \sum_{(i,q) \in D(i)} Y_{i,q}(j,i) \tag{A.47}$$

We can therefore write the resumption of flow equation in a general form using an auxiliary parameter F as

$$r_d(j,i) = r_i F + K_4 \frac{r_d(j,i)}{p_d(j,i)} \tag{A.48}$$

with

$$F = 1 \tag{A.49}$$

$$K_4 = \sum_{(i,q)\in D(i)} (r_d(i,q) - r_i)\mathbf{p}[(i,q); N(i,q)10]\frac{1}{E(j,i)} \tag{A.50}$$

A.2 Resumption of Flow Equations: Merge System

A.2.1 Upstream Machine

We derive the resumption of flow equation in the usual way using the definition of virtual machine states:

$$r_u(i,q) = \text{prob}\left[\{\alpha_u[(i,q),t+1]=1\}\Big|\right. \tag{A.51}$$
$$\left.\{\alpha_u[(i,q),t]=0\} \text{ and } \{n[(i,q),t-1]<N(i,q)\}\right]$$

$$= \text{prob}\left[\{\alpha_u[(i,q),t+1]=1\}\Big|\right. \tag{A.52}$$
$$\left\{\{\alpha_i(t)=0\} \text{ or }\right.$$
$$\left.\{n[(j,i),t-1]=0,\forall j\in\{j_1,j_2\}\}\right\} \text{ and }$$
$$\left.\{n[(i,q),t-1]<N(i,q)\}\right]$$

Since the different events that force Machine $M_u(i,q)$ down are mutually disjoint, we can break equation (A.52) down by decomposing the conditioning event to find

$$r_u(i,q) = A(i,q)X(i,q) + B(i,q)Y(i,q) \tag{A.53}$$

where we define

$$A(i,q) = \text{prob}\left[\{\alpha_u[(i,q),t+1]=1\}\Big|\right. \tag{A.54}$$
$$\left.\{\alpha_i(t)=0\} \text{ and } \{n[(i,q),t-1]<N(i,q)\}\right]$$

$$X(i,q) = \text{prob}\left[\{\alpha_i(t)=0\} \text{ and } \{n[(i,q),t-1]<N(i,q)\}\Big|\right. \tag{A.55}$$
$$\left.\{\alpha_u[(i,q),t]=0\} \text{ and } \{n[(i,q),t-1]<N(i,q)\}\right]$$

$$B(i,q) = \text{prob}\left[\{\alpha_u[(i,q),t+1]=1\}\Big| \right. \tag{A.56}$$
$$\{n[(j,i),t-1]=0, \forall j \in \{j_1,j_2\}\} \text{ and}$$
$$\left. \{n[(i,q),t-1] < N(i,q)\}\right]$$

$$Y(i,q) = \text{prob}\left[\{n[(j,i),t-1]=0, \forall j \in \{j_1,j_2\}\} \text{ and} \right. \tag{A.57}$$
$$\{n[(i,q),t-1] < N(i,q)\}\Big|$$
$$\left. \{\alpha_u[(i,q),t]=0\} \text{ and } \{n[(i,q),t-1] < N(i,q)\}\right]$$

In (A.54), $A(i,q)$ is the probability that Machine M_i is repaired, i.e.,

$$A(i,q) = r_i, \tag{A.58}$$

whereas $B(i,q)$ in (A.56) is the probability that at least one virtual upstream machine is repaired. It is expressed in terms of the probability that none of the upstream machines is repaired, i.e.,

$$B(i,q) = 1 - \prod_{\forall j \in \{j_1,j_2\}} (1 - r_u(j,i)), \tag{A.59}$$

since repairs are independent, as they do not depend on buffer levels or the states of other machines.

The probability that Machine M_i is starved given that Machine $M_u(i,q)$ is down is

$$Y(i,q) = \text{prob}\left[\{n[(j,i),t-1]=0, \forall j \in \{j_1,j_2\}\} \text{ and} \right. \tag{A.60}$$
$$\left. \{n[(i,q),t-1] < N(i,q)\}\right]:$$
$$\text{prob}\left[\{\alpha_u[(i,q),t]=0\} \text{ and } \{n[(i,q),t-1] < N(i,q)\}\right]$$

We approximate the numerator by assuming independence:

$$Y(i,q) \approx \prod_{\forall j \in \{j_1,j_2\}} \text{p}[(j,i);001]\frac{r_u(i,q)}{p_u(i,q)E(i,q)} \tag{A.61}$$

The probability $X(i,q)$ that a failure of Machine $M_u(i,q)$ is due to a failure of Machine M_i itself is expressed in terms of the probability of the single other reason why Machine $M_u(i,q)$ can be down:

$$X(i,q) = 1 - Y(i,q)$$
$$= 1 - \prod_{\forall j \in \{j_1,j_2\}} \text{p}[(j,i);001]\frac{r_u(i,q)}{p_u(i,q)E(i,q)} \tag{A.62}$$

We can hence express the resumption of flow equation for Machine $M_u(i,q)$ using the routing probability $d_{i,q} = 1$ as

$$r_u(i,q) = \left[r_i + K_3 \frac{r_u(i,q)}{p_u(i,q)} \right] d_{i,q} \tag{A.63}$$

with

$$d_{i,q} = 1 \tag{A.64}$$

$$K_3 = \left(1 - \prod_{\forall j \in \{j_1,j_2\}} (1 - r_u(j,i)) - r_i \right) \prod_{\forall j \in \{j_1,j_2\}} \mathbf{p}[(j,i); 001] \frac{1}{E(i,q)} \tag{A.65}$$

A.2.2 Downstream Machine in the Priority Two Line

The different reasons for Machine $M_d(j_2, i)$ to be down as shown in (4.42) are not disjoint. Table A.3 gives the six mutually exclusive and collectively exhaustive reasons for Machine $M_d(j_2, i)$ to be down at time t. In three out of the six cases, it is not possible that Machine $M_d(j_2, i)$ is up at time $t + 1$.

In Case 1, M_i is down and not blocked while the priority one buffer is empty. Since M_i might be repaired, it is possible that the virtual machine $M_d(j_2, i)$ that is seen by an observer in the priority two buffer is up at time $t + 1$.

Case 2 is a situation where $M_d(j_2, i)$ is down for two reasons: M_i is down and the priority one buffer $B_{j_1,i}$ is non-empty. Even if Machine M_i were repaired at time $t + 1$, the priority one buffer would still be non-empty as nothing was removed at time t. For this reason, it is not possible that Machine $M_d(j_2, i)$ is up at time $t + 1$.

Machine M_i is blocked at time t in Case 3 and the priority one buffer is empty. Since the blocking machine $M_d(i, q)$ might be repaired, it is possible that Machine $M_d(j_2, i)$ is up at time $t + 1$.

In Case 4, however, Machine M_i is blocked and the priority one buffer is non-empty. Even if Machine $M_d(i, q)$ were repaired at time $t + 1$, Machine

Case	Machine M_i at time t	Buffer $B_{i,q}$ at time $t-1$	Buffer $B_{j_1,i}$ at time $t-1$	$M_d(j_2,i)$ up at time $t+1$?
1	down	not full	empty	possible
2	down	not full	not empty	impossible
3	up	full	empty	possible
4	up	full	not empty	impossible
5	up	not full	$n[(j_1,i), t-1] = 1$	possible
6	up	not full	$n[(j_1,i), t-1] > 1$	impossible

Table A.3. Disjoint Reasons for Machine $M_d(j_2, i)$ to be Down

$M_d(j_2, i)$ would still be down due to the non-empty priority one buffer in the previous period.

The priority one buffer is almost empty in Case 5, i.e. $n[(j_1, i), t - 1] = 1$, and Machine M_i can operate as it is up and not blocked. Since the virtual upstream machine related to the priority one buffer might be down or fail, it is possible that the priority one buffer is empty at time $t + 1$. If this happens, $M_d(j_2, i)$ is up at time $t + 1$.

In Case 6, there is more than one part in the priority one buffer $B_{j_1, i}$ at time t and therefore $B_{j_1, i}$ cannot be empty at time $t + 1$. For this reason, $M_d(j_2, i)$ cannot be up at time $t + 1$.

The resumption of flow probability $r_d(j_2, i)$ describing repairs of Machine $M_d(j_2, i)$ is defined as

$$r_d(j_2, i) = \text{prob}\Big[\{\alpha_d[(j_2, i), t + 1] = 1\} \mid$$
$$\{\alpha_d[(j_2, i), t] = 0\} \text{ and } \{n[(j_2, i), t - 1] > 0\}\Big] \quad (A.66)$$

and is evaluated by decomposing the conditioning event. However, in the decomposition we omit Cases 2, 4, and 6 in Table A.3 since in these cases it is not possible that Machine $M_d(j_2, i)$ is up at time $t + 1$:

$$r_d(j_2, i) = A(j_2, i)X(j_2, i) + B(j_2, i)Y(j_2, i) + C(j_2, i)Z(j_2, i) \quad (A.67)$$

where we define

$$A(j_2, i) = \text{prob}\Big[\{\alpha_d[(j_2, i), t + 1] = 1\}\Big| \quad (A.68)$$
$$\{\alpha_i(t) = 0\} \text{ and } \{n[(j_1, i), t - 1] = 0\} \text{ and }$$
$$\{n[(j_2, i), t - 1] > 0\}\Big]$$

$$X(j_2, i) = \text{prob}\Big[\{\alpha_i(t) = 0\} \text{ and } \{n[(j_1, i), t - 1] = 0\} \text{ and } \quad (A.69)$$
$$\{n[(j_2, i), t - 1] > 0\}\Big|$$
$$\{\alpha_d[(j_2, i), t] = 0\} \text{ and } \{n[(j_2, i), t - 1] > 0\}\Big]$$

$$B(j_2, i) = \text{prob} \left[\{\alpha_d[(j_2, i), t+1] = 1\} \right|$$ (A.70)
$$\{n[(i, q), t-1] = N(i, q)\} \text{ and } \{n[(j_1, i), t-1] = 0\}$$
$$\text{and } \{n[(j_2, i), t-1] > 0\} \right]$$

$$Y(j_2, i) = \text{prob} \left[\{n[(i, q), t-1] = N(i, q)\} \text{ and} \right.$$ (A.71)
$$\{n[(j_1, i), t-1] = 0\} \text{ and } \{n[(j_2, i), t-1] > 0\} \right|$$
$$\left. \{\alpha_d[(j_2, i), t] = 0\} \text{ and } \{n[(j_2, i), t-1] > 0\} \right]$$

$$C(j_2, i) = \text{prob} \left[\{\alpha_d[(j_2, i), t+1] = 1\} \right|$$ (A.72)
$$\{\alpha_i(t) = 1\} \text{ and } \{n[(i, q), t-1] < N(i, q)\} \text{ and}$$
$$\{n[(j_1, i), t-1] = 1\} \text{ and } \{n[(j_2, i), t-1] > 0\} \right]$$

$$Z(j_2, i) = \text{prob} \left[\{\alpha_i(t) = 1\} \text{ and } \{n[(i, q), t-1] < N(i, q)\} \text{ and} \right.$$ (A.73)
$$\{n[(j_1, i), t-1] = 1\} \text{ and } \{n[(j_2, i), t-1] > 0\} \right|$$
$$\left. \{\alpha_d[(j_2, i), t] = 0\} \text{ and } \{n[(j_2, i), t-1] > 0\} \right]$$

Probabilities $A(j_2, i)$ in (A.68) and $X(j_2, i)$ in (A.69) are related to Case 1 in Table A.3. Assuming that repairs of Machines M_i and $M_u(j_1, i)$ are independent, we find that $A(j_2, i)$ is the repair probability of Machine M_i times the probability that Machine $M_u(j_1, i)$ is not repaired, i.e.

$$A(j_2, i) = r_i(1 - r_u(j_1, i))$$ (A.74)

since the reason for the buffer of the priority one Line $L(j_1, i)$ to be empty at time t is a failure of Machine $M_u(j_1, i)$. Note that this is a term of a new kind that reflects the priorities assigned to the two buffers upstream of a merge machine: One machine (M_i) has to be repaired, while a another machine ($M_u(j_1, i)$) must not be repaired: If Machine $M_u(j_1, i)$ were repaired, the priority one buffer would become non-empty and Machine $M_d(j_2, i)$ would stay down, but now for a different reason.

A similar reasoning applies for the probability $B(j_2, i)$ in (A.70) that is related to Case 3 in Table A.3. The virtual downstream machine $M_d(i, q)$ must be repaired and Machine $M_u(j_1, i)$ must not be repaired in order for the flow to resume out of the priority two buffer:

$$B(j_2, i) = r_d(i, q)(1 - r_u(j_1, i))$$ (A.75)

In (A.71), $Y(j_2, i)$ is the probability that the buffer of Line $L(i, q)$ is full and the buffer of Line $L(j_1, i)$ empty given that Machine $M_d(j_2, i)$ is down. Assuming independence for buffer levels in Lines $L(j_1, i)$ and $L(i, q)$, we find, after a derivation similar to that leading to (A.19),

$$Y(j_2, i) \approx \mathbf{p}[(j_1, i); 001]\mathbf{p}[(i, q); N(i, q)10]\frac{r_d(j_2, i)}{p_d(j_2, i)E(j_2, i)}. \qquad (A.76)$$

We next analyze the probability $Z(j_2, i)$ in (A.73) related to Case 5 in Table A.3. Since the event $\{n[(j_1, i), t-1] = 1\}$ implies that Machine $M_d(j_2, i)$ is down at time t, we can use the definition of conditional probability and write

$$
\begin{aligned}
Z(j_2, i) &= \text{prob}\left[\{\alpha_i(t) = 1\} \text{ and } \{n[(i, q), t-1] < N(i, q)\} \text{ and}\right.\\
&\qquad\qquad \left.\{n[(j_1, i), t-1] = 1\} \text{ and } \{n[(j_2, i), t-1] > 0\}\right] : \\
&\qquad \text{prob}\left[\{\alpha_d[(j_2, i), t] = 0\} \text{ and } \{n[(j_2, i), t-1] > 0\}\right] \\
Z(j_2, i) &= \text{prob}[\{\text{ Case 5 }\}] : \\
&\qquad \text{prob}\left[\{\alpha_d[(j_2, i), t] = 0\} \text{ and } \{n[(j_2, i), t-1] > 0\}\right] \\
&= \text{prob}[\{\text{ Case 5 }\}]\frac{r_d(j_2, i)}{E(j_2, i)p_d(j_2, i)}. \qquad (A.77)
\end{aligned}
$$

It is not necessary to derive the probability of Case 5 explicitly, since this quantity can be canceled out in the term $C(j_2, i)Z(j_2, i)$ of the resumption of flow equation (A.67) after the derivation of $C(j_2, i)$.

To derive an expression for the conditional probability $C(j_2, i)$ in (A.72), we list in Table A.4 the implications of Case (Table A.3 on Page 215), i.e. of having one part in the buffer of the priority one Line $L(j_1, i)$ while Machine M_i is up and not blocked. These implications are used to decompose the conditional probability $C(j_2, i)$ in (A.72).

The first entry in Table A.4 is related to Line $L(i, q)$. In Case 5, Machine M_i is up, not starved and not blocked. This implies that the buffer level $n[(i, q), t-1]$ is in the interval $[0, N(i, q) - 1]$ and that Machine $M_u(i, q)$ is up. However, nothing is implied with respect to Machine $M_d(i, q)$. It can be up or down.

The second entry in Table A.4 describes the situation in Line $L(j_1, i)$. Case 5 states explicitly the buffer level $n[(j_1, i), t-1] = 1$. Since Machine M_i is up and not blocked, Machine $M_d(j_1, i)$ is also up. Nothing is implied with respect to Machine $M_u(j_1, i)$.

To reduce the notational effort when decomposing the conditioning event of $C(j_2, i)$, the auxiliary events a_1, a_2, a_3, as well as b_1 and b_2 are defined in Table A.5.

The mutually exclusive and collectively exhaustive events a_1, a_2, and a_3 describe the possible states of Line $L(i, q)$ implied by Case 5. In a similar manner, the auxiliary events b_1 and b_2 describe those of Line $L(j_1, i)$.

Line	Buffer Level	Machine State
$L(i,q)$	$0 \leq n[(i,q),t-1] \leq N(i,q)-1$	$\alpha_u[(i,q),t] = 1,$ $\alpha_d[(i,q),t] \in \{0,1\}$
$L(j_1,i)$	$n[(j_1,i),t-1] = 1$	$\alpha_u[(j_1,i),t] \in \{0,1\},$ $\alpha_d[(j_1,i),t] = 1$

Table A.4. Implications of Case 5

$$
\begin{aligned}
a_1 &= \{0 \leq n[(i,q),t-1] \leq N(i,q)-2\} \text{ and} \\
&\quad \{\alpha_u[(i,q),t] = 1\} \text{ and } \{\alpha_d[(i,q),t] \in \{0,1\}\} \\
a_2 &= \{n[(i,q),t-1] = N(i,q)-1\} \text{ and} \\
&\quad \{\alpha_u[(i,q),t] = 1\} \text{ and } \{\alpha_d[(i,q),t] = 1\} \\
a_3 &= \{n[(i,q),t-1] = N(i,q)-1\} \text{ and} \\
&\quad \{\alpha_u[(i,q),t] = 1\} \text{ and } \{\alpha_d[(i,q),t] = 0\} \\[6pt]
b_1 &= \{n[(j_1,i),t-1] = 1\} \text{ and} \\
&\quad \{\alpha_u[(j_1,i),t] = 1\} \text{ and } \{\alpha_d[(j_1,i),t] = 1\} \\
b_2 &= \{n[(j_1,i),t-1] = 1\} \text{ and} \\
&\quad \{\alpha_u[(j_1,i),t] = 0\} \text{ and } \{\alpha_d[(j_1,i),t] = 1\} \\[6pt]
c &= \{\alpha_d[(j_2,i),t+1] = 1\} \\
d &= \{n[(j_2,i),t-1] > 0\}
\end{aligned}
$$

Table A.5. Definition of Auxiliary Events

Given the definition of the auxiliary variables, we find

$$\text{prob}[\{ \text{ Case 5 } \}] = \text{prob}[\{a_1 \text{ or } a_2 \text{ or } a_3\} \text{ and } \{b_1 \text{ or } b_2\}]$$

and therefore

$$
\begin{aligned}
C(j_2,i) &= \text{prob}\Big[c \,\Big|\, \{a_1 \text{ or } a_2 \text{ or } a_3\} \text{ and } \{b_1 \text{ or } b_2\} \text{ and } d\Big] \\
&= \text{prob}\Big[c \,\Big|\, a_1\,b_1\,d \text{ or } a_1\,b_2\,d \text{ or} \\
&\qquad\qquad\quad a_2\,b_1\,d \text{ or } a_2\,b_2\,d \text{ or} \\
&\qquad\qquad\quad a_3\,b_1\,d \text{ or } a_3\,b_2\,d \text{ or }\Big]
\end{aligned}
$$

Each of the events $a_k\,b_l\,d$ with $k \in \{1,2,3\}$, $l \in \{1,2\}$ implies Case 5 and this implies that the priority two buffer is not empty, i.e. event d. The last implication is due to the assumption of operation dependent failures. As an approximation, we assume that the events related to different lines are independent. In this case $C(j_2, i)$ can be decomposed to find

$$
\begin{aligned}
C(j_2, i) \;=\; \Big[& \text{prob}[\, c \mid a_1 \, b_1 \,] \; \text{prob}[\, a_1 \,] \; \text{prob}[\, b_1 \,] + \\
& \text{prob}[\, c \mid a_1 \, b_2 \,] \; \text{prob}[\, a_1 \,] \; \text{prob}[\, b_2 \,] + \\
& \text{prob}[\, c \mid a_2 \, b_1 \,] \; \text{prob}[\, a_2 \,] \; \text{prob}[\, b_1 \,] + \\
& \text{prob}[\, c \mid a_2 \, b_2 \,] \; \text{prob}[\, a_2 \,] \; \text{prob}[\, b_2 \,] + \\
& \text{prob}[\, c \mid a_3 \, b_1 \,] \; \text{prob}[\, a_3 \,] \; \text{prob}[\, b_1 \,] + \\
& \text{prob}[\, c \mid a_3 \, b_2 \,] \; \text{prob}[\, a_3 \,] \; \text{prob}[\, b_2 \,] \Big] \cdot
\end{aligned}
$$

$$
\frac{1}{\text{prob}[\{\text{ Case 5 }\}]}
\tag{A.78}
$$

The conditional probabilities in (A.78) can be determined using the information in Table A.5. The first conditional probability, $\text{prob}[\, c \mid a_1 \, b_1 \,]$, is the probability that Machine $M_d(j_2, i)$ is up at time $t + 1$, given that the buffer level in Line $L(i, q)$ that receives the part at time t, is below $N(i, q) - 1$ at time $t - 1$ (event a_1), that the priority one buffer is almost empty, i.e. $n[(j, i), t - 1] = 1$, and that the Machine $M_u(j_1, i)$ is up at time t (event b_1). The priority one buffer looses a part at time t that is processed by Machine M_i. If Machine $M_u(j_1, i)$ fails at time t, the buffer becomes empty which is necessary for Machine $M_d(j_2, i)$ to be up at time $t + 1$.

Thus, the following has to happen, in order for Machine $M_d(j_2, i)$ to be up at time $t + 1$:

- Machine M_i in the real system must not fail, with probability $(1 - p_i)$.
- The virtual upstream machine $M_u(j_1, i)$ related to the priority one buffer must fail, with probability $p_u(j_1, i)$.

Given the buffer level in Line $L(i, q)$, Machine M_i cannot be blocked at time $t + 1$. Since failures of M_i and $M_u(j_2, i)$ are assumed to be independent, we find

$$
\text{prob}[\, c \mid a_1 \, b_1 \,] = (1 - p_i) \, p_u(j_1, i).
\tag{A.79}
$$

In the second conditional probability in (A.78), $\text{prob}[\, c \mid a_1 \, b_2 \,]$, event b_2 says that Buffer $B_{j,i}$, which looses a part at time t, is almost empty, i.e. $n[(j, i), t - 1] = 1$, and that Machine $M_u(j_1, i)$ is down at time t. In this situation, the following has to happen in order for Machine $M_d(j_2, i)$ to be up at time $t + 1$:

- Machine M_i in the real system must not fail, with probability $(1 - p_i)$.
- Machine $M_u(j_1, i)$ must not be repaired (because otherwise Buffer $B_{j_2, i}$ would become non-empty), with probability $(1 - r_u(j_1, i))$.

The second conditional probability is therefore

$$\text{prob}[\,c \mid a_1\, b_2\,] = (1 - p_i)\,(1 - r_u(j_1, i)). \tag{A.80}$$

In the third conditional probability in (A.78), $\text{prob}[\,c \mid a_2\, b_1\,]$, Buffer $B_{i,q}$ is almost full and Machine $M_d(i, q)$ is up (event a_2). Furthermore, buffer $B_{j_1,i}$ is almost empty and Machine $M_u(j_1, i)$ is also up.

The following has to happen in order for Machine $M_d(j_2, i)$ to be up at time $t + 1$:

- Machine M_i in the real system must not fail, with probability $(1 - p_i)$.
- The virtual downstream machine $M_d(i, q)$ must not fail to avoid blocking of M_i, with probability $(1 - p_d(i, q))$.
- The virtual upstream machine $M_u(j_1, i)$ must fail for the priority one buffer $B_{j_1,i}$ to become empty, with probability $p_u(j_1, i)$.

We find

$$\text{prob}[\,c \mid a_2\, b_1\,] = (1 - p_i)\,(1 - p_d(i, q))\,p_u(j_1, i). \tag{A.81}$$

The derivation for the three remaining conditional probabilities is completely analogous and we find:

$$
\begin{aligned}
\text{prob}[\,c \mid a_2\, b_2\,] &= (1 - p_i)\,(1 - p_d(i, q))\,(1 - r_u(j_1, i)) & \text{(A.82)} \\
\text{prob}[\,c \mid a_3\, b_1\,] &= (1 - p_i)\,r_d(i, q)\,p_u(j_1, i) & \text{(A.83)} \\
\text{prob}[\,c \mid a_3\, b_2\,] &= (1 - p_i)\,r_d(i, q)\,(1 - r_u(j_1, i)) & \text{(A.84)}
\end{aligned}
$$

The special structure of these six conditional probabilities allows to factor (A.78). This yields

$$
\begin{aligned}
C(&j_2, i) \\
= \;& (1 - p_i) \\
& \Big[\, \text{prob}[\,a_1\,] + (1 - p_d(i, q))\, \text{prob}[\,a_2\,] + r_d(i, q))\, \text{prob}[\,a_3\,] \Big] \cdot \\
& \Big[\, p_u(j_1, i)\, \text{prob}[\,b_1\,] + (1 - r_u(j_1, i))\, \text{prob}[\,b_2\,] \Big] \cdot \\
& \frac{1}{\text{prob}[\{\,\text{Case 5}\,\}]}.
\end{aligned}
\tag{A.85}
$$

Using the information in Table A.5, the unconditional probabilities can be expressed as

$$\text{prob}[\, a_1 \,] \;\; = \;\; \sum_{n=0}^{N(i,q)-2} \Big(\mathbf{p}[(i,q); n11] + \mathbf{p}[(i,q); n10] \Big)$$

$$\text{prob}[\, a_2 \,] \;\; = \;\; \mathbf{p}[(i,q); N(i,q) - 1, 11]$$

$$\text{prob}[\, a_3 \,] \;\; = \;\; \mathbf{p}[(i,q); N(i,q) - 1, 10]$$

$$\text{prob}[\, b_1 \,] \;\; = \;\; \mathbf{p}[(j_1,i); 111]$$

$$\text{prob}[\, b_2 \,] \;\; = \;\; \mathbf{p}[(j_1,i); 101]$$

and we find

$$
\begin{aligned}
C(j_2, i) \;\; = \;\; & (1 - p_i) \cdot \\
& \left[\sum_{n=0}^{N(i,q)-2} \Big(\mathbf{p}[(i,q); n11] + \mathbf{p}[(i,q); n10] \Big) \right. \\
& \quad + (1 - p_d(i,q)\mathbf{p}[(i,q); N(i,q) - 1, 11] \\
& \left. \quad + r_d(i,q)\mathbf{p}[(i,q); N(i,q) - 1, 10]) \right] \cdot \\
& \Big[p_u(j_1, i)\mathbf{p}[(j_1, i); 011] \\
& \quad + (1 - r_u(j_1, i))\mathbf{p}[(j_1, i); 001] \Big] \cdot \\
& \frac{1}{\text{prob}[\{ \text{ Case 5 }\}]}
\end{aligned}
\tag{A.86}
$$

Multiplying $C(j_2, i)$ and $Z(j_2, i)$ as required in (A.67) yields

$$
\begin{aligned}
C(j_2, i)Z(j_2, i) \;\; = \;\; & (1 - p_i) \cdot \\
& \left[\sum_{n=0}^{N(i,q)-2} \Big(\mathbf{p}[(i,q); n11] + \mathbf{p}[(i,q); n10] \Big) \right. \\
& \quad + (1 - p_d(i,q)\mathbf{p}[(i,q); N(i,q) - 1, 11] \\
& \left. \quad + r_d(i,q)\mathbf{p}[(i,q); N(i,q) - 1, 10]) \right] \cdot \\
& \Big[p_u(j_1, i)\mathbf{p}[(j_1, i); 011] \\
& \quad + (1 - r_u(j_1, i))\mathbf{p}[(j_1, i); 001] \Big] \cdot \\
& \frac{r_d(j_2, i)}{E(j_2, i)p_d(j_2, i)},
\end{aligned}
\tag{A.87}
$$

i.e. the probability of Case 5 can be canceled out.

The last probability we have to determine is $X(j_2, i)$ in (A.69), i.e. the conditional probability that Machine M_i is down and Buffer $B_{j_1,i}$ is empty given that Machine $M_d(j_2, i)$ is down. In Table A.3 on Page 215, the situation that Machine M_i is down and the buffer of the priority one line is empty is referred to as Case 1. Since Table A.3 contains all the disjoint reasons for Machine $M_d(j_2, i)$ to be down, we can write

$$\text{prob}\Big[\{ \text{ Case 1}\}\Big|\{\alpha_d[(j_2, i), t] = 0\} \text{ and } \{n[(j_2, i), t-1] > 0\}\Big]$$

$$= \quad 1 - \text{prob} \ \Big[\{ \text{ Case 2 or 3 or 4 or 5 or 6}\}\Big| \qquad \text{(A.88)}$$
$$\{\alpha_d[(j_2, i), t] = 0\} \text{ and } \{n[(j_2, i), t-1] > 0\}\Big]$$

$$= \quad 1 - \frac{\text{prob}\Big[\{\text{Case 2 or 5 or 6}\} \text{ or } \{ \text{ Case 3 or 4}\}\Big]}{\text{prob}[\alpha_d[(j_2, i), t] = 0 \text{ and } n[(j_2, i), t-1] > 0]} \qquad \text{(A.89)}$$

$$= \quad 1 - \frac{\text{prob}\Big[\{\text{Case 2 or 5 or 6}\}\Big] + \text{prob}\Big[\{ \text{ Case 3 or 4}\}\Big]}{\text{prob}[\alpha_d[(j_2, i), t] = 0 \text{ and } n[(j_2, i), t-1] > 0]} \qquad \text{(A.90)}$$

since all cases imply that Machine $M_d(j_2, i)$ is down. We see in Table A.3 that the probability of the $\{$Case 2 or 5 or 6$\}$ is the probability of having a non-empty buffer in the priority one line $L(j_1, i)$ and a non-full buffer in Line $L(i, q)$. As an approximation, we assume again that these buffer levels are independent, or

$$\text{prob} \ \Big[\{ \text{ Case 2 or 5 or 6 } \} \ \Big]$$
$$\approx \quad (1 - \mathbf{p}[(i, q); N(i, q)10])(1 - \mathbf{p}[(j_1, i); 001]) \qquad \text{(A.91)}$$

and we also find that

$$\text{prob} \ [\ \{\text{Case 3 or 4 } \} \ ] = \mathbf{p}[(i, q); N(i, q)10] \qquad \text{(A.92)}$$

since in Cases 3 and 4, Machine M_i must be up (as a blocked machine cannot fail). We hence conclude that

$$X(j_2, i) \quad = \quad \text{prob}\Big[\ \{\text{Case 1}\} \mid \alpha_d[(j_2, i), t] = 0 \text{ and } n[(j_2, i), t-1] > 0\Big]$$

$$= \quad 1 - \Big[(1 - \mathbf{p}[(i, q); N(i, q)10])(1 - \mathbf{p}[(j_1, i); 001])$$

$$+\mathbf{p}[(i, q); N(i, q)10]\Big] \frac{r_d(j_2, i)}{p_d(j_2, i)E(j_2, i)}. \qquad \text{(A.93)}$$

The resumption of flow equation can thus be expressed as

$$r_d(j_2, i) \;\; = \;\; r_i F + K_4 \frac{r_d(j_2, i)}{p_d(j_2, i)} \tag{A.94}$$

where we define

$$F \;\; = \;\; (1 - r_u(j_1, i)) \tag{A.95}$$

$$
\begin{aligned}
K_4 \;\; = \;\; & \Bigg[r_d(i, q) F \mathbf{p}[(i, q); N(i, q)10] \mathbf{p}[(j_1, i); 001] + \\
& (1 - p_i) \cdot \\
& \Bigg[\sum_{n=0}^{N(i,q)-2} \mathbf{p}[(i, q); n10] + \mathbf{p}[(i, q); n11] \\
& + (1 - p_d(i, q) \mathbf{p}[(i, q); N(i, q) - 1, 11] \\
& + r_d(i, q) \mathbf{p}[(i, q); N(i, q) - 1, 10] \Bigg] \\
& \Big[p_u(j_1, i) \mathbf{p}[(j_1, i); 111] + (1 - r_u(j_1, i)) \mathbf{p}[(j_1, i); 101] \Big] \\
& - r_i F \Big[(1 - \mathbf{p}[(i, q); N(i, q)10])(1 - \mathbf{p}[(j_1, i); 001]) \\
& + \mathbf{p}[(i, q); N(i, q)10] \Big] \Bigg] \frac{1}{E(j_2, i)}
\end{aligned}
\tag{A.96}
$$

B. Derivation for the Continuous Material Flow Line

This appendix presents the complete derivation of the resumption and interruption of flow equations for the continuous material model of a flow line with split and merge operations and machine-specific deterministic processing times presented in Section 5.

B.1 Resumption of Flow Equations: Split System

B.1.1 Upstream Machine

Since the different events that force Machine $M_u(i, m)$ down are mutually disjoint, we can break equation (5.12) down by decomposing the conditioning event to find:

$$
\begin{aligned}
r_u(i,m)\delta t \;=\; & A(i,m)W(i,m) + B(i,m)X(i,m) \\
& + \sum_{(i,q)\in D(i),q\neq m} C_{i,q}(i,m)Y_{i,q}(i,m)
\end{aligned}
\tag{B.1}
$$

where we define

$$
A(i,m) \;=\; \text{prob}\Big[\{\alpha_u[(i,m),t+\delta t] = 1\} \mid \{\alpha_i(t) = 0\}\Big]
\tag{B.2}
$$

$$
W(i,m) \;=\; \text{prob}\Big[\{\alpha_i(t) = 0\} \mid \{\alpha_u[(i,m),t] = 0\}\Big]
\tag{B.3}
$$

$$B(i,m) \;=\; \text{prob}\Big[\{\alpha_u[(i,m),t+\delta t]=1\}\,|$$
$$\{n[(j,i),t]=0\} \text{ and } \{\alpha_u[(j,i),t]=0\}\Big] \qquad \text{(B.4)}$$

$$X(i,m) \;=\; \text{prob}\Big[\{n[(j,i),t]=0\} \text{ and } \{\alpha_u[(j,i),t]=0\}\,|$$
$$\{\alpha_u[(i,m),t]=0\}\Big] \qquad \text{(B.5)}$$

$$C_{i,q}(i,m) \;=\; \text{prob}\Big[\{\alpha_u[(i,m),t+\delta t]=1\}\,|$$
$$\{n[(i,q),t]=N(i,q)\} \text{ and } \{\alpha_d[(i,q),t]=0\}\Big] \qquad \text{(B.6)}$$

$$Y_{i,q}(i,m) \;=\; \text{prob}\Big[\{n[(i,q),t]=N(i,q)\} \text{ and } \{\alpha_d[(i,q),t]=0\}\,|$$
$$\{\alpha_u[(i,m),t]=0\}\Big] \qquad \text{(B.7)}$$

We now determine the conditional probabilities. Probabilities $A(i,m)$ and $W(i,m)$ deal with a possible failure of Machine M_i in the real system. In (B.2), $A(i,m)$ is the probability that flow has resumed into Buffer $B_{i,m}$ at time $t+\delta t$ given that Machine M_i was down at time t. For this to happen, Machine M_i must be repaired (with probability $r_i \delta t$), i.e.

$$A(i,m) = r_i \delta t \qquad \text{(B.8)}$$

In (B.3), $W(i,m)$ is the probability that a failure of the virtual machine $M_u(i,m)$ is due to a failure of Machine M_i. It can be expressed in terms of the conditional probabilities of all the other events that can lead to a failure of Machine $M_u(i,m)$

$$W(i,m) = 1 - X(i,m) - \sum_{(i,q)\in D(i),q\neq m} Y_{i,q}(i,m) \qquad \text{(B.9)}$$

since all these conditional probabilities add up to one.

Probabilities $B(i,m)$ and $X(i,m)$ account for starvation of Machine M_i. In (B.4), $B(i,m)$ is the probability that flow has resumed into Buffer $B_{i,m}$ at time $t+\delta t$, given that Machine M_i was starved at time t. The reason for starvation is an upstream machine failure. Therefore, the virtual upstream machine $M_u(j,i)$ must be repaired, so

$$B(i,m) = r_u(j,i)\delta t \qquad \text{(B.10)}$$

In (B.5), $X(i, m)$ is the probability that Machine M_i is starved given that Machine $M_u(i, m)$ is down. Since $\{n[(j, i), t] = 0\}$ and $\{\alpha_u[(j, i), t] = 0\}$ implies $\{\alpha_u[(i, m), t] = 0\}$, we can write

$$X(i, m) \quad = \quad \text{prob}\Big[\{n[(j, i), t] = 0\} \text{ and } \{\alpha_u[(j, i), t] = 0\}\} \\ \text{and } \{\alpha_u[(i, m), t] = 0\} \mid \\ \{\alpha_u[(i, m), t] = 0\}\Big] \tag{B.11}$$

Using the definition of conditional probability

$$\text{prob}[A \mid B] = \frac{\text{prob}[A \text{ and } B]}{\text{prob}[B]}, \tag{B.12}$$

we can write (B.11) as a quotient, i.e.

$$X(i, m) \quad = \quad \text{prob}\left[\begin{array}{c} \{n[(j, i), t] = 0\} \text{ and } \{\alpha_u[(j, i), t] = 0\} \\ \text{and } \{\alpha_u[(i, m), t] = 0\} \end{array}\right] : \\ \text{prob}\Big[\{\alpha_u[(i, m), t] = 0\}\Big] \tag{B.13}$$

The downstream machine $M_d(j, i)$ in Line $L(j, i)$ cannot fail if it is starved. Thus, $\{n[(j, i), t] = 0\}$ and $\{\alpha_u[(j, i), t] = 0\}$ implies that Machine $M_d(j, i)$ is up at time t, i.e. event $\{\alpha_d[(j, i), t] = 1\}$. For this reason, the numerator in (B.13) is approximately $\mathbf{p}[(j, i); 001]$ where $\mathbf{p}[(j, i); n\alpha_u\alpha_d]$ denotes the probability of finding the virtual upstream machine $M_u(j, i)$ in Line $L(j, i)$ in state α_u, the downstream machine $M_d(j, i)$ in state α_d and the buffer at level n.

As there must be exactly one repair for each failure, the following equation for the two-machine model (Gershwin, 1994, p 125) holds exactly as in the continuous material transfer line decomposition (Burman, 1995, p 69)

$$r_u(i, m) \, \text{prob}\Big[\{\alpha_u[(i, m), t] = 0\}\Big] \\ = \quad p_u(i, m) \, \text{prob}\Big[\{\alpha_u[(i, m), t] = 1\}\Big] \\ = \quad p_u(i, m) E_u(i, m) \tag{B.14}$$

where $E_u(i, m)$ is the fraction of time that Machine $M_u(i, m)$ is operational with $PR(i, m) = E_u(i, m)\mu_u(i, m)$. The denominator in (B.13) can hence be written as:

$$\text{prob}\Big[\{\alpha_u[(i, m), t] = 0\}\Big] = \frac{p_u(i, m) PR(i, m)}{r_u(i, m)\mu_u(i, m)} \tag{B.15}$$

Using these expressions for the numerator and denominator of (B.11), we find:

$$X(i,m) = \frac{\mathbf{p}[(j,i);001]r_u(i,m)\mu_u(i,m)}{p_u(i,m)PR(i,m)} \tag{B.16}$$

In (B.6), $C_{i,q}(i,m)$ is the probability that flow resumes into Buffer $B_{i,m}$ after a blocking of Machine M_i due to a failure of Machine $M_d(i,q)$. It is the repair probability of $M_d(i,q)$:

$$C_{i,q}(i,m) = r_d(i,q) \tag{B.17}$$

The expression for $Y_{i,q}(i,m)$ is derived like the one for $X(i,m)$ to find:

$$Y_{i,q}(i,m) = \frac{\mathbf{p}[(i,q); N(i,q)10]r_u(i,m)\mu_u(i,m)}{p_u(i,m)PR(i,m)} \tag{B.18}$$

The resumption of flow equation for the virtual upstream machine $M_u(i,m)$ can hence be written as

$$r_u(i,m) \quad = \quad \left[r_i + K_3\frac{r_u(i,m)\mu_u(i,m)}{p_u(i,m)}\right] \tag{B.19}$$

where we define

$$K_3 \quad = \quad \left[(r_u(j,i) - r_i)\mathbf{p}[(j,i);001]\right.$$
$$+ \quad \sum_{(i,q)\in D(i),q\neq m}(r_d(i,q) - r_i)\mathbf{p}[(i,q); N(i,q)10]\left.\right]\frac{1}{PR(i,m)} \tag{B.20}$$

B.1.2 Downstream Machine

The resumption of flow probability $r_d(j,i)$ is defined as

$$r_d(j,i) = \mathrm{prob}\left[\{\alpha_d[(j,i),t+\delta t] = 1\} \mid \{\alpha_d[(j,i),t] = 0\}\right] \tag{B.21}$$

and can again be evaluated by decomposing the conditioning event

$$r_d(j,i) = A(j,i)X(j,i) + \sum_{(i,q)\in D(i)}B_{i,q}(j,i)Y_{i,q}(j,i) \tag{B.22}$$

where we define

$$A(j,i) \;=\; \mathrm{prob}\Big[\{\alpha_d[(j,i),t+\delta t]=1\}\mid\{\alpha_i(t)=0\}\Big] \tag{B.23}$$

$$X(j,i) \;=\; \mathrm{prob}\Big[\{\alpha_i(t)=0\}\mid\{\alpha_d[(j,i),t]=0\}\Big] \tag{B.24}$$

$$B_{i,q}(j,i) \;=\; \mathrm{prob}\,\Big[\{\alpha_d[(j,i),t+\delta t]=1\}\mid \tag{B.25}$$
$$\{n[(i,q),t]=N(i,q)\}\text{ and }\{\alpha_d[(i,q),t]=0\}\Big]$$

$$Y_{i,q}(j,i) \;=\; \mathrm{prob}\,\Big[\{n[(i,q),t]=N(i,q)\}\text{ and }\{\alpha_d[(i,q),t]=0\}\mid$$
$$\{\alpha_d[(j,i),t]=0\}\Big]$$
$$\tag{B.26}$$

We find that $A(j,i)$ in (B.23) is the repair probability of Machine M_i, i.e.

$$A(j,i) = r_i\delta t \tag{B.27}$$

and $B_{i,q}(j,i)$ in (B.25) is the repair probability of the blocking machine $M_d(i,q)$, i.e.

$$B_{i,q}(j,i) = r_d(i,q)\delta t \tag{B.28}$$

After a derivation similar to that leading to (B.18), we find

$$Y_{i,q}(j,i) = \frac{\mathbf{P}[(i,q);N(i,q)10]r_d(j,i)\mu_d(j,i)}{p_d(j,i)PR(j,i)} \tag{B.29}$$

and

$$X(j,i) = 1 - \sum_{(i,q)\in D(i)} Y_{i,q}(j,i) \tag{B.30}$$

We can therefore write the resumption of flow equation as

$$r_d(j,i) = r_i + K_4\frac{r_d(j,i)\mu_d(j,i)}{p_d(j,i)} \tag{B.31}$$

with

$$K_4 = \sum_{(i,q)\in D(i)} (r_d(i,q)-r_i)\mathbf{P}[(i,q);N(i,q)10]\frac{1}{PR(j,i)} \tag{B.32}$$

B.2 Resumption of Flow Equations: Merge System

B.2.1 Upstream Machine

Since the different events that force Machine $M_u(i, q)$ down are mutually disjoint, we can break equation (5.23) down by decomposing the conditioning event to find

$$r_u(i, q) = A(i, q)X(i, q) + B(i, q)Y(i, q) \tag{B.33}$$

where we define

$$A(i, q) = \text{prob}\left[\{\alpha_u[(i, q), t + \delta t] = 1\} \big| \{\alpha_i(t) = 0\}\right] \tag{B.34}$$

$$X(i, q) = \text{prob}\left[\{\alpha_i(t) = 0\} \big| \{\alpha_u[(i, q), t] = 0\}\right] \tag{B.35}$$

$$B(i, q) = \text{prob}\left[\{\alpha_u[(i, q), t + \delta t] = 1\} \big| \right. \tag{B.36}$$
$$\left. \left\{\{n[(j, i), t] = 0\} \text{ and } \{\alpha_u[(j, i), t] = 0\}, \right.\right.$$
$$\left.\left. \forall j \in \{j_1, j_2\}\right\}\right]$$

$$Y(i, q) = \text{prob}\left[\left\{\{n[(j, i), t] = 0\} \text{ and } \{\alpha_u[(j, i), t] = 0\},\right.\right. \tag{B.37}$$
$$\left.\left. \forall j \in \{j_1, j_2\}\right\} \big| \{\alpha_u[(i, q), t] = 0\}\right]$$

In (B.34), $A(i, q)$ is the probability that Machine M_i itself is repaired, i.e

$$A(i, q) = r_i \delta t, \tag{B.38}$$

whereas $B(i, q)$ in (B.36) is the probability that one of the two virtual upstream machine is repaired.[1] It is expressed in terms of the probability that none of the upstream machines is repaired, i.e.

$$B(i, q) = 1 - \prod_{j \in \{j_1, j_2\}} (1 - r_u(j, i)\delta t) \tag{B.39}$$

$$\approx 1 - 1 + \sum_{j \in \{j_1, j_2\}} r_u(j, i)\delta t \tag{B.40}$$

$$= \sum_{j \in \{j_1, j_2\}} r_u(j, i)\delta t \tag{B.41}$$

---

[1] The probability of a repair of both machines leads to a second order term that can be omitted for a small δt.

since repairs are independent as they do not depend on buffer levels or the states of other machines. In the approximation in (B.40), all second order terms have been omitted.

The probability that Machine M_i is starved given that Machine $M_u(i,q)$ is down is

$$Y(i,q) \quad = \quad \text{prob} \left[\left\{ \{n[(j,i),t] = 0\} \text{ and } \{\alpha_u[(j,i),t] = 0\}, \right. \right. \qquad \text{(B.42)}$$
$$\left. \left. \forall j \in \{j_1, j_2\} \right\} \right] :$$
$$\text{prob} \left[\{\alpha_u[(i,q),t] = 0\} \right] \qquad \text{(B.43)}$$

We approximate the numerator by assuming independence of the input buffer levels and determine the denominator as in (B.15) to find:

$$Y(i,q) \approx \prod_{j \in \{j_1, j_2\}} \mathbf{p}[(j,i); 001] \frac{r_u(i,q)\mu_u(i,q)}{p_u(i,q)PR(i,q)} \qquad \text{(B.44)}$$

The probability $X(i,q)$ that a failure of Machine $M_u(i,q)$ is due to a failure of Machine M_i itself is expressed in terms of the probability of the single other reason why Machine $M_u(i,q)$ can be down:

$$X(i,q) \quad = \quad 1 - Y(i,q)$$
$$= \quad 1 - \prod_{j \in \{j_1, j_2\}} \mathbf{p}[(j,i); 001] \frac{r_u(i,q)\mu_u(i,q)}{p_u(i,q)PR(i,q)} \qquad \text{(B.45)}$$

We can hence state the resumption of flow equation for Machine $M_u(i,q)$ as

$$r_u(i,q) = r_i + K_3 \frac{r_u(i,q)\mu_u(i,q)}{p_u(i,q)} \qquad \text{(B.46)}$$

with

$$K_3 = \left(\sum_{j \in \{j_1, j_2\}} r_u(j,i) - r_i \right) \prod_{j \in \{j_1, j_2\}} \mathbf{p}[(j,i); 001] \frac{1}{PR(i,q)} \qquad \text{(B.47)}$$

B.2.2 Downstream Machine in the Priority Two Line

To decompose the resumption of flow equation on the conditioning event, we need to derive disjoint reasons for Machine $M_d(j_2, i)$ to be down at time t.

Case	$\alpha_i(t)$	$\alpha_u[(j_1, i), t]$	$n[(j_1, i), t]$	M_i blocked at time t	$M_d(j_2, i)$ up at time $t + \delta t$
1	0	0	$= 0$	no	possible
2	0	0 or 1	> 0	no	impossible
3	1	0	$> \mu_d(j_1, i)\delta t$	no	impossible
4	1	0	> 0, but $\leq \mu_d(j_1, i)\delta t$	no	possible
5	1	1	$> (\mu_d(j_1, i) - \mu_u(j_1, i))\delta t$	no	impossible
6	1	1	> 0, but $\leq (\mu_d(j_1, i) - \mu_u(j_1, i))\delta t$	no	possible
7	1	0	$= 0$	yes	possible
8	1	0 or 1	> 0	yes	impossible

Table B.1. Disjoint Cases in which Machine $M_d(j_2, i)$ is Down

This is done in Table B.1. We consider eight different cases. In Cases 1 to 6, Machine M_i is not blocked.

In Case 1 in Table B.1, Machine M_i is down while the priority one input buffer is empty due to a failure of Machine $M_u(j_1, i)$. If Machine M_i is repaired before Machine $M_u(j_1, i)$ is repaired, the flow of material out of the priority two buffer can resume. It is therefore possible that Machine $M_d(j_2, i)$ is up at time $t + \delta t$. This is the first out of four cases that actually allow for a resumption of flow out of the priority two buffer.

Case 2 has an inventory level $n[(j_1, i), t]$ for the priority one buffer that is positive. Immediately after a repair of Machine M_i, material will be taken out of the non-empty priority one buffer only. For this reason, Machine $M_d(j_2, i)$ cannot be up at time $t + \delta t$, irrespective of the state of Machine $M_u(j_1, i)$ at time t.

In Cases 3 and 4, the virtual machine $M_u(j_1, i)$ upstream of the priority one buffer is down at time t. No material flows into the priority one buffer since Machine $M_u(j_1, i)$ is down. If the buffer level $n[(j_1, i), t]$ of the priority one buffer is larger than the amount of material $\mu_d(j_1, i)\delta t$ that can be taken out of the buffer during the interval $[t, t + \delta t]$, there will still be material left in the priority one buffer at time $t + \delta t$. In this situation, Machine M_i

continues to empty the priority one buffer. Thus, Machine $M_d(j_2, i)$ cannot be up at time $t + \delta t$ in Case 3. However, if in Case 4 the priority one buffer level is positive but lower than $\mu_d(j_1, i)\delta t$, the priority one buffer can become empty during the interval $[t, t + \delta t]$. If this happens, flow out of the priority two buffer resumes and Machine $M_d(j_2, i)$ can be up at time $t + \delta t$.

The reasoning for Cases 5 and 6 is very similar, except that now the virtual machine upstream of the priority one buffer, $M_u(j_1, i)$, is up. For this reason we have to take the amount $\mu_u(j_1, i)\delta t$ of material into account that is produced by Machine $M_u(j_1, i)$ during a time interval of length δt. The state of Machine $M_d(j_2, i)$ can only change in Case 6.

In Cases 7 and 8, Machine M_i is blocked at time t. The priority one buffer level $n[(j_1, i), t]$ in Case 7 is 0 while Machine M_i is blocked. This implies that Machine $M_u(j_1, i)$ is down. If it were up, the buffer would become non-empty immediately. In this situation it is possible that Machine $M_d(j_2, i)$ is up at time $t + \delta t$ if the blocking downstream machine, $M_d(i, q)$, is repaired. Finally, in Case 8 the priority one buffer level $n[(j_1, i), t]$ is positive and therefore Machine $M_d(j_2, i)$ will still be down at time $t + \delta t$, even if the blocking downstream machine, $M_d(i, q)$, should be repaired at time $t + \delta t$.

The resumption of flow probability $r_d(j_2, i)$ is defined as

$$r_d(j_2, i) = \mathrm{prob}\Big[\{\alpha_d[(j_2, i), t + \delta t] = 1\} \mid \{\alpha_d[(j_2, i), t] = 0\}\Big] \qquad \text{(B.48)}$$

and is evaluated by decomposing the conditioning event. However, in the decomposition we omit Cases 2, 3, 5, and 8 in Table B.1 since in these cases a transition such that Machine $M_d(j_2, i)$ is up at time $t + \delta t$ is not possible:

$$
\begin{aligned}
r_d(j_2, i) &= A(j_2, i)W(j_2, i) + B(j_2, i)X(j_2, i) \\
&+ C(j_2, i)Y(j_2, i) + D(j_2, i)Z(j_2, i)
\end{aligned}
\qquad \text{(B.49)}
$$

where we define

$$A(j_2, i) \quad = \quad \text{prob}\Big[\{\alpha_d[(j_2, i), t + \delta t] = 1\}\big|\{\text{Case 1}\}\Big] \qquad \text{(B.50)}$$

$$W(j_2, i) \quad = \quad \text{prob}\Big[\{\text{Case 1}\}\big|\{\alpha_d[(j_2, i), t] = 0\}\Big] \qquad \text{(B.51)}$$

$$B(j_2, i) \quad = \quad \text{prob}\Big[\{\alpha_d[(j_2, i), t + \delta t] = 1\}\big|\{\text{Case 4}\}\Big] \qquad \text{(B.52)}$$

$$X(j_2, i) \quad = \quad \text{prob}\Big[\{\text{Case 4}\}\big|\{\alpha_d[(j_2, i), t] = 0\}\Big] \qquad \text{(B.53)}$$

$$C(j_2, i) \quad = \quad \text{prob}\Big[\{\alpha_d[(j_2, i), t + \delta t] = 1\}\big|\{\text{Case 6}\}\Big] \qquad \text{(B.54)}$$

$$Y(j_2, i) \quad = \quad \text{prob}\Big[\{\text{Case 6}\}\big|\{\alpha_d[(j_2, i), t] = 0\}\Big] \qquad \text{(B.55)}$$

$$D(j_2, i) \quad = \quad \text{prob}\Big[\{\alpha_d[(j_2, i), t + \delta t] = 1\}\big|\{\text{Case 7}\}\Big] \qquad \text{(B.56)}$$

$$Z(j_2, i) \quad = \quad \text{prob}\Big[\{\text{Case 7}\}\big|\{\alpha_d[(j_2, i), t] = 0\}\Big] \qquad \text{(B.57)}$$

Probabilities $A(j_2, i)$ in (B.50) and $W(j_2, i)$ in (B.51) are related to Case 1 in Table B.1. We find that $A(j_2, i)$ is the repair probability of Machine M_i times the probability that Machine $M_u(j_1, i)$ is not repaired, i.e.

$$A(j_2, i) = r_i \delta t (1 - r_u(j_1, i)\delta t) \approx r_i \delta t, \qquad \text{(B.58)}$$

which is the repair probability of Machine M_i if the the second order term is omitted.

Probability $W(j_2, i)$ in (B.51) can be expressed in terms of all the other reasons that can force Machine $M_d(j_2, i)$ down:

$$
\begin{aligned}
W(j_2, i) \quad = \quad & \text{prob}\Big[\{\text{Case 1}\}\big|\{\alpha_d[(j_2, i), t] = 0\}\Big] \\
= \quad & 1 - \text{prob}\Big[\{\text{Case 2 or 3 or 4 or 5 or 6 or 7 or 8 }\}\big| \\
= \quad & \qquad \{\alpha_d[(j_2, i), t] = 0\}\Big] \\
= \quad & 1 - \text{prob}\Big[\{\text{Case 2 or 3 or 4 or 5 or 6 }\}\big|\{\alpha_d[(j_2, i), t] = 0\}\Big] \\
& - \text{prob}\Big[\{\text{Case 7 or 8 }\}\big|\{\alpha_d[(j_2, i), t] = 0\}\Big] \qquad \text{(B.59)}
\end{aligned}
$$

since these cases disjoint. From Table B.1 we see that

$$\text{prob}\Big[\{\text{Case 7 or 8 }\}\big|\{\alpha_d[(j_2, i), t] = 0\}\Big] \qquad \text{(B.60)}$$

$$= \quad \text{prob}\Big[\{ \text{Machine } M_i \text{ is blocked at time } t\}\big|\{\alpha_d[(j_2, i), t] = 0\}\Big]$$

Equation (B.60) can be reformulated using the definition of conditional probability and transforming the numerator as in equation (B.15), i.e.

$$\text{prob}\Big[\{ \text{ Machine } M_i \text{ is blocked at time } t\}\big|\{\alpha_d[(j_2,i),t]=0\}\Big]$$

$$= \text{prob}\Big[\{ \text{ Machine } M_i \text{ is blocked at time } t\}\Big] :$$

$$\text{prob}\Big[\{\alpha_d[(j_2,i),t]=0\}\Big]$$

$$= \mathbf{p}[(i,q); N(i,q)10] :$$

$$\text{prob}\Big[\{\alpha_d[(j_2,i),t]=0\}\Big]$$

$$= \mathbf{p}[(i,q); N(i,q)10]\frac{r_d(j_2,i)\mu_d(j_2,i)}{p_d(j_2,i)PR(j_2,i)} \tag{B.61}$$

A similar reasoning applies for the subtrahend related to Cases 2 to 6 in (B.59). In all these cases, the priority one buffer level $n[(j_1,i),t]$ is positive and Machine M_i is not blocked, so

$$\text{prob}\Big[\{\text{Case 2 or 3 or 4 or 5 or 6 }\}\big|\{\alpha_d[(j_2,i),t]=0\}\Big]$$

$$= \text{prob}\Big[\{n[(j_1,i),t]>0\} \text{ and } \{M_i \text{ not blocked at time } t\}\big|$$

$$\{\alpha_d[(j_2,i),t]=0\}\Big]$$

$$= \text{prob}\Big[\{n[(j_1,i),t]>0\} \text{ and } \{M_i \text{ not blocked at time } t\}\Big] :$$

$$\text{prob}\Big[\{\alpha_d[(j_2,i),t]=0\}\Big]$$

$$\approx (1-\mathbf{p}[(j_1,i);001]-\mathbf{p}[(j_1,i);011])(1-\mathbf{p}[(i,q); N(i,q)10]) :$$

$$\text{prob}\Big[\{\alpha_d[(j_2,i),t]=0\}\Big]$$

$$= (1-\mathbf{p}[(j_1,i);001]-\mathbf{p}[(j_1,i);011])(1-\mathbf{p}[(i,q); N(i,q)10])$$

$$\frac{r_d(j_2,i)\mu_d(j_2,i)}{p_d(j_2,i)PR(j_2,i)}$$

if we assume as an approximation that the two-machine lines $L(j_1,i)$ and $L(i,q)$ are independent. The result for $W(j_2,i)$ is therefore

$$W(j_2,i)$$

$$= 1 - \Big[(1-\mathbf{p}[(j_1,i);001]-\mathbf{p}[(j_1,i);011])(1-\mathbf{p}[(i,q); N(i,q)10]) +$$

$$\mathbf{p}[(i,q); N(i,q)10]\Big]\frac{r_d(j_2,i)\mu_d(j_2,i)}{p_d(j_2,i)PR(j_2,i)} \tag{B.62}$$

The next probability to determine is $B(j_2, i)$ in (B.52). It is related to Case 4 in Table B.1, i.e. the situation that Machine M_i is not blocked, the machine upstream of the priority one buffer is down, and the priority one buffer is almost empty, i.e. $0 < n[(j_1, i), t] < \mu_d(j_1, i)\delta t$. In this situation, the priority one buffer can become empty during a time interval of length δt, if Machine M_i does not fail and the virtual upstream machine $M_u(j_1, i)$ is not repaired, so

$$B(j_2, i) = 1 - p_i\delta t - r_u(j_1, i)\delta t \tag{B.63}$$

The conditional probability $X(j_2, i)$ in (B.53) is the probability that a failure of $M_d(j_2, i)$ is due to Case 4, given that $M_d(j_2, i)$ is down. Using the information related to Case 4 in Table B.1, we find

$$
\begin{aligned}
X(j_2, i) \quad = \quad & \text{prob}\Big[\{\text{Case 4}\}\Big|\{\alpha_d[(j_2, i), t] = 0\}\Big] && \text{(B.64)}\\[1mm]
= \quad & \text{prob}\Big[\ \{M_i \text{ up at time } t\ \} \text{ and} && \text{(B.65)}\\
& \quad \{M_u(j_1, i) \text{ down at time } t\ \} \text{ and}\\
& \quad \{0 < n[(j_1, i), t] < \mu_d(j_1, i)\delta t\} \text{ and}\\
& \quad \{M_i \text{ not blocked at time } t\ \}\Big|\\
& \quad \{\alpha_d[(j_2, i), t] = 0\}\Big]
\end{aligned}
$$

Since Case 4 implies $\{\alpha_d[(j_2, i), t] = 0\}$, the definition of conditional probability yields

$$
\begin{aligned}
X(j_2, i) \quad = \quad & \text{prob}\Big[\ \{M_i \text{ up at time } t\ \} \text{ and}\\
& \quad \{M_u(j_1, i) \text{ down at time } t\ \} \text{ and}\\
& \quad \{M_d(j_1, i) \text{ up at time } t\ \} \text{ and}\\
& \quad \{0 < n[(j_1, i), t] < \mu_d(j_1, i)\delta t\} \text{ and}\\
& \quad \{M_i \text{ not blocked at time } t\ \}\Big] :\\[1mm]
& \text{prob}\Big[\{\alpha_d[(j_2, i), t] = 0\}\Big] && \text{(B.66)}
\end{aligned}
$$

As an approximation, we assume that Lines $L(j_1, i)$ and $L(i, q)$ are independent to find

$$X(j_2, i) \approx \text{prob}\Big[\{M_i \text{ up at time } t \} \text{ and}$$
$$\{M_u(j_1, i) \text{ down at time } t \} \text{ and}$$
$$\{0 < n[(j_1, i), t] < \mu_d(j_1, i)\delta t\}\Big] \cdot$$
$$\text{prob}\Big[\{M_i \text{ up at time } t \} \text{ and}$$
$$\{M_i \text{ not starved at time } t \} \text{ and}$$
$$\{M_i \text{ not blocked at time } t \}\Big] :$$
$$\text{prob}\Big[\{\alpha_d[(j_2, i), t] = 0\}\Big] \tag{B.67}$$

The first probability in (B.67) is for a small time interval of length δt

$$\text{prob}\Big[\{M_u(j_1, i) \text{ down at time } t \} \text{ and}$$
$$\{M_d(j_1, i) \text{ up at time } t \} \text{ and}$$
$$\{0 < n[(j_1, i), t] < \mu_d(j_1, i)\delta t\}\Big]$$
$$= \int_0^{\mu_d(j_1, i)\delta t} f_{(j_1, i)}(x, 0, 1) dx$$
$$= f_{(j_1, i)}(0, 0, 1)\mu_d(j_1, i)\delta t \tag{B.68}$$

where $f_{(j_1, i)}(n, \alpha_1, \alpha_2)$ is the probability density function related to buffer level n, upstream machine state α_1, and downstream machine state α_2 in Line $L(j_1, i)$.[2]

The second probability in (B.67) is

$$\text{prob}\Big[\{M_i \text{ up at time } t \} \text{ and}$$
$$\{M_i \text{ not starved at time } t \} \text{ and}$$
$$\{M_i \text{ not blocked at time } t \}\Big]$$
$$= \text{prob}\Big[\{M_u(i, q) \text{ up at time } t \} \text{ and}$$
$$\{M_u(i, q) \text{ not blocked at time } t \}\Big]$$
$$= \int_0^{N(i,q)} (f_{(i,q)}(x, 1, 0) + f_{(i,q)}(x, 1, 1)) dx$$
$$+\mathbf{p}[(i, q); 011] + \mathbf{p}[(i, q); N(i, q), 11] \tag{B.69}$$

Using equations (3.224), (3.247), and (3.238) in Gershwin (1994), it can be shown that the following equation holds:

---

[2] (Gershwin, 1994, p.118)

$$\int_0^{N(i,q)} (f_{(i,q)}(x,1,0) + f_{(i,q)}(x,1,1))dx + \mathbf{p}[(i,q);011]$$

$$= e_u(i,q)(1 - \mathbf{p}[(i,q);N(i,q),10] - \mathbf{p}[(i,q);N(i,q),11]) +$$

$$(e_u(i,q) - 1)\frac{\mu_d(i,q)}{\mu_u(i,q)}\mathbf{p}[(i,q);N(i,q),11] \qquad (\text{B.70})$$

The second probability in (B.67) is therefore

$$\text{prob}\bigg[\begin{array}{l} \{M_i \text{ up at time } t \} \text{ and} \\ \{M_i \text{ not starved at time } t \} \text{ and} \\ \{M_i \text{ not blocked at time } t \} \end{array}\bigg]$$

$$= e_u(i,q)(1 - \mathbf{p}[(i,q);N(i,q),10] - \mathbf{p}[(i,q);N(i,q),11]) +$$

$$(e_u(i,q) - 1)\frac{\mu_d(i,q)}{\mu_u(i,q)}\mathbf{p}[(i,q);N(i,q),11]$$

$$+\mathbf{p}[(i,q);N(i,q),11] \qquad (\text{B.71})$$

The numerator in (B.67) is transformed as in (B.15) to find

$$X(j_2,i) = \text{prob}\bigg[\text{Case } 4 \bigg| \{\alpha_d[(j_2,i),t] = 0\}\bigg]$$

$$= f_{(j_1,i)}(0,0,1)\mu_d(j_1,i)\delta t \cdot$$

$$\bigg[e_u(i,q)(1 - \mathbf{p}[(i,q);N(i,q),10] - \mathbf{p}[(i,q);N(i,q),11]) +$$

$$(e_u(i,q) - 1)\frac{\mu_d(i,q)}{\mu_u(i,q)}\mathbf{p}[(i,q);N(i,q),11] + \mathbf{p}[(i,q);N(i,q),11]\bigg] \cdot$$

$$\frac{r_d(j_2,i)\mu_d(j_2,i)}{p_d(j_2,i)PR(j_2,i)} \qquad (\text{B.72})$$

Note that both the expression for $B(j_2,i)$ in (B.63) and for $X(j_2,i)$ in (B.72) contain the term δt. In the resumption of flow equation (B.49), $B(j_2,i)$ and $X(j_2,i)$ are multiplied. For small time intervals δt, the second order terms can be omitted to find

$$B(j_2, i) X(j_2, i)$$

$$= (1 - p_i \delta t - r_u(j_1, i) \delta t) f_{(j_1, i)}(0, 0, 1) \mu_d(j_1, i) \delta t \cdot$$

$$\left[e_u(i, q) \left(1 - \mathbf{p}[(i, q); N(i, q), 10] - \mathbf{p}[(i, q); N(i, q), 11]\right) + \right.$$

$$\left. (e_u(i, q) - 1) \frac{\mu_d(i, q)}{\mu_u(i, q)} \mathbf{p}[(i, q); N(i, q), 11] + \mathbf{p}[(i, q); N(i, q), 11] \right] \cdot$$

$$\frac{r_d(j_2, i) \mu_d(j_2, i)}{p_d(j_2, i) PR(j_2, i)} \tag{B.73}$$

$$= f_{(j_1, i)}(0, 0, 1) \mu_d(j_1, i) \delta t \cdot$$

$$\left[e_u(i, q) \left(1 - \mathbf{p}[(i, q); N(i, q), 10] - \mathbf{p}[(i, q); N(i, q), 11]\right) + \right.$$

$$\left. (e_u(i, q) - 1) \frac{\mu_d(i, q)}{\mu_u(i, q)} \mathbf{p}[(i, q); N(i, q), 11] + \mathbf{p}[(i, q); N(i, q), 11] \right] \cdot$$

$$\frac{r_d(j_2, i) \mu_d(j_2, i)}{p_d(j_2, i) PR(j_2, i)} \tag{B.74}$$

The next terms of the resumption of flow equation (B.49) that need to be evaluated are $C(j_2, i)$ and $Y(j_2, i)$ related to Case 6 in Table B.1, i.e. the situation that Machine M_i is up and not blocked, the virtual machine $M_u(j_1, i)$ upstream of the priority one buffer is up, and the priority one buffer is almost empty. In this situation, the priority one buffer can become empty during a time interval of length δt, which in turn might result in a flow out of the priority two buffer, i.e. a repair of the virtual machine $M_d(j_2, i)$ seen by an observer in this priority two buffer. However, the probability of these events depends on the speed of the virtual machines up- and downstream of the priority one buffer.

$C(j_2, i)$ in (B.54) is the probability that an almost empty priority one buffer becomes empty and that the downstream machine starts to take material out of the priority two buffer. To evaluate $C(j_2, i)$, we first assume that the downstream machine is faster, i.e. $\mu_d(j_1, i) > \mu_u(j_1, i)$. If the downstream machine is faster than the upstream machine, a possible failure of the virtual upstream machine $M_u(j_1, i)$ during the time interval $[t, t + \delta t]$ is irrelevant: The priority one buffer becomes empty unless Machine M_i itself fails. We find

$$C(j_2, i) = 1 - p_i \delta t \quad \text{if} \quad \mu_d(j_1, i) > \mu_u(j_1, i). \tag{B.75}$$

However, if the downstream machine is slower than the upstream machine, the buffer level increases if both virtual machine operate. Therefore, the virtual machine $M_u(j_1, i)$ upstream of the priority one buffer must fail while Machine M_i must not fail in order for the priority one buffer to become empty and the flow out of the priority two buffer to resume. $C(j_2, i)$ is the probability that the faster virtual upstream machine fails and Machine M_i does not fail, so

$$C(j_2, i) = p_u(j_1, i)\delta t(1 - p_i\delta t) \approx p_u(j_1, i)\delta t \quad \text{if } \mu_d(j_1, i) \leq \mu_u(j_1, i) \text{ (B.76)}$$

since the second order terms can be omitted for small time intervals of length δt.

In (B.55), $Y(j_2, i)$ is the probability that a failure of Machine $M_d(j_2, i)$ is due to Case 6 in Table B.1, i.e. an almost empty priority one buffer. The derivation is similar to those for $X(j_2, i)$. Using the information related to Case 6 in Table B.1, we find

$$
\begin{aligned}
Y(j_2, i) \quad = \quad & \text{prob}\Big[\{\text{Case 6}\}\big|\{\alpha_d[(j_2, i), t] = 0\}\Big] & \text{(B.77)} \\[2mm]
= \quad & \text{prob}\Big[\ \{M_i \text{ up at time } t \} \text{ and} & \text{(B.78)} \\
& \{M_u(j_1, i) \text{ up at time } t \} \text{ and} \\
& \{0 < n[(j_1, i), t] < (\mu_d(j_1, i) - \mu_u(j_1, i))\delta t\} \text{ and} \\
& \{M_i \text{ not blocked at time } t \}\Big| \\
& \{\alpha_d[(j_2, i), t] = 0\}\Big]
\end{aligned}
$$

Since Case 6 implies $\{\alpha_d[(j_2, i), t] = 0\}$, the definition of conditional probability yields

$$
\begin{aligned}
Y(j_2, i) \quad = \quad & \text{prob}\Big[\ \{M_i \text{ up at time } t \} \text{ and} \\
& \{M_u(j_1, i) \text{ up at time } t \} \text{ and} \\
& \{M_d(j_1, i) \text{ up at time } t \} \text{ and} \\
& \{0 < n[(j_1, i), t] < (\mu_d(j_1, i) - \mu_u(j_1, i))\delta t\} \text{ and} \\
& \{M_i \text{ not blocked at time } t \}\Big] : \\[2mm]
& \text{prob}\Big[\{\alpha_d[(j_2, i), t] = 0\}\Big] & \text{(B.79)}
\end{aligned}
$$

As an approximation, we assume that Lines $L(j_1, i)$ and $L(i, q)$ are independent to find

$$Y(j_2, i) \approx \text{prob}\Big[\{M_i \text{ up at time } t \} \text{ and}$$
$$\{M_u(j_1, i) \text{ up at time } t \} \text{ and}$$
$$\{0 < n[(j_1, i), t] < (\mu_d(j_1, i) - \mu_u(j_1, i))\delta t\}\Big] \cdot$$

$$\text{prob}\Big[\{M_i \text{ up at time } t \} \text{ and}$$
$$\{M_i \text{ not starved at time } t \} \text{ and}$$
$$\{M_i \text{ not blocked at time } t \}\Big] :$$

$$\text{prob}\Big[\{\alpha_d[(j_2, i), t] = 0\}\Big] \qquad (B.80)$$

The first probability in (B.80) is for a small time interval of length δt

$$\text{prob}\Big[\{M_u(j_1, i) \text{ up at time } t \} \text{ and}$$
$$\{M_d(j_1, i) \text{ up at time } t \} \text{ and}$$
$$\{0 < n[(j_1, i), t] < (\mu_d(j_1, i) - \mu_u(j_1, i))\delta t\}\Big]$$

$$= \int_0^{(\mu_d(j_1,i)-\mu_u(j_1,i))\delta t} f_{(j_1,i)}(x, 1, 1)dx$$

$$= f_{(j_1,i)}(0, 1, 1)(\mu_d(j_1, i) - \mu_u(j_1, i))\delta t \qquad (B.81)$$

The other two probabilities in (B.80) are the same as in (B.67). The resulting term for $Y(j_2, i)$ is therefore

$$Y(j_2, i) = \text{prob}\Big[\{\text{Case } 6\}\big|\{\alpha_d[(j_2, i), t] = 0\}\Big]$$
$$= f_{(j_1,i)}(0, 1, 1)(\mu_d(j_1, i) - \mu_u(j_1, i))\delta t \cdot$$
$$\Big[e_u(i, q)\,(1 - \mathbf{p}[(i, q); N(i, q), 10] - \mathbf{p}[(i, q); N(i, q), 11]) +$$
$$(e_u(i, q) - 1)\frac{\mu_d(i, q)}{\mu_u(i, q)}\mathbf{p}[(i, q); N(i, q), 11] + \mathbf{p}[(i, q); N(i, q), 11]\Big] \cdot$$
$$\frac{r_d(j_2, i)\mu_d(j_2, i)}{p_d(j_2, i)PR(j_2, i)} \qquad (B.82)$$

In the resumption of flow equation (B.49), $C(j_2, i)$ in (B.75), as well as (B.76) and $Y(j_2, i)$ in (B.82) are multiplied. The product $C(j_2, i)Y(j_2, i)$ approaches 0 and can hence be omitted for $\mu_d(j_1, i) \leq \mu_u(j_1, i)$ as it only has second order terms with respect to δt. This product is non-zero only if $\mu_d(j_1, i) > \mu_u(j_1, i)$, i.e. if the downstream machine is faster than the upstream machine. We therefore find

$C(j_2, i)Y(j_2, i)$

$$= (1 - p_i\delta t)f_{(j_1,i)}(0, 1, 1)(\mu_d(j_1, i) - \mu_u(j_1, i))\delta t \cdot$$

$$\left[e_u(i, q)(1 - \mathbf{p}[(i, q); N(i, q), 10] - \mathbf{p}[(i, q); N(i, q), 11]) + \right.$$

$$\left. (e_u(i, q) - 1)\frac{\mu_d(i, q)}{\mu_u(i, q)}\mathbf{p}[(i, q); N(i, q), 11] + \mathbf{p}[(i, q); N(i, q), 11] \right] \cdot$$

$$\frac{r_d(j_2, i)\mu_d(j_2, i)}{p_d(j_2, i)PR(j_2, i)} \qquad \text{if } \mu_d(j_1, i) > \mu_u(j_1, i)$$

$$= f_{(j_1,i)}(0, 1, 1)(\mu_d(j_1, i) - \mu_u(j_1, i))\delta t \cdot \qquad (B.83)$$

$$\left[e_u(i, q)(1 - \mathbf{p}[(i, q); N(i, q), 10] - \mathbf{p}[(i, q); N(i, q), 11]) + \right.$$

$$\left. (e_u(i, q) - 1)\frac{\mu_d(i, q)}{\mu_u(i, q)}\mathbf{p}[(i, q); N(i, q), 11] + \mathbf{p}[(i, q); N(i, q), 11] \right] \cdot$$

$$\frac{r_d(j_2, i)\mu_d(j_2, i)}{p_d(j_2, i)PR(j_2, i)} \qquad \text{if } \mu_d(j_1, i) > \mu_u(j_1, i)$$

The sum $B(j_2, i)\, X(j_2, i) + C(j_2, i)\, Y(j_2, i)$ in (B.49) can be simplified using some equations for the two-machine model given in Gershwin (1994). The required equations depend on which of the two virtual machines in Line $L(j_1, i)$ is faster. The first equation (see (3.229), (3.212), and (3.188) in (Gershwin, 1994, p. 116-123))

$$f_{(j_1,i)}(0, 0, 1)\mu_d(j_1, i)\delta t + f_{(j_1,i)}(0, 1, 1)(\mu_d(j_1, i) - \mu_u(j_1, i))\delta t$$
$$= f_{(j_1,i)}(0, 1, 0)\mu_u(j_1, i)\delta t$$
$$= \mathbf{p}[(j_1, i); 011]p_d(j_1, i)\delta t\frac{\mu_u(j_1, i)}{\mu_d(j_1, i)} \qquad (B.84)$$

is used if $\mu_d(j_1, i) > \mu_u(j_1, i)$. The second equation ((3.205) in (Gershwin, 1994, p. 120))

$$f_{(j_1,i)}(0, 0, 1)\mu_d(j_1, i)\delta t = \mathbf{p}[(j_1, i); 001]r_u(j_1, i)\delta t \qquad (B.85)$$

is used if $\mu_d(j_1, i) \leq \mu_u(j_1, i)$.

We first determine $B(j_2, i)X(j_2, i) + C(j_2, i)Y(j_2, i)$ in (B.49) for the case $\mu_d(j_1, i) > \mu_u(j_1, i)$. From (B.74) and (B.83) we see that

$$B(j_2, i)X(j_2, i) + C(j_2, i)Y(j_2, i) \tag{B.86}$$

$$
\begin{aligned}
= \quad & \left[f_{(j_1, i)}(0, 0, 1)\mu_d(j_1, i)\delta t + f_{(j_1, i)}(0, 1, 1)(\mu_d(j_1, i) - \mu_u(j_1, i))\delta t \right] \cdot \\
& \left[e_u(i, q)\left(1 - \mathbf{p}[(i, q); N(i, q), 10] - \mathbf{p}[(i, q); N(i, q), 11]\right) + \right. \\
& \left. (e_u(i, q) - 1)\frac{\mu_d(i, q)}{\mu_u(i, q)}\mathbf{p}[(i, q); N(i, q), 11] + \mathbf{p}[(i, q); N(i, q), 11] \right] \cdot \\
& \frac{r_d(j_2, i)\mu_d(j_2, i)}{p_d(j_2, i)PR(j_2, i)} \qquad \text{if } \mu_d(j_1, i) > \mu_u(j_1, i)
\end{aligned}
$$

Using (B.84), this can be transformed into

$$B(j_2, i)X(j_2, i) + C(j_2, i)Y(j_2, i) \tag{B.87}$$

$$
\begin{aligned}
= \quad & \left[\mathbf{p}[(j_1, i); 011]p_d(j_1, i)\delta t \frac{\mu_u(j_1, i)}{\mu_d(j_1, i)} \right] \cdot \\
& \left[e_u(i, q)\left(1 - \mathbf{p}[(i, q); N(i, q), 10] - \mathbf{p}[(i, q); N(i, q), 11]\right) + \right. \\
& \left. (e_u(i, q) - 1)\frac{\mu_d(i, q)}{\mu_u(i, q)}\mathbf{p}[(i, q); N(i, q), 11] + \mathbf{p}[(i, q); N(i, q), 11] \right] \cdot \\
& \frac{r_d(j_2, i)\mu_d(j_2, i)}{p_d(j_2, i)PR(j_2, i)} \qquad \text{if } \mu_d(j_1, i) > \mu_u(j_1, i)
\end{aligned}
$$

This expression depends only on parameters and performance measures of the two-machine models that will be computed in the course of the iterative solution of the decomposition equations.

If the virtual upstream machine related to Line $L(j_1, i)$ is faster than the downstream machine, the $C(j_2, i)$ is zero and the sum is

$$B(j_2, i)X(j_2, i) + C(j_2, i)Y(j_2, i) = B(j_2, i)X(j_2, i)$$

$$
\begin{aligned}
= \quad & \left[f_{(j_1, i)}(0, 0, 1)\mu_d(j_1, i)\delta t \right] \cdot \\
& \left[e_u(i, q)\left(1 - \mathbf{p}[(i, q); N(i, q), 10] - \mathbf{p}[(i, q); N(i, q), 11]\right) + \right. \\
& \left. (e_u(i, q) - 1)\frac{\mu_d(i, q)}{\mu_u(i, q)}\mathbf{p}[(i, q); N(i, q), 11] + \mathbf{p}[(i, q); N(i, q), 11] \right] \cdot \\
& \frac{r_d(j_2, i)\mu_d(j_2, i)}{p_d(j_2, i)PR(j_2, i)} \qquad \text{if } \mu_d(j_1, i) \leq \mu_u(j_1, i)
\end{aligned}
$$

Using (B.85), this can be transformed into

$$B(j_2, i) X(j_2, i)$$

$$= \left[\mathbf{p}[(j_1, i); 001] r_u(j_1, i)\delta t \right] \cdot \qquad (B.88)$$

$$\left[e_u(i, q) \left(1 - \mathbf{p}[(i, q); N(i, q), 10] - \mathbf{p}[(i, q); N(i, q), 11]\right) + \right.$$

$$\left. (e_u(i, q) - 1)\frac{\mu_d(i, q)}{\mu_u(i, q)} \mathbf{p}[(i, q); N(i, q), 11] + \mathbf{p}[(i, q); N(i, q), 11] \right] \cdot$$

$$\frac{r_d(j_2, i)\mu_d(j_2, i)}{p_d(j_2, i) PR(j_2, i)} \qquad \text{if } \mu_d(j_1, i) \leq \mu_u(j_1, i)$$

This is again an expression that depends only on parameters and performance measures of the two-machine lines.

To complete the derivation of the resumption of flow equation for the virtual downstream machine of the priority two buffer, Case 7 in Table B.1 has to be analyzed. In this situation, the virtual machine $M_d(j_2, i)$ is down at time t because the real machine M_i is blocked while the priority one buffer is empty due to a failure of the virtual machine upstream of the priority one buffer. The flow out of Buffer $B_{j_2, i}$ can only resume if the blocking machine downstream of Machine M_i is repaired and the virtual machine $M_u(j_1, i)$ is not repaired. If Machine $M_u(j_1, i)$ were repaired, the priority one buffer would become non-empty and the virtual machine $M_d(j_2, i)$ would still be down, even if the blocking machine $M_d(i, q)$ were repaired. For this reason we find that

$$D(j_2, i) = r_d(i, q)\delta t(1 - r_u(j_1, i)\delta t) \approx r_d(i, q)\delta t \qquad (B.89)$$

since second order terms can be omitted.

The conditional probability $Z(j_2, i)$ in (B.57) is the probability that a failure of $M_d(j_2, i)$ is due to Case 7, given that $M_d(j_2, i)$ is down. Using the information related to Case 7 in Table B.1, we find

$$Z(j_2, i) = \text{prob}\left[\{\text{Case 7}\} \big| \{\alpha_d[(j_2, i), t] = 0\} \right] \qquad (B.90)$$

$$= \text{prob}\left[\begin{array}{l} \{M_i \text{ up at time } t\} \text{ and} \\ \{M_u(j_1, i) \text{ down at time } t\} \text{ and} \\ \{n[(j_1, i), t] = 0\} \text{ and} \\ \{M_i \text{ blocked at time } t\} \end{array} \right| \qquad (B.91)$$

$$\{\alpha_d[(j_2, i), t] = 0\} \right]$$

Since Case 7 implies $\{\alpha_d[(j_2, i), t] = 0\}$, the definition of conditional probability yields

$$
\begin{aligned}
Z(j_2, i) \quad = \quad \text{prob} \Big[& \{M_i \text{ up at time } t \} \text{ and} \\
& \{M_u(j_1, i) \text{ down at time } t \} \text{ and} \\
& \{n[(j_1, i), t] = 0\} \text{ and} \\
& \{M_i \text{ blocked at time } t \} \Big] : \\
& \text{prob} \Big[\{\alpha_d[(j_2, i), t] = 0\}\Big] \quad\quad\quad\text{(B.92)}
\end{aligned}
$$

From the perspective of an observer in an empty priority one buffer, there is no way to tell whether the virtual downstream machine is up or down, if its virtual upstream machine is down and if the buffer is empty. This is because no material is flowing into the buffer. If no material is flowing into the buffer at time t, no material can flow out of the buffer at time t. If no material is flowing out of the priority one buffer, its virtual downstream machine looks like it is up and starved. However, at this very moment Machine M_i in the real system may be blocked while trying to serve the priority two buffer, which is invisible to the observer in the priority one buffer. We can therefore add the event $\{M_d(j_1, i) \text{ up at time } t \}$ to find

$$
\begin{aligned}
Z(j_2, i) \quad = \quad \text{prob} \Big[& \{M_i \text{ up at time } t \} \text{ and} \\
& \{M_u(j_1, i) \text{ down at time } t \} \text{ and} \\
& \{M_d(j_1, i) \text{ up at time } t \} \text{ and} \\
& \{n[(j_1, i), t] = 0\} \text{ and} \\
& \{M_i \text{ blocked at time } t \} \Big] : \\
& \text{prob} \Big[\{\alpha_d[(j_2, i), t] = 0\}\Big] \quad\quad\quad\text{(B.93)}
\end{aligned}
$$

As an approximation, we assume that Lines $L(j_1, i)$ and $L(i, q)$ are independent to find

$$
\begin{aligned}
Z(j_2, i) \quad \approx \quad \text{prob} \Big[& \{M_u(j_1, i) \text{ down at time } t \} \text{ and} \\
& \{M_d(j_1, i) \text{ up at time } t \} \text{ and} \\
& \{n[(j_1, i), t] = 0\}\Big] \cdot \\
\text{prob} \Big[& \{M_i \text{ blocked at time } t \} \Big] : \\
& \text{prob} \Big[\{\alpha_d[(j_2, i), t] = 0\}\Big] \quad\quad\quad\text{(B.94)}
\end{aligned}
$$

The first probability is $\mathbf{p}[(j_1, i); 001]$ and the second is $\mathbf{p}[(i, q); N(i, q)10]$. We therefore find

$$Z(j_2, i) = \mathbf{p}[(j_1, i); 001]\mathbf{p}[(i, q); N(i, q)10]\frac{r_d(j_2, i)\mu_d(j_2, i)}{p_d(j_2, i)PR(j_2, i)} \quad \text{(B.95)}$$

Now all conditional probabilities have been determined and we can write the resumption of flow equation for the downstream machine of a priority two buffer in a form that is useful for the decomposition algorithm:

$$r_d(j_2, i) = \left[r_i + K_4 \frac{r_d(j_2, i)\mu_d(j_2, i)}{p_d(j_2, i)} \right] \quad \text{(B.96)}$$

where we define

$$K_4 = \left[\left[\begin{array}{ll} \mathbf{p}[(j_1, i); 011]p_d(j_1, i)\frac{\mu_u(j_1, i)}{\mu_d(j_1, i)} & \text{if } \mu_d(j_2, i) > \mu_u(j_2, i) \\ \mathbf{p}[(j_1, i); 001]r_u(j_1, i) & \text{if } \mu_d(j_2, i) \leq \mu_u(j_2, i) \end{array} \right] \right.$$

$$\left[e_u(i, q)\Big(1 - \mathbf{p}[(i, q); N(i, q), 10] - \mathbf{p}[(i, q); N(i, q), 11]\Big) + \right.$$

$$\left. \Big(e_u(i, q) - 1\Big)\frac{\mu_d(i, q)}{\mu_u(i, q)}\mathbf{p}[(i, q); N(i, q), 11] + \mathbf{p}[(i, q); N(i, q), 11] \right] +$$

$$\Big[r_d(i, q)\mathbf{p}[(j_1, i); 001]\mathbf{p}[(i, q); N(i, q)10] \Big] -$$

$$r_i\left[\Big(1 - \mathbf{p}[(j_1, i); 001] - \mathbf{p}[(j_1, i); 011]\Big)\Big(1 - \mathbf{p}[(i, q); N(i, q)10]\Big) \right.$$

$$\left. + \mathbf{p}[(i, q); N(i, q)10] \right] \left. \frac{1}{PR(j_2, i)} \quad \text{(B.97)}$$

B.3 Interruption of Flow Equations: Upstream Machine of a Split System

To reduce the notational effort, we define the auxiliary expression $U_u(i, m, t)$ for the event that Machine $M_u(i, m)$ is up and that Buffer $B_{i,m}$ is not full at time t, i.e.

$$U_u(i, m, t) := \{\alpha_u[(i, m), t] = 1\} \text{ and } \{n[(i, m), t] < N(i, m)\} \quad \text{(B.98)}$$

and can hence write

$$p_u(i,m)\delta t = \text{prob}\left[\{\alpha_u[(i,m),t+\delta t]=0\}\Big|U_u(i,m,t)\right] \tag{B.99}$$

$$= \text{prob}\left[\{\alpha_i(t+\delta t)=0\}\text{ or} \tag{B.100}\right.$$
$$\{n[(j,i),t+\delta t]=0\}\text{ and }\{\alpha_u[(j,i),t+\delta t]=0\}\text{ or}$$
$$\Big\{\{n[(i,q),t+\delta t]=N(i,q)\}\text{ and }\{\alpha_d[(i,q),t+\delta t]=0\}$$
$$\left.\text{for some }(i,q)\in D(i),q\neq m\Big\}\Big|U_u(i,m,t)\right]$$

Since the events that can force Machine M_i down are disjoint, this can be written as

$$p_u(i,m)\delta t \tag{B.101}$$
$$= \text{prob}\left[\{\alpha_i(t+\delta t)=0\}\Big|U_u(i,m,t)\right]\ +$$
$$\text{prob}\left[\{n[(j,i),t+\delta t]=0\}\text{ and }\{\alpha_u[(j,i),t+\delta t]=0\}\Big|U_u(i,m,t)\right]\ +$$
$$\sum_{\substack{(i,q)\in D(i)\\q\neq m}}\text{prob}\left[\{n[(i,q),t+\delta t]=N(i,q)\}\text{ and }\{\alpha_d[(i,q),t+\delta t]=0\}\Big|\right.$$
$$\left. U_u(i,m,t)\right]$$

The derivation for the interruption of flow equation generalizes the one by (Burman, 1995, p. 55-67) for the linear arrangement of machines in a flow line. The difference is that we have to take the multiple output buffers of a split machine into account.

In order to derive an expression for the first term on the right hand side of (B.101), we first derive an auxiliary equation. It says that the input buffer $B_{j,i}$ can be either empty or non-empty while either all downstream buffers $B_{i,q}, q\neq m$ are non-full or at most one of them is full:

$$\text{prob}\left[\Big\{\{n[(j,i),t]=0\}\text{ or }\{n[(j,i),t]>0\}\Big\}\text{ and}\right.$$
$$\Big\{\{n[(i,q),t]<N(i,q),\forall(i,q)\in D(i),q\neq m\}\text{ or}$$
$$\left.\{n[(i,q),t]=N(i,q),\text{ for one }(i,q)\in D(i),q\neq m\Big\}\right]$$
$$= \text{prob}\left[\{n[(j,i),t]=0\}\text{ and}\right.$$
$$\{n[(i,q),t]<N(i,q),\forall(i,q)\in D(i),q\neq m\}\text{ or}$$

$$\{n[(j,i),t]>0\}\text{ and}$$
$$\{n[(i,q),t]<N(i,q),\forall(i,q)\in D(i),q\neq m\}\text{ or}$$

$$\{n[(j,i),t]>0\}\text{ and}$$
$$\left.\{n[(i,q),t]=N(i,q),\text{ for one }(i,q)\in D(i),q\neq m\}\right]$$
$$= 1 \tag{B.102}$$

In the continuous material model, the probability of Buffer $B_{j,i}$ being empty while at least one buffer $B_{i,q}$ is full is zero.[3] For a similar reason, the probability that multiple output buffers of a split machine are full at the same time is zero as well. Since the probabilities of the disjoint events in (B.102) add up to one, we can use this equation to decompose the first term on the right hand side of (B.101):

$$\text{prob}\left[\{\alpha_i(t+\delta t) = 0\}\Big| U_u(i,m,t)\right]$$

$$\begin{aligned}
= \quad & \text{prob}\left[\{\alpha_i(t+\delta t) = 0\} \text{ and } \{n[(j,i),t] = 0\} \text{ and} \qquad\qquad \text{(B.103)}\right. \\
& \left.\{n[(i,q),t] < N(i,q), \forall(i,q) \in D(i), q \neq m\}\Big| U_u(i,m,t)\right] \\
+ \quad & \text{prob}\left[\{\alpha_i(t+\delta t) = 0\} \text{ and } \{n[(j,i),t] > 0\} \text{ and} \right. \\
& \left.\{n[(i,q),t] < N(i,q), \forall(i,q) \in D(i), q \neq m\}\Big| U_u(i,m,t)\right] \\
+ \quad & \sum_{\substack{(i,q)\in D(i) \\ q \neq m}} \text{prob}\left[\{\alpha_i(t+\delta t) = 0\} \text{ and } \{n[(j,i),t] > 0\} \text{ and} \right. \\
& \left.\{n[(i,q),t] = N(i,q)\}\Big| U_u(i,m,t)\right]
\end{aligned}$$

From the definition of conditional probability the following identity can be derived:

$$\begin{aligned}
\text{prob}[A \text{ and } B | C] \quad &= \quad \text{prob}[A | B \text{ and } C] \, \frac{\text{prob}[B \text{ and } C]}{\text{prob}[C]} \\
&= \quad \text{prob}[A | B \text{ and } C] \, \text{prob}[B | C] \qquad\qquad \text{(B.104)}
\end{aligned}$$

Using this identity, the first term on the right hand side of (B.103) can be expressed as

$$\text{prob}\left[\{\alpha_i(t+\delta t) = 0\} \text{ and } \{n[(j,i),t] = 0\} \text{ and} \right.$$
$$\left.\{n[(i,q),t] < N(i,q), \forall(i,q) \in D(i), q \neq m\}\Big| U_u(i,m,t)\right]$$

$$= \quad \text{prob}\left[\{\alpha_i(t+\delta t) = 0\}\Big| \right.$$
$$\{n[(j,i),t] = 0\} \text{ and}$$
$$\{n[(i,q),t] < N(i,q), \forall(i,q) \in D(i), q \neq m\} \text{ and}$$
$$\left. U_u(i,m,t)\right] \cdot$$

$$\text{prob}\left[\{n[(j,i),t] = 0\} \text{ and} \qquad\qquad\qquad\qquad \text{(B.105)}\right.$$
$$\{n[(i,q),t] < N(i,q), \forall(i,q) \in D(i), q \neq m\}\Big|$$
$$\left. U_u(i,m,t)\right]$$

---

[3] Dallery et al. (1989)

The first factor in (B.105) is the probability that Machine M_i fails, given that its upstream buffer is empty. It can only be empty if the virtual upstream machine $M_u(j, i)$ is slower than or equally fast as Machine M_i since none of M_i's output buffers is full. If the Machine $M_u(j, i)$ is slower than M_i, operation dependent failures of M_i occur less frequently than when Machine M_i operates in isolation. The decrease of the failure rate is proportional to the relative speed of $M_u(j, i)$ and M_i ((Burman, 1995, p. 58)), so

$$\text{prob} \left[\{\alpha_i(t + \delta t) = 0\} \middle| \right.$$
$$\{n[(j, i), t] = 0\} \text{ and}$$
$$\{n[(i, q), t] < N(i, q), \forall(i, q) \in D(i), q \neq m\} \text{ and}$$
$$\left. U_u(i, m, t) \right].$$

$$= \begin{cases} \frac{\mu_u(j,i)}{\mu_i} p_i \delta t & \text{if } \mu_u(j, i) \leq \mu_i \\ 0 & \text{otherwise.} \end{cases}$$

$$\approx \begin{cases} \frac{\mu_u(j,i)}{\mu_i} p_i \delta t & \text{if } \mu_u(j, i) \leq \mu_d(j, i) \\ 0 & \text{otherwise.} \end{cases} \quad (B.106)$$

where μ_i has been replaced in (B.106) by $\mu_d(j, i)$ as in (Burman, 1995, p. 58).

The second factor in (B.105) is the probability that Buffer $B_{j,i}$ is empty, given that Machine $M_u(i, m)$ is up and not blocked. Due to the definition of conditional probability, the following holds:

$$\text{prob} \left[\{n[(j, i), t] = 0\} \text{ and } \{n[(i, q), t] < N(i, q), \forall(i, q) \in D(i), q \neq m\} \middle| U_u(i, m, t) \right]$$

$$= \text{prob} \left[\{n[(j, i), t] = 0\} \text{ and } \{n[(i, q), t] < N(i, q), \forall(i, q) \in D(i), q \neq m\} \text{ and } U_u(i, m, t) \middle| U_u(i, m, t) \right]$$

$$= \text{prob} \left[\{n[(j, i), t] = 0\} \text{ and } \{n[(i, q), t] < N(i, q), \forall(i, q) \in D(i), q \neq m\} \text{ and } U_u(i, m, t) \right] : \text{prob}[U_u(i, m, t)]$$

$$(B.107)$$

In the numerator in (B.107), an empty buffer $B_{j,i}$, i.e. event $\{n[(j, i), t] = 0\}$, implies that no buffer $B_{i,q}$ is full, i.e. event $\{n[(i, q), t] < N(i, q), \forall(i, q) \in D(i), q \neq m\}$. Given that Buffer $B_{j,i}$ is empty at time t, event $U_u(i, m, t)$ implies that both Machines $M_u(j, i)$ and $M_d(j, i)$ are up. The first factor is therefore $\mathbf{p}[(j, i); 011]$.[4]

---

[4] Burman (1995, p. 59)

The denominator in (B.107) is the probability that Machine $M_u(i, m)$ is up and Buffer $B_{i,m}$ is not full. It is related to the fraction of time $E_u(i, m)$ that Machine $M_u(i, m)$ is operating (Burman (1995, p. 59)) via

$$
\begin{aligned}
E_u(i, m) \;=\;& \text{prob}[\{\alpha_u[(i, m), t] = 1\} \text{ and } \{n[(i, m), t] < N(i, m)\}] + \\
& \mathbf{p}[(i, m); N(i, m), 11]\frac{\mu_d(i, m)}{\mu_u(i, m)} \\
\;=\;& \text{prob}[U_u(i, m, t)] + \mathbf{p}[(i, m); N(i, m), 11]\frac{\mu_d(i, m)}{\mu_u(i, m)} \quad \text{(B.108)}
\end{aligned}
$$

which takes into account that when Buffer $B_{i,m}$ is full and both machines are up, Machine $M_u(i, m)$ is operating at a reduced speed.[5] We can therefore write (B.107) as

$$
\begin{aligned}
&\text{prob}\Big[\{n[(j, i), t] = 0\} \text{ and } \{n[(i, q), t] < N(i, q), \forall (i, q) \in D(i), q \neq m\} \\
&\qquad\qquad \text{and } U_u(i, m, t)\big|U_u(i, m, t)\Big] \\
&= \frac{\mathbf{p}[(j, i); 011]}{E_u(i, m) - \mathbf{p}[(i, m); N(i, m), 11]\frac{\mu_d(i,m)}{\mu_u(i,m)}} \qquad\qquad \text{(B.109)}
\end{aligned}
$$

The expression for the first term on the right hand side of (B.103) is therefore

$$
\begin{aligned}
&\text{prob}\Big[\{\alpha_i(t + \delta t) = 0\} \text{ and } U_u(i, m, t) \text{ and} \qquad\qquad\qquad \text{(B.110)}\\
&\quad \{n[(j, i), t] = 0\} \text{ and} \\
&\quad \{n[(i, q), t] < N(i, q), \forall (i, q) \in D(i), q \neq m\}\big|U_u(i, m, t)\Big] \\
&= \frac{\mu_u(j, i)}{\mu_i}p_i\delta t\frac{\mathbf{p}[(j, i); 011]}{E_u(i, m) - \mathbf{p}[(i, m); N(i, m), 11]\frac{\mu_d(i,m)}{\mu_u(i,m)}} \qquad \text{(B.111)}
\end{aligned}
$$

Note that $\mathbf{p}[(j, i); 011]$ is zero if Machine $M_u(j, i)$ is faster than $M_d(j, i)$. This takes care of the two cases in (B.106).

The third term on the right hand side of (B.103) can be analyzed in an almost identical way to find

$$
\begin{aligned}
&\text{prob}\Big[\{\alpha_i(t + \delta t) = 0\} \text{ and } \{n[(j, i), t] > 0\} \text{ and} \\
&\qquad\qquad \{n[(i, q), t] = N(i, q)\}\big|U_u(i, m, t)\Big] \\
&= \frac{\frac{\mu_d(i, q)}{d_{i,q}}}{\mu_i}p_i\delta t\frac{\mathbf{p}[(i, q); N(i, q); 11]}{E_u(i, m) - \mathbf{p}[(i, m); N(i, m), 11]\frac{\mu_d(i,m)}{\mu_u(i,m)}} \qquad \text{(B.112)}
\end{aligned}
$$

---

[5] Burman (1995, p. 45)

Note that this last term takes into account that only a fraction $d_{i,q}$ of material flows from Machine M_i to M_q. If Buffer $B_{i,q}$ is full while the downstream machine $M_d(i,q)$ processes material at rate $\mu_d(i,q)$, the rate at which Machine M_i processes material is $\mu_d(i,q)/d_{i,q}$ as all the other downstream machines also receive material. The adjusted processing rate $\mu_d(i,q)/d_{i,q}$ has to be considered to determine the modified failure rate.

For the second term of (B.103), we find

$$
\begin{aligned}
&\text{prob} \Big[\{\alpha_i(t+\delta t)=0\} \text{ and} \\
&\quad \{n[(j,i),t]>0\} \text{ and} \\
&\quad \{n[(i,q),t]<N(i,q), \forall(i,q)\in D(i), q\neq m\} \Big| U_u(i,m,t) \Big] \\
&= \text{prob} \Big[\{\alpha_i(t+\delta t)=0\} \Big| \\
&\quad \{n[(j,i),t]>0\} \text{ and} \\
&\quad \{n[(i,q),t]<N(i,q), \forall(i,q)\in D(i), q\neq m\} \text{ and} \\
&\quad U_u(i,m,t) \Big] \cdot \\
&\quad \text{prob} \Big[\{n[(j,i),t]>0\} \text{ and} \\
&\quad \{n[(i,q),t]<N(i,q), \forall(i,q)\in D(i), q\neq m\} \Big| U_u(i,m,t) \Big]
\end{aligned}
\tag{B.113}
$$

The first factor in (B.113) is the probability that Machine M_i fails during a time interval of length δt, given that its input buffer is non-empty and all its output buffers are non-full. This is the failure probability of Machine M_i in the real system, so

$$
\begin{aligned}
&\text{prob} \Big[\{\alpha_i(t+\delta t)=0\} \Big| \\
&\quad \{n[(j,i),t]>0\} \text{ and} \\
&\quad \{n[(i,q),t]<N(i,q), \forall(i,q)\in D(i), q\neq m\} \text{ and} \\
&\quad U_u(i,m,t) \Big] \\
&= p_i \delta t
\end{aligned}
\tag{B.114}
$$

For the second factor in (B.113) we find that

$$
\text{prob} \Big[\{n[(j,i),t]>0\} \text{ and} \tag{B.115}
$$
$$
\{n[(i,q),t]<N(i,q), \forall(i,q)\in D(i), q\neq m\} \Big| U_u(i,m,t) \Big]
$$
$$
= 1 - \text{prob} \Big[\{n[(j,i),t]=0\} \text{ or} \tag{B.116}
$$
$$
\{n[(i,q),t]<N(i,q), \text{ some } (i,q)\in D(i), q\neq m\} \Big| U_u(i,m,t) \Big]
$$
$$
= 1 - \frac{\Big[\mathbf{p}[(j,i);011] + \sum_{\substack{(i,q)\in D(i) \\ q\neq m}} \mathbf{p}[(i,q);N(i,q),11] \Big]}{E_u(i,m) - \mathbf{p}[(i,m);N(i,m),11]\frac{\mu_d(i,m)}{\mu_u(i,m)}}
\tag{B.117}
$$

The second term on the right hand side of (B.103) is therefore

$$
\text{prob} \left[\{\alpha_i(t + \delta t) = 0\} \text{ and} \right. \tag{B.118}
$$
$$
\left. \{n[(j, i), t] > 0\} \text{ and } \{n[(i, q), t] = N(i, q)\} \middle| U_u(i, m, t) \right]
$$
$$
= p_i \delta t \left(1 - \frac{\left[\mathbf{p}[(j, i); 011] + \sum_{\substack{(i,q) \in D(i) \\ q \neq m}} \mathbf{p}[(i, q); N(i, q), 11] \right]}{E_u(i, m) - \mathbf{p}[(i, m); N(i, m), 11]\frac{\mu_d(i,m)}{\mu_u(i,m)}} \right)
$$

Now all components of (B.103) can be assembled

$$
\text{prob} \left[\{\alpha_i(t + \delta t) = 0\} \middle| U_u(i, m, t) \right] \tag{B.119}
$$
$$
= \frac{\mu_u(j, i)}{\mu_i} p_i \delta t \frac{\mathbf{p}[(j, i); 011]}{E_u(i, m) - \mathbf{p}[(i, m); N(i, m), 11]\frac{\mu_d(i,m)}{\mu_u(i,m)}}
$$
$$
+ p_i \delta t \left(1 - \frac{\left[\mathbf{p}[(j, i); 011] + \sum_{\substack{(i,q) \in D(i) \\ q \neq m}} \mathbf{p}[(i, q); N(i, q), 11] \right]}{E_u(i, m) - \mathbf{p}[(i, m); N(i, m), 11]\frac{\mu_d(i,m)}{\mu_u(i,m)}} \right)
$$
$$
+ \sum_{\substack{(i,q) \in D(i) \\ q \neq m}} \frac{\frac{\mu_d(i,q)}{d_{i,q}}}{\mu_i} p_i \delta t \frac{\mathbf{p}[(i, q); N(i, q); 11]}{E_u(i, m) - \mathbf{p}[(i, m); N(i, m), 11]\frac{\mu_d(i,m)}{\mu_u(i,m)}}
$$

and the last two components of (B.101) remain to be determined. The second term can be approximated as in Burman (1995, p. 65) for a flow line to find

$$
\text{prob} \left[\{n[(j, i), t + \delta t] = 0\} \text{ and } \{\alpha_u[(j, i), t + \delta t] = 0\} \middle| U_u(i, m, t) \right]
$$
$$
\approx r_u(j, i) \delta t \frac{\mathbf{p}[(j, i); 001]}{E_u(i, m) - \mathbf{p}[(i, m); N(i, m), 11]\frac{\mu_d(i,m)}{\mu_u(i,m)}} \tag{B.120}
$$

and an almost identical derivation results in

$$
\text{prob} \left[\{n[(i, q), t + \delta t] = N(i, q)\} \text{ and } \{\alpha_d[(i, q), t + \delta t] = 0\} \middle| U_u(i, m, t) \right]
$$
$$
\approx r_d(i, q) \delta t \frac{\mathbf{p}[(i, q); N(i, q), 10]}{E_u(i, m) - \mathbf{p}[(i, m); N(i, m), 11]\frac{\mu_d(i,m)}{\mu_u(i,m)}} \tag{B.121}
$$

for the third component of (B.101). The interruption of flow equation (B.101) can hence be written as

$$p_u(i,m)\delta t$$

$$= \frac{\mu_u(j,i)}{\mu_i}p_i\delta t\frac{\mathbf{p}[(j,i);011]}{E_u(i,m) - \mathbf{p}[(i,m);N(i,m),11]\frac{\mu_d(i,m)}{\mu_u(i,m)}}$$

$$+ \quad p_i\delta t\left(1 - \frac{\left[\mathbf{p}[(j,i);011] + \sum\limits_{\substack{(i,q)\in D(i)\\q\neq m}}\mathbf{p}[(i,q);N(i,q),11]\right]}{E_u(i,m) - \mathbf{p}[(i,m);N(i,m),11]\frac{\mu_d(i,m)}{\mu_u(i,m)}}\right)$$

$$+ \quad \sum\limits_{\substack{(i,q)\in D(i)\\q\neq m}}\frac{\frac{\mu_d(i,q)}{d_{i,q}}}{\mu_i}p_i\delta t\frac{\mathbf{p}[(i,q);N(i,q);11]}{E_u(i,m) - \mathbf{p}[(i,m);N(i,m),11]\frac{\mu_d(i,m)}{\mu_u(i,m)}}$$

$$+ \quad r_u(j,i)\delta t\frac{\mathbf{p}[(j,i);001]}{E_u(i,m) - \mathbf{p}[(i,m);N(i,m),11]\frac{\mu_d(i,m)}{\mu_u(i,m)}}$$

$$+ \quad \sum\limits_{\substack{(i,q)\in D(i)\\q\neq m}}r_d(i,q)\delta t\frac{\mathbf{p}[(i,q);N(i,q),10]}{E_u(i,m) - \mathbf{p}[(i,m);N(i,m),11]\frac{\mu_d(i,m)}{\mu_u(i,m)}} \quad (\text{B.122})$$

The interruption of flow equation given in (Burman, 1995, p. 66) is a special case of (B.122). Equation (B.122) says that the failure rate $p_u(i,m)$ that an observer in Buffer $B_{i,m}$ downstream of a split machine M_i sees is a weighted average. The first three components of (B.122) account for failures of Machine M_i. The first term deals with failures of Machine M_i when M_i is slowed down by upstream machines. In a very similar way, the third term models failures of Machine M_i when M_i is slowed down by one of its virtual downstream machines $M_d(i,q), q \neq m$. Machine M_i can also fail if it is not slowed down by up- or downstream machines. This leads to the second term. If Buffer $B_{j,i}$ is empty, Machine $M_u(i,m)$ can fail due to failure of Machine $M_u(j,i)$. The rate at which these failures occur is equal to the rate $r_u(j,i)\mathbf{p}[(j,i);001]$ at which the corresponding repairs occur. This leads to the fourth component of (B.122). The last group of terms in (B.122) deals in a similar way with failures of downstream machines $M_d(i,q)$ when Buffer $B_{i,q}$ is full.

In order to get the iterative algorithm that is based on the decomposition equations to converge, we made the following substitutions as proposed in (Burman, 1995, p. 76):

$$\frac{\mu_u(j,i)}{\mu_i} \implies \frac{\mu_u(j,i)}{\mu_d(j,i)} \qquad \text{(B.123)}$$

$$\frac{\frac{\mu_d(i,q)}{d_{i,q}}}{\mu_i} \implies \frac{\mu_d(i,q)}{\mu_u(i,q)} \qquad \text{(B.124)}$$

$$E_u(i,m) - \mathbf{p}[(i,m); N(i,m), 11]\frac{\mu_d(i,m)}{\mu_u(i,m)} \implies E_u(i,m)$$

$$= \frac{PR(i,m)}{\mu_u(i,m)} \quad \text{(B.125)}$$

Based on these substitutions, (B.122) can be expressed in a form that is more useful for algorithmic purposes:

$$p_u(i,m) = p_i + K_5\mu_u(i,m) \qquad \text{(B.126)}$$

with

$$K_5 = \left[p_i\left(\frac{\mu_u(j,i)}{\mu_d(j,i)} - 1\right)\mathbf{p}[(j,i); 011] + r_u(j,i)\mathbf{p}[(j,i); 001] + \right.$$

$$\sum_{\substack{(i,q)\in D(i) \\ q\neq m}}\left(p_i\left(\frac{\mu_d(i,q)}{\mu_u(i,q)} - 1\right)\mathbf{p}[(i,q); N(i,q), 11] + \right.$$

$$\left. r_d(i,q)\mathbf{p}[(i,q); N(i,q), 10]\right)\right]\frac{1}{PR(i,m)} \qquad \text{(B.127)}$$

B.4 Interruption of Flow Equations: Merge System

B.4.1 Upstream Machine

We define the auxilliary expression $U_u(i,q,t)$ for the event that Machine $M_u(i,q)$ is up and Buffer $B_{i,q}$ not full at time t, i.e.

$$U_u(i,q,t) := \{\alpha_u[(i,q),t] = 1\} \text{ and } \{n[(i,q),t] < N(i,q)\} \qquad \text{(B.128)}$$

and can hence write

$$p_u(i,q)\delta t = \text{prob}\left[\{\alpha_u[(i,q),t+\delta t] = 0\}\big|U_u(i,q,t)\right] \qquad \text{(B.129)}$$

$$= \text{prob}\left[\{\alpha_i(t+\delta t) = 0\} \text{ or } \left\{\{n[(j,i),t+\delta t] = 0\} \text{ and } \right.\right.$$

$$\left.\left.\{\alpha_u[(j,i),t+\delta t] = 0\}, \forall j \in \{j_1, j_2\}\right\}\big|U_u(i,q,t)\right]$$

Since the events that can force Machine M_i down are disjoint, this can be written as

$$p_u(i,q)\delta t \qquad\qquad\qquad\qquad (\text{B.130})$$

$$= \quad \text{prob}\ \left[\{\alpha_i(t+\delta t)=0\}\Big|U_u(i,q,t)\right]\ +$$

$$\text{prob}\ \left[\Big\{\{n[(j,i),t+\delta t]=0\}\ \text{and}\right.$$
$$\left.\{\alpha_u[(j,i),t+\delta t]=0\}, \forall j\in\{j_1,j_2\}\Big\}\Big|U_u(i,q,t)\right]$$

The first term on the right hand side of (B.130) deals with failures of Machine M_i and the second with those of virtual upstream machines $M_u(j_1,i)$ and $M_u(j_2,i)$.

The failure rate of Machine M_i depends on whether M_i is slowed down by upstream machines. This can only happen if both input buffers are empty. To evaluate the first term on the right hand side of (B.130), we will consider the two cases that either both input buffers are empty or that at least one of them is non-empty, i.e

$$\text{prob}\ \left[\{\alpha_i(t+\delta t)=0\}\Big|U_u(i,q,t)\right]$$

$$= \quad \text{prob}\ \left[\{\alpha_i(t+\delta t)=0\}\ \text{and}\right.$$
$$\Big\{\{n[(j_1,i),t]=0\}\ \text{and}\ \{n[(j_2,i),t]=0\}\ \text{or}$$
$$\left.\Big\{\{n[(j_1,i),t]>0\}\ \text{or}\ \{n[(j_2,i),t]>0\}\Big\}\Big\}\Big|U_u(i,q,t)\right]$$

$$= \quad \text{prob}\ \left[\{\alpha_i(t+\delta t)=0\}\ \text{and}\right.$$
$$\left.\{n[(j_1,i),t]=0\}\ \text{and}\ \{n[(j_2,i),t]=0\}\Big|U_u(i,q,t)\right]$$

$$+ \quad \text{prob}\ \left[\{\alpha_i(t+\delta t)=0\}\ \text{and}\right. \qquad\qquad (\text{B.131})$$
$$\left.\Big\{\{n[(j_1,i),t]>0\}\ \text{or}\ \{n[(j_2,i),t]>0\}\Big\}\Big|U_u(i,q,t)\right]$$

The first term on the right hand side of (B.131) deals with failures of Machine M_i when it is slowed down by its upstream machines. The conditioning event $U_u(i,q,t)$ says that Machine $M_u(i,q)$ is up at time t. Given that both input buffers are empty at time t, this implies that at least one of the two machines $M_u(j_1,i)$ and $M_u(j_2,i)$ is up at time t. The rate at which Machine M_i fails when both input buffers are empty depends on the state and speed of the Machines $M_u(j_1,i)$ and $M_u(j_2,i)$. To determine the first term on the right hand side of (B.131), we consider the different possible combinations of machine states:

$$= \text{ prob } \Big[\{\alpha_i(t + \delta t) = 0\} \text{ and}$$
$$\{n[(j_1, i), t] = 0\} \text{ and } \{n[(j_2, i), t] = 0\} \text{ and}$$
$$\Big\{ \{\alpha_u[(j_1, i), t] = 1\} \text{ and } \{\alpha_u[(j_2, i), t] = 1\} \text{ or}$$
$$\{\alpha_u[(j_1, i), t] = 0\} \text{ and } \{\alpha_u[(j_2, i), t] = 1\} \text{ or}$$
$$\{\alpha_u[(j_1, i), t] = 1\} \text{ and } \{\alpha_u[(j_2, i), t] = 0\} \Big\} \Big| U_u(i, q, t) \Big]$$

$$= \text{ prob } \Big[\{\alpha_i(t + \delta t) = 0\} \text{ and}$$
$$\{n[(j_1, i), t] = 0\} \text{ and } \{n[(j_2, i), t] = 0\} \text{ and}$$
$$\{\alpha_u[(j_1, i), t] = 1\} \text{ and } \{\alpha_u[(j_2, i), t] = 1\} \Big| U_u(i, q, t) \Big]$$

$$+ \text{ prob } \Big[\{\alpha_i(t + \delta t) = 0\} \text{ and}$$
$$\{n[(j_1, i), t] = 0\} \text{ and } \{n[(j_2, i), t] = 0\} \text{ and}$$
$$\{\alpha_u[(j_1, i), t] = 0\} \text{ and } \{\alpha_u[(j_2, i), t] = 1\} \Big| U_u(i, q, t) \Big]$$

$$+ \text{ prob } \Big[\{\alpha_i(t + \delta t) = 0\} \text{ and} \hspace{3cm} \text{(B.132)}$$
$$\{n[(j_1, i), t] = 0\} \text{ and } \{n[(j_2, i), t] = 0\} \text{ and}$$
$$\{\alpha_u[(j_1, i), t] = 1\} \text{ and } \{\alpha_u[(j_2, i), t] = 0\} \Big| U_u(i, q, t) \Big]$$

The first term on the right hand side of (B.132) reflects failures of Machine M_i when both input buffers are empty and both upstream machines operate. Using the definition of conditional probablity and (B.104), this first term can be expressed as

$$\text{prob } \Big[\{\alpha_i(t + \delta t) = 0\} \text{ and}$$
$$\{n[(j_1, i), t] = 0\} \text{ and } \{n[(j_2, i), t] = 0\} \text{ and}$$
$$\{\alpha_u[(j_1, i), t] = 1\} \text{ and } \{\alpha_u[(j_2, i), t] = 1\} \Big| U_u(i, q, t) \Big]$$

$$= \text{ prob } \Big[\{\alpha_i(t + \delta t) = 0\} \Big| \{n[(j_1, i), t] = 0\} \text{ and } \{n[(j_2, i), t] = 0\} \text{ and}$$
$$\{\alpha_u[(j_1, i), t] = 1\} \text{ and } \{\alpha_u[(j_2, i), t] = 1\} \text{ and } U_u(i, q, t) \Big] \cdot$$

$$\text{prob } \Big[\{n[(j_1, i), t] = 0\} \text{ and } \{n[(j_2, i), t] = 0\} \text{ and}$$
$$\{\alpha_u[(j_1, i), t] = 1\} \text{ and } \{\alpha_u[(j_2, i), t] = 1\} \text{ and } U_u(i, q, t) \Big] :$$

$$\text{prob } \Big[U_u(i, q, t) \Big] \hspace{3cm} \text{(B.133)}$$

The first factor on the right hand side of (B.133) is the probability that a merge machine M_i fails during a time interval of length δt given that both input buffers are empty and both virtual upstream machines $M_u(j_1, i)$ and $M_u(j_2, i)$ operate. Since Machine M_i is slowed down, the failure rate of the operation dependent failures has to be adjusted, where the speed at which M_i is operating is $\mu_u(j_1, i) + \mu_u(j_2, i)$, so

$$\text{prob}\left[\{\alpha_i(t+\delta t)=0\}\big|\{n[(j_1,i),t]=0\}\text{ and }\{n[(j_2,i),t]=0\}\text{ and }\right.$$
$$\left.\{\alpha_u[(j_1,i),t]=1\}\text{ and }\{\alpha_u[(j_2,i),t]=1\}\text{ and }U_u(i,q,t)\right]$$

$$=\begin{cases}\frac{\mu_u(j_1,i)+\mu_u(j_2,i)}{\mu_i}p_i\delta t & \text{if }\mu_u(j_1,i)\le\mu_d(j_1,i)\text{ and}\\ & \mu_u(j_2,i)\le\mu_d(j_2,i)\qquad\text{(B.134)}\\ 0 & \text{otherwise.}\end{cases}$$

as in both input buffers the virtual upstream machines must be slower than the virtual downstream machines. If at least one virtual upstream machine related to one of the two input buffers is faster than the respective downstream machine, this buffer is never empty while the virtual upstream machine is up. If one of the two input buffers is non-empty, Machine M_i is not slowed down.

In the second factor on the right hand side of (B.133), the event $U_u(i,q,t)$ says that the virtual machine $M_u(i,q)$ is up and not blocked at time t. This implies that the virtual downstream machine of the priority one buffer is also up. Since the priority one buffer is empty at time t, it also implies that the virtual downstream machine of the priority two buffer is up at time t, so

$$\text{prob}\left[\{n[(j_1,i),t]=0\}\text{ and }\{n[(j_2,i),t]=0\}\text{ and }\right.$$
$$\left.\{\alpha_u[(j_1,i),t]=1\}\text{ and }\{\alpha_u[(j_2,i),t]=1\}\text{ and }U_u(i,q,t)\right]$$

$$=\ \text{prob}\left[\{n[(j_1,i),t]=0\}\text{ and }\right.$$
$$\{\alpha_u[(j_1,i),t]=1\}\text{ and }\{\alpha_d[(j_1,i),t]=1\}\text{ and }$$
$$\{n[(j_2,i),t]=0\}\text{ and }$$
$$\left.\{\alpha_u[(j_2,i),t]=1\}\text{ and }\{\alpha_d[(j_2,i),t]=1\}\text{ and }\ U_u(i,q,t)\right]$$

$$=\ \text{prob}\left[\{n[(j_1,i),t]=0\}\text{ and }\right.\qquad\qquad\text{(B.135)}$$
$$\{\alpha_u[(j_1,i),t]=1\}\text{ and }\{\alpha_d[(j_1,i),t]=1\}\text{ and }$$
$$\{n[(j_2,i),t]=0\}\text{ and }$$
$$\left.\{\alpha_u[(j_2,i),t]=1\}\text{ and }\{\alpha_d[(j_2,i),t]=1\}\right]$$

since $\{n[(j_1,i),t]=0\}$ and $\{\alpha_u[(j_1,i),t]=1\}$ and $\{\alpha_d[(j_1,i),t]=1\}$ implies that $M_u(i,q)$ is up and $B_{i,q}$ is not full, i.e. event $U_u(i,q,t)$. As an approximation, we assume that the two lines $L(j_1,i)$ and $L(j_2,i)$ are independent, so

$$\text{prob}\ \Big[\{n[(j_1,i),t]=0\}\ \text{and}$$
$$\{\alpha_u[(j_1,i),t]=1\}\ \text{and}\ \{\alpha_d[(j_1,i),t]=1\}\ \text{and}$$
$$\{n[(j_2,i),t]=0\}\ \text{and}$$
$$\{\alpha_u[(j_2,i),t]=1\}\ \text{and}\ \{\alpha_d[(j_2,i),t]=1\}\Big]$$

$$\approx\ \text{prob}\ \Big[\{n[(j_1,i),t]=0\}\ \text{and}$$
$$\{\alpha_u[(j_1,i),t]=1\}\ \text{and}\ \{\alpha_d[(j_1,i),t]=1\}\Big]\cdot$$

$$\text{prob}\ \Big[\{n[(j_1,i),t]=0\}\ \text{and}$$
$$\{\alpha_u[(j_1,i),t]=1\}\ \text{and}\ \{\alpha_d[(j_1,i),t]=1\}\Big]$$

$$=\ \mathbf{p}[(j_1,i);011]\mathbf{p}[(j_2,i);011] \tag{B.136}$$

The probability $\text{prob}[U_u(i,q,t)]$ that Machine $M_u(i,q)$ is up and Buffer $B_{i,q}$ is not full is determined as in (B.108) to find

$$\text{prob}[U_u(i,q,t)] = E_u(i,q) - \mathbf{p}[(i,q);N(i,q),11]\frac{\mu_d(i,q)}{\mu_u(i,q)} \tag{B.137}$$

We therefore find for the first term on the right hand side of (B.132)

$$=\ \text{prob}\ \Big[\{\alpha_i(t+\delta t)=0\}\ \text{and} \tag{B.138}$$
$$\{n[(j_1,i),t]=0\}\ \text{and}\ \{n[(j_2,i),t]=0\}\ \text{and}$$
$$\{\alpha_u[(j_1,i),t]=1\}\ \text{and}\ \{\alpha_u[(j_2,i),t]=1\}\Big|U_u(i,q,t)\Big]$$

$$=\ \frac{\mu_u(j_1,i)+\mu_u(j_2,i)}{\mu_i}p_i\delta t\frac{\mathbf{p}[(j_1,i);011]\mathbf{p}[(j_2,i);011]}{E_u(i,q)-\mathbf{p}[(i,q);N(i,q),11]\frac{\mu_d(i,q)}{\mu_u(i,q)}}$$

The second and third term of (B.132) treat the two cases that both input buffers are empty and one of the virtual upstream machines $M_u(j_1,i)$ and $M_u(j_2,i)$ is up while the other one is down. A similar derivation yields

$$\text{prob}\ \Big[\{\alpha_i(t+\delta t)=0\}\ \text{and}$$
$$\{n[(j_1,i),t]=0\}\ \text{and}\ \{n[(j_2,i),t]=0\}\ \text{and}$$
$$\{\alpha_u[(j_1,i),t]=0\}\ \text{and}\ \{\alpha_u[(j_2,i),t]=1\}\Big|U_u(i,q,t)\Big]$$

$$=\ \frac{\mu_u(j_2,i)}{\mu_i}p_i\delta t\frac{\mathbf{p}[(j_1,i);001]\mathbf{p}[(j_2,i);011]}{E_u(i,q)-\mathbf{p}[(i,q);N(i,q),11]\frac{\mu_d(i,q)}{\mu_u(i,q)}} \tag{B.139}$$

for the second term where Machine $M_u(j_1,i)$ is down and

$$\text{prob}\ \Big[\{\alpha_i(t+\delta t)=0\}\ \text{and}$$
$$\{n[(j_1,i),t]=0\}\ \text{and}\ \{n[(j_2,i),t]=0\}\ \text{and}$$
$$\{\alpha_u[(j_1,i),t]=1\}\ \text{and}\ \{\alpha_u[(j_2,i),t]=0\}\big|U_u(i,q,t)\Big]$$

$$=\ \frac{\mu_u(j_1,i)}{\mu_i}p_i\delta t\frac{\mathbf{p}[(j_1,i);011]\mathbf{p}[(j_2,i);001]}{E_u(i,q)-\mathbf{p}[(i,q);N(i,q),11]\frac{\mu_d(i,q)}{\mu_u(i,q)}}\qquad\text{(B.140)}$$

for the third term where Machine $M_u(j_2,i)$ is down.

The second term on the right hand side of (B.131) deals with failures of Machine M_i when it is not slowed down by its upstream machines, as at least one of the two input buffers is non-empty.

Using the definition of conditional probablity and (B.104), this second term on the right hand side of (B.131) can be expressed as

$$\text{prob}\ \Big[\{\alpha_i(t+\delta t)=0\}\ \text{and}\ \big\{\{n[(j_1,i),t]>0\}\ \text{or}\ \{n[(j_2,i),t]>0\}\big\}\big|$$
$$U_u(i,q,t)\Big]$$

$$=\ \text{prob}\ \Big[\{\alpha_i(t+\delta t)=0\}\big| \qquad\qquad\qquad\text{(B.141)}$$
$$\big\{\{n[(j_1,i),t]>0\}\ \text{or}\ \{n[(j_2,i),t]>0\}\big\}\ \text{and}\ U_u(i,q,t)\Big]\cdot$$
$$\text{prob}\ \Big[\{n[(j_1,i),t]>0\}\ \text{or}\ \{n[(j_2,i),t]>0\}\ \text{and}\ U_u(i,q,t)\big|U_u(i,q,t)\Big]$$

The first term on the right hand side of (B.141) is the probability of a failure of Machine M_i when it is not slowed down, i.e.

$$\text{prob}\ \Big[\{\alpha_i(t+\delta t)=0\}\big|$$
$$\big\{\{n[(j_1,i),t]>0\}\ \text{or}\ \{n[(j_2,i),t]>0\}\big\}\ \text{and}\ U_u(i,q,t)\Big]$$

$$=\ p_i\delta t \qquad\qquad\qquad\qquad\qquad\qquad\qquad\text{(B.142)}$$

The second term on the right hand side of (B.141) is the probability that at least one input buffer of Machine M_i is non-empty and the virtual machine $M_u(i,q)$ is up. It can be expressed in terms of the approximated probability to have both input buffers empty and Machine $M_u(i,q)$ up

$$\text{prob} \left[\{ n[(j_1, i), t] > 0 \} \text{ or } \{ n[(j_2, i), t] > 0 \} \text{ and } U_u(i, q, t) \Big| U_u(i, q, t) \right]$$

$$= 1 - \text{prob} \left[\{ n[(j_1, i), t] = 0 \} \text{ and } \{ n[(j_2, i), t] = 0 \} \text{ and } U_u(i, q, t) \Big| \right.$$
$$\left. U_u(i, q, t) \right]$$

$$= 1 -$$
$$\left[\mathbf{p}[(j_1, i); 011] \mathbf{p}[(j_2, i); 011] + \right.$$
$$\mathbf{p}[(j_1, i); 001] \mathbf{p}[(j_2, i); 011] +$$
$$\left. \mathbf{p}[(j_1, i); 011] \mathbf{p}[(j_2, i); 001] \right] :$$
$$\left[E_u(i, m) - \mathbf{p}[(i, m); N(i, m), 11] \frac{\mu_d(i, m)}{\mu_u(i, m)} \right] \tag{B.143}$$

if we assume that Lines $L(j_1, i)$ and $L(j_2, i)$ are independent.

In the next step, the second term on the right hand side of (B.130) has to be analyzed. It is the probability that both input buffers $B_{j_1, i}$ and $B_{j_2, i}$ are empty and both virtual upstream machine $M_u(j_1, i)$ and $M_u(j_2, i)$ are down at time $t + \delta t$, given that at least one input buffer was non-empty or at least one virtual upstream machine was up at time t.

This probability is determined in a way identical to those used for the second and third component on the right hand side of (B.101). We ask for the possible previous states at time t and their probabilities and for the probabilities of the transitions from the different possible states at time t to the state at time $t + \delta t$. If both input buffers are empty at time $t + \delta t$, they must have been either empty or almost empty at time t, as only a small amount of material can be processed during a small time interval of length δt. Furthermore, since M_i is up and not blocked at time t, both virtual downstream machines $M_d(j_1, i)$ and $M_d(j_2, i)$ are up at time t. None of the lines $L(j, i), j \in \{ j_1, j_2 \}$ can therefore be in any state at time t other than one of the following three:

- Buffer empty, upstream machine down, downstream machine up:
  $(n[(j, i), t] = 0, \alpha_u[(j, i), t] = 0, \alpha_d[(j, i), t] = 1)$
- Buffer *almost* empty, upstream machine down,[6] downstream machine up:
  $(0 < n[(j, i), t] = 0 < \mu_d(j, i) \delta t, \alpha_u[(j, i), t] = 0, \alpha_d[(j, i), t] = 1)$
- Buffer empty, upstream machine up, downstream machine up:
  $(n[(j, i), t] = 0, \alpha_u[(j, i), t] = 1, \alpha_d[(j, i), t] = 1)$

As there are two lines $L(j_1, i)$ and $L(j_2, i)$, this leads to a total of $3 \cdot 3 = 9$ possible previous states and respective transition probabilities. The following approximation expresses the state at time $t + \delta t$ in terms of those at time t:

---

[6] If the upstream machine is up, the corresponding probability times the transition probability leads to a second order expression $(\delta t)^2$ and can hence be omitted.

prob $\left[\left\{\{n[(j,i),t+\delta t]=0\}\text{ and}\right.\right.$

$\qquad\left.\left.\{\alpha_u[(j,i),t+\delta t]=0\},j\in\{j_1,j_2\}\right\}\Big|U_u(i,q,t)\right]$

$\approx\quad(1-r_u(j_1,i)\delta t-r_u(j_2,i)\delta t)\cdot$

$\qquad$ prob $\left[\left\{\{n[(j,i),t]=0\}\text{ and}\right.\right.$

$\qquad\qquad\left.\left.\{\alpha_u[(j,i),t]=0\},j\in\{j_1,j_2\}\right\}\Big|U_u(i,q,t)\right]$

$+\quad(1-r_u(j_1,i)\delta t-p_d(j_1,i)\delta t-r_u(j_2,i)\delta t)\cdot$

$\qquad\dfrac{\int_0^{\mu_d(j_1,i)\delta t}f_{(j_1,i)}(x,0,1)\delta x\ \mathbf{p}[(j_2,i);001]}{\text{prob}[U_u(i,q,t)]}$

$+\quad p_u(j_1,i)\delta t\dfrac{\mathbf{p}[(j_1,i);011]\ \mathbf{p}[(j_2,i);001]}{\text{prob}[U_u(i,q,t)]}$

$+\quad(1-r_u(j_1,i)\delta t-r_u(j_2,i)\delta t-p_d(j_2,i)\delta t)\cdot$

$\qquad\dfrac{\mathbf{p}[(j_1,i);001]\ \int_0^{\mu_d(j_2,i)\delta t}f_{(j_2,i)}(x,0,1)\delta x}{\text{prob}[U_u(i,q,t)]}$

$+\quad(1-r_u(j_1,i)\delta t-p_d(j_1,i)\delta t-r_u(j_2,i)\delta t-p_d(j_2,i)\delta t)\cdot$

$\qquad\dfrac{\int_0^{\mu_d(j_1,i)\delta t}f_{(j_1,i)}(x,0,1)\delta x\ \int_0^{\mu_d(j_2,i)\delta t}f_{(j_2,i)}(x,0,1)\delta x}{\text{prob}[U_u(i,q,t)]}$

$+\quad p_u(j_1,i)\delta t\dfrac{\mathbf{p}[(j_1,i);011]\ \int_0^{\mu_d(j_2,i)\delta t}f_{(j_2,i)}(x,0,1)\delta x}{\text{prob}[U_u(i,q,t)]}$

$+\quad p_u(j_2,i)\delta t\dfrac{\mathbf{p}[(j_1,i);001]\ \mathbf{p}[(j_2,i);011]}{\text{prob}[U_u(i,q,t)]}$

$+\quad p_u(j_2,i)\delta t\dfrac{\int_0^{\mu_d(j_1,i)\delta t}f_{(j_1,i)}(x,0,1)\delta x\ \mathbf{p}[(j_2,i);011]}{\text{prob}[U_u(i,q,t)]}$

$+\quad p_u(j_1,i)p_u(j_2,i)(\delta t)^2\dfrac{\mathbf{p}[(j_1,i);011]\ \mathbf{p}[(j_2,i);011]}{\text{prob}[U_u(i,q,t)]}\qquad\text{(B.144)}$

This is an approximation as we treat Lines $L(j_1,i)$ and $L(j_2,i)$ as if they were independent. The first three terms on the right hand side of (B.144) deal with the first possible state of Line $L(j_2,i)$ as described above, i.e. Machine $M_u(j_2,i)$ down, $M_d(j_2,i)$ up, and Buffer $B_{j_2,i}$ empty at time t. The second three terms reflect the second and the last three terms the third possible state of Line $L(j_2,i)$ at time t.

Some of these terms can be omitted for small time intervals δt. Consider the first term on the right hand side of (B.144): If the virtual machine $M_u(i,q)$

is up at time t due to the conditioning event of the second factor, it is not possible that both machines $M_u(j_1, i)$ and $M_u(j, 2)$ are down and both input buffers $B_{j_1, i}$ and $B_{j_2, i}$ are empty at time t. For this reason, the first term is 0.

To evaluate the second term, the integral over the probability density function $f_{(j_1, i)}(x, 0, 1)$ has to be determined. Due to the small amount of material that can be processed during a time interval of length δt, the following equation holds:

$$\int_0^{\mu_d(j, i)\delta t} f_{(j, i)}(x, 0, 1)\delta x = \mu_d(j, i)\delta t f_{(j, i)}(0, 0, 1), \quad j \in \{j_1, j_2\} \quad (B.145)$$

Note the term δt on the right hand side of (B.145). In the case of the second term on the right hand side of (B.144), multiplication of this term δt with the transisition probability $(1 - r_u(j_1, i)\delta t - p_d(j_1, i)\delta t - r_u(j_2, i)\delta t)$ leads to several second order terms that can be omitted. Using (B.145) and omitting the second order terms, we can write for the second term on the right hand side of (B.144):

$$(1 - r_u(j_1, i)\delta t - p_d(j_1, i)\delta t - r_u(j_2, i)\delta t) \cdot$$
$$\frac{\int_0^{\mu_d(j_1, i)\delta t} f_{(j_1, i)}(x, 0, 1)\delta x \; \mathbf{p}[(j_2, i); 001]}{\text{prob}[U_u(i, q, t)]}$$

$$= \frac{\mu_d(j, i)\delta t f_{(j, i)}(0, 0, 1) \; \mathbf{p}[(j_2, i); 001]}{\text{prob}[U_u(i, q, t)]} \quad (B.146)$$

The fourth, fifth, sixth, and eighth term on the right hand side of (B.144) can be evaluated in the same way. The fifth, sixth, eighth, and nineth term are second order terms with respect to δt and can hence be omitted and we find

$$\text{prob}\left[\left\{\{n[(j, i), t + \delta t] = 0\} \text{ and}\right.\right.$$
$$\left.\left.\{\alpha_u[(j, i), t + \delta t] = 0\}, j \in \{j_1, j_2\}\right\}\middle| U_u(i, q, t)\right]$$

$$= \left[\mu_d(j_1, i)\delta t \; f_{(j_1, i)}(0, 0, 1) \; \mathbf{p}[(j_2, i); 001]\right.$$

$$+ p_u(j_1, i)\delta t \; \mathbf{p}[(j_1, i); 011] \; \mathbf{p}[(j_2, i); 001]$$

$$+ \mathbf{p}[(j_1, i); 001] \; \mu_d(j_2, i)\delta t \; f_{(j_2, i)}(0, 0, 1)$$

$$+ p_u(j_2, i)\delta t \; \mathbf{p}[(j_1, i); 001] \; \mathbf{p}[(j_2, i); 011]\left]\frac{1}{\text{prob}[U_u(i, q, t)]} \right. \quad (B.147)$$

Using the equation (3.205) in (Gershwin, 1994, p. 120), we find

$$\text{prob} \left[\left\{\{n[(j,i),t+\delta t]=0\} \text{ and} \atop \{\alpha_u[(j,i),t+\delta t]=0\}, j \in \{j_1,j_2\}\right\} \Big| U_u(i,q,t)\right]$$

$$= \frac{(r_u(j_1,i)\delta t + r_u(j_2,i)\delta t)\, \mathbf{p}[(j_1,i);001]\, \mathbf{p}[(j_2,i);001]}{\text{prob}[\{\alpha_u[(i,q),t]=1\} \text{ and } \{n[(i,q),t] < N(i,q)\}]} \quad (B.148)$$

which is more compact and readable than the original equation (B.144).

This last equation can be interpreted in the following way: The rate of failures of Machine $M_u(i,q)$ that are due to failures of both upstream machines is equal to the rate of repairs of upstream machines if they are both down. It is the two-input-buffer version of the "one repair for each failure" equation.

Similar to the procedure in (Burman, 1995, p. 76), we make the following substitutions to get the algorithm to converge:

$$\frac{\mu_u(j,i)}{\mu_i} \implies \frac{\mu_u(j,i)}{\mu_d(j,i)}, j \in \{j_1,j_2\} \quad (B.149)$$

$$\frac{\mu_u(j_1,i)+\mu_u(j_2,i)}{\mu_i} \implies \frac{\mu_u(j_1,i)+\mu_u(j_2,i)}{\mu_d(j_1,i)+\mu_d(j_2,i)} \quad (B.150)$$

$$E_u(i,q) - \mathbf{p}[(i,q);N(i,q),11]\frac{\mu_d(i,q)}{\mu_u(i,q)} \implies E_u(i,q)$$

$$= \frac{PR(i,q)}{\mu_u(i,q)} \quad (B.151)$$

The final form of the interruption of flow equation for Machine $M_u(i,q)$ in a merge system is therefore

$$p_u(i,q) = p_i + K_5\, \mu_u(i,q) \quad (B.152)$$

with

$$K_5 = \left[p_i \left(\frac{\mu_u(j_1, i) + \mu_u(j_2, i)}{\mu_d(j_1, i) + \mu_d(j_2, i)} - 1 \right) \mathbf{p}[(j_1, i); 011] \mathbf{p}[(j_1, i); 011] + \right.$$

$$p_i \left(\frac{\mu_u(j_1, i)}{\mu_d(j_1, i)} - 1 \right) \mathbf{p}[(j_1, i); 011] \mathbf{p}[(j_2, i); 001] + \qquad \text{(B.153)}$$

$$p_i \left(\frac{\mu_u(j_2, i)}{\mu_d(j_2, i)} - 1 \right) \mathbf{p}[(j_1, i); 001] \mathbf{p}[(j_2, i); 011] +$$

$$\left. (r_u(j_1, i) + r_u(j_2, i)) \, \mathbf{p}[(j_1, i); 001] \mathbf{p}[(j_2, i); 001] \right] \frac{1}{PR(i, q)}$$

B.4.2 Downstream Machine of the Priority Two Line

To derive the interruption of flow equation, it is convenient to write the definition of Machine $M_d(j_2, i)$ being down as such that the different conditions are disjoint. Using Table B.1 on Page 232, we find

$$\{\alpha_d[(j_2, i), t] = 0\} \quad \text{iff} \quad \{ \text{Case 1 or 2} \} \text{ or}$$
$$\{ \text{Case 3 or 4 or 5 or 6} \} \text{ or}$$
$$\{ \text{Case 7 or 8} \} \qquad \text{(B.154)}$$

which is equivalent to

$$\{\alpha_d[(j_2, i), t] = 0\} \text{ iff } \{\alpha_i(t) = 0\} \text{ or} \qquad \text{(B.155)}$$
$$\left\{ \{n[(j_1, i), t] > 0\} \text{ and} \right.$$
$$\{\alpha_i(t) = 1\} \text{ and}$$
$$\left\{ \{n[(i, q), t] < N(i, q)\} \text{ or} \right.$$
$$\left. \{n[(i, q), t] = N(i, q)\} \text{ and } \{\alpha_d[(i, q), t] = 1\} \right\} \right\} \text{ or}$$
$$\{n[(i, q), t] = N(i, q)\} \text{ and } \{\alpha_d[(i, q), t] = 0\}$$

Using the definition of virtual machine states (5.29) and (B.156), the interruption of flow equation for the virtual machine $M_d(j_2, i)$ is

$$p_d(j_2, i)\delta t = \text{prob} \left[\{\alpha_d[(j_2, i), t + \delta t] = 0\} \right| \qquad \text{(B.156)}$$
$$\left. \{\alpha_d[(j_2, i), t] = 1\} \text{ and } \{n[(j_2, i), t] > 0\} \right]$$

We define the auxilliary expression $U_d(j_2, i, t)$ for the event that Machine $M_d(j_2, i)$ is up and Buffer $B_{j_2, i}$ not empty at time t, i.e.

$$U_d(j_2, i, t) := \{\alpha_d[(j_2, i), t] = 1\} \text{ and } \{n[(j_2, i), t] > 0\} \qquad \text{(B.157)}$$

and can hence decompose the interruption of flow equation

$$p_d(j_2, i)\delta t = \text{prob} \left[\{\alpha_d[(j_2, i), t + \delta t] = 0\} \big| U_d(j_2, i, t) \right] \qquad \text{(B.158)}$$

$$= \text{prob} \left[\{\alpha_i(t + \delta t) = 0\} \text{ or} \right. \qquad \text{(B.159)}$$

$$\left\{ \{n[(j_1, i), t + \delta t] > 0\} \text{ and} \right.$$
$$\{\alpha_i(t + \delta t) = 1\} \text{ and}$$
$$\left\{ \{n[(i, q), t + \delta t] < N(i, q)\} \text{ or} \right.$$
$$\{n[(i, q), t + \delta t] = N(i, q)\} \text{ and}$$
$$\left. \{\alpha_d[(i, q), t + \delta t] = 1\} \right\} \right\} \text{ or}$$
$$\left\{ \{n[(i, q), t + \delta t] = N(i, q)\} \text{ and} \right.$$
$$\left. \left. \{\alpha_d[(i, q), t + \delta t] = 0\} \right\} \big| U_d(j_2, i, t) \right]$$

$$= \text{prob} \left[\{\alpha_i(t + \delta t) = 0\} \big| U_d(j_2, i, t + \delta t) \right] \qquad \text{(B.160)}$$

$$+ \text{prob} \left[\{n[(j_1, i), t + \delta t] > 0\} \text{ and} \right.$$
$$\{\alpha_i(t + \delta t) = 1\} \text{ and}$$
$$\left\{ \{n[(i, q), t + \delta t] < N(i, q)\} \text{ or} \right.$$
$$\left\{ \{n[(i, q), t + \delta t] = N(i, q)\} \text{ and} \right.$$
$$\left. \left. \{\alpha_d[(i, q), t + \delta t] = 1\} \right\} \right\} \big| U_d(j_2, i, t) \right]$$

$$+ \text{prob} \left[\{n[(i, q), t + \delta t] = N(i, q)\} \text{ and } \{\alpha_d[(i, q), t + \delta t] = 0\} \right|$$
$$\left. U_d(j_2, i, t) \right]$$

The first term on the right hand side of (B.160) deals with failures of Machine M_i, the second with priority one buffers becoming non-empty and the third with failures of machines downstream of M_i.

The failure rate of Machine M_i depends on whether M_i is slowed down by downstream machines. This can only happen if the output buffer $B_{i,q}$ is full. To evaluate the first term on the right hand side of (B.160), we will consider the two cases that Buffer $B_{i,q}$ is either full or non-full at time t, i.e

$$\text{prob}\ \Big[\{\alpha_i(t+\delta t)=0\}\big|U_d(i,q,t)\Big]$$

$$=\ \text{prob}\ \Big[\{\alpha_i(t+\delta t)=0\}\ \text{and}$$
$$\Big\{\{n[(i,q),t]=N(i,q)\}\ \text{or}\ \{n[(i,q),t]<N(i,q)\}\Big\}\big|U_d(j_2,i,t)\Big]$$

$$=\ \text{prob}\ \Big[\{\alpha_i(t+\delta t)=0\}\ \text{and}\ \{n[(i,q),t]=N(i,q)\}\big|U_d(j_2,i,t)\Big]$$

$$+\ \text{prob}\ \Big[\{\alpha_i(t+\delta t)=0\}\ \text{and}\ \{n[(i,q),t]<N(i,q)\}\big|U_d(j_2,i,t)\Big]\ \text{(B.161)}$$

The first term on the right hand side of (B.161) deals with failures of Machine M_i when M_i is slowed down by the virtual downstream machine $M_d(i,q)$ and the priority one buffer $B_{j_1,i}$ is empty. Using (B.104), the following decomposition is possible:

$$\text{prob}\ \Big[\{\alpha_i(t+\delta t)=0\}\ \text{and}\ \{n[(i,q),t]=N(i,q)\}\big|U_d(j_2,i,t)\Big]$$

$$=\ \text{prob}\ \Big[\{\alpha_i(t+\delta t)=0\}\big|\{n[(i,q),t]=N(i,q)\}\ \text{and}\ U_d(j_2,i,t)\Big]\cdot$$

$$\text{prob}\ \Big[\{n[(i,q),t]=N(i,q)\}\ \text{and}\ U_d(j_2,i,t)\Big]:$$

$$\text{prob}\ \Big[U_d(j_2,i,t)\Big] \hspace{3cm} \text{(B.162)}$$

It is analyzed in the same way as (B.105). The first term on the right hand side of (B.162) is the failure rate of Machine M_i when it is slowed down by $M_d(i,q)$, i.e.

$$\text{prob}\ \Big[\{\alpha_i(t+\delta t)=0\}\big|\{n[(i,q),t]=N(i,q)\}\ \text{and}\ U_d(j_2,i,t)\Big]$$

$$=\ \begin{cases} \dfrac{\mu_d(i,q)}{\mu_i}p_i\delta t & \text{if}\ \mu_u(i,q)\geq\mu_d(i,q) \\ 0 & \text{otherwise.} \end{cases} \hspace{2cm} \text{(B.163)}$$

In the second term on the right hand side of (B.162), the event $U_d(j_2,i,t)$ implies that Machine $M_u(i,q)$ is up at time t. If Buffer $B_{i,q}$ is full at time t, $U_d(j_2,i,t)$ further implies that the virtual downstream machine $M_d(i,q)$ is up at time t. With respect to the priority one line $L(j_1,i)$, the event $U_d(j_2,i,t)$ implies that the downstream machine $M_d(j_1,i)$ is up and Buffer $B_{j_1,i}$ is empty at time t. However, nothing is implied with respect to the virtual upstream machine of the priority one line $L(j_1,i)$. This leads to the following approximation

$$\text{prob}\ \Big[\{n[(i,q),t] = N(i,q)\}\ \text{and}\ U_d(j_2,i,t)\Big]$$

$$= \ \text{prob}\ \Big[\{n[(i,q),t] = N(i,q)\}\ \text{and}$$
$$\{\alpha_u[(i,q),t] = 1\}\ \text{and}\ \{\alpha_d[(i,q),t] = 1\}\ \text{and}$$
$$\{n[(j_1,i),t] = 0\}\ \text{and}$$
$$\Big\{\{\alpha_u[(j_1,i),t] = 1\}\ \text{or}\ \{\alpha_u[(j_1,i),t] = 0\}\Big\}\ \text{and}$$
$$\{\alpha_d[(j_1,i),t] = 1\}\Big]$$

$$\approx\ \mathbf{p}[(i,q);N(i,q),11]\ \Big(\mathbf{p}[(j_1,i);001] + \mathbf{p}[(j_1,i);011]\Big) \qquad (B.164)$$

if we treat Lines $L(j_1,i)$ and $L(i,q)$ as if they were independent.

The third term on the right hand side of (B.162) is the probability that Machine $M_d(j_2,i)$ is up and not starved at time t. It is derived as the denominator in (B.107):

$$\text{prob}\Big[U_d(j_2,i,t)\Big] = E_d(j_2,i) - \mathbf{p}[(j_2,i);011]\frac{\mu_u(j_2,i)}{\mu_d(j_2,i)} \qquad (B.165)$$

The first term on the right hand side of (B.161) can therefore be expressed as follows:

$$\text{prob}\ \Big[\{\alpha_i(t+\delta t) = 0\}\ \text{and}\ \{n[(i,q),t] = N(i,q)\}\Big|U_d(j_2,i,t)\Big]$$

$$= \ \frac{\mu_d(i,q)}{\mu_i}p_i\delta t\frac{\mathbf{p}[(i,q);N(i,q),11]\ \Big(\mathbf{p}[(j_1,i);001] + \mathbf{p}[(j_1,i);011]\Big)}{E_d(j_2,i) - \mathbf{p}[(j_2,i);011]\frac{\mu_u(j_2,i)}{\mu_d(j_2,i)}}(B.166)$$

The second term on the right hand side of (B.161) is determined like (B.113) to find

$$\text{prob}\ \Big[\{\alpha_i(t+\delta t) = 0\}\ \text{and}\ \{n[(i,q),t] < N(i,q)\}\Big|U_d(j_2,i,t)\Big]$$

$$= \ p_i\delta t\left(1 - \frac{\mathbf{p}[(i,q);N(i,q),11]\ \Big(\mathbf{p}[(j_1,i);001] + \mathbf{p}[(j_1,i);011]\Big)}{E_d(j_2,i) - \mathbf{p}[(j_2,i);011]\frac{\mu_u(j_2,i)}{\mu_d(j_2,i)}}\right)(B.167)$$

The second term on the right hand side of (B.160) is the probability of a priority one buffer $B(j_1,i)$ being non-empty at time $t + \delta t$, given that the buffer was empty and Machine $M_d(j_1,i)$ up at time t, both due to the conditioning event $U_d(j_2,i,t)$.

Buffer $B_{j_1,i}$ can only be empty at time t and non-empty at time $t + \delta t$ if two conditions hold: Machine $M_u(j_1,i)$ is faster than $M_d(j_1,i)$ and Machine

$M_u(j_1, i)$ is down at time t. The reasoning is as follows: If Machine $M_u(j_2, i)$ is slower than Machine $M_d(j_2, i)$, Buffer $B_{j_1, i}$ can only become non-empty if $M_d(j_1, i)$ fails during the time interval $[t, t + \delta t]$. This does not happen. This implies that $M_u(j_1, i)$ is faster than $M_d(j_1, i)$. If Machine $M_u(j_1, i)$ is up at time t and is faster than Machine $M_d(j_1, i)$, Buffer $B_{j_1, i}$ becomes non-empty immediately. Since Buffer $B_{j_1, i}$ is empty at time t, Machine $M_u(j_1, i)$ must therefore be down at time t.

Using equation (B.104), the second term on the right hand side of (B.160) can be reformulated as

$$
\text{prob} \left[\left\{ \left\{ n[(j_1, i), t + \delta t] > 0 \right\} \text{ and} \right. \right.
$$
$$
\left\{ \alpha_i(t + \delta t) = 1 \right\} \text{ and}
$$
$$
\left\{ \left\{ n[(i, q), t + \delta t] < N(i, q) \right\} \text{ or} \right.
$$
$$
\left. \left. \left\{ n[(i, q), t + \delta t] = N(i, q) \right\} \text{ and } \left\{ \alpha_d[(i, q), t + \delta t] = 1 \right\} \right\} \right\} \text{ and}
$$
$$
\left. \left\{ \alpha_u[(j_1, i), t] = 0 \right\} \Big| U_d(j_2, i, t) \right]
$$
$$
= \text{prob} \left[\left\{ \left\{ n[(j_1, i), t + \delta t] > 0 \right\} \text{ and} \right. \right.
$$
$$
\left\{ \alpha_i(t + \delta t) = 1 \right\} \text{ and}
$$
$$
\left\{ \left\{ n[(i, q), t + \delta t] < N(i, q) \right\} \text{ or} \right.
$$
$$
\left. \left. \left\{ n[(i, q), t + \delta t] = N(i, q) \right\} \text{ and } \left\{ \alpha_d[(i, q), t + \delta t] = 1 \right\} \right\} \right\} \Big|
$$
$$
\left. \left\{ \alpha_u[(j_1, i), t] = 0 \right\} \text{ and } U_d(j_2, i, t) \right] \cdot
$$
$$
\text{prob} \left[\left\{ \alpha_u[(j_1, i), t] = 0 \right\} \text{ and } U_d(j_2, i, t) \right] :
$$
$$
\text{prob} \left[U_d(j_2, i, t) \right] \tag{B.168}
$$

since $M_u(j_1, i)$ must be down at time t. The first term on the right side of (B.168) is the probability that Buffer $B_{j_1, i}$ is non-empty at time $t + \delta t$ given that it is empty and Machine $M_u(j_1, i)$ is down at time t. This is the repair probability of Machine $M_u(j_1, i)$ if it is faster than $M_d(j_1, i)$ and zero otherwise, i.e.

$$\text{prob} \left[\left\{ \{n[(j_1,i),t+\delta t] > 0\} \text{ and} \right. \right.$$
$$\{\alpha_i(t+\delta t) = 1\} \text{ and}$$
$$\left\{ \{n[(i,q),t+\delta t] < N(i,q)\} \text{ or} \right.$$
$$\left. \left. \{n[(i,q),t+\delta t] = N(i,q)\} \text{ and } \{\alpha_d[(i,q),t+\delta t] = 1\} \right\} \right\} \right|$$
$$\left. \{\alpha_u[(j_1,i),t] = 0\} \text{ and } U_d(j_2,i,t) \right]$$

$$= \begin{cases} r_u(j_1,i)\delta t & \text{if } \mu_u(j_1,i) > \mu_d(j_1,i) \\ 0 & \text{otherwise} \end{cases} \qquad (B.169)$$

The second and third term on the right hand side of (B.168) can be approximated as

$$\frac{\text{prob}[\{\alpha_u[(j_1,i),t] = 0\} \text{ and } U_d(j_2,i,t)]}{\text{prob}[U_d(j_2,i,t)]}$$

$$= \frac{\mathbf{p}[(j_1,i);001]}{E_d(j_2,i) - \mathbf{p}[(j_2,i);011]\frac{\mu_u(j_2,i)}{\mu_d(j_2,i)}} \qquad (B.170)$$

if we assume that the probability that both input buffers of a merge machine are empty at the same time is small.

The second term on the right hand side of (B.160) is therefore

$$\text{prob} \left[\left\{ \{n[(j_1,i),t+\delta t] > 0\} \text{ and} \right. \right.$$
$$\{\alpha_i(t+\delta t) = 1\} \text{ and}$$
$$\left\{ \{n[(i,q),t+\delta t] < N(i,q)\} \text{ or} \right.$$
$$\left. \left. \{n[(i,q),t+\delta t] = N(i,q)\} \text{ and } \{\alpha_d[(i,q),t+\delta t] = 1\} \right\} \right\} \text{ and}$$
$$\left. \{\alpha_u[(j_1,i),t] = 0\} \middle| U_d(j_2,i,t) \right]$$

$$= \frac{r_u(j_1,i)\delta t \, \mathbf{p}[(j_1,i);001]}{E_d(j_2,i) - \mathbf{p}[(j_2,i);011]\frac{\mu_u(j_2,i)}{\mu_d(j_2,i)}} \qquad \text{if } \mu_u(j_1,i) > \mu_d(j_1,i) \quad (B.171)$$

and 0 otherwise.

The third term on the right hand side of (B.160) deals with failures of the virtual machine $M_d(j_2,i)$ that are due to a blocking of Machine M_i at time $t + \delta t$, given that M_i was not blocked at time t. There are two possible mutually exclusive states that can lead to blocking of Machine M_i at time $t + \delta t$:

- Buffer $B_{j_1,i}$ is empty, $M_d(j_1,i)$ is up, Buffer $B_{i,q}$ is almost full ($N(i,q) - \mu_u(i,q)\delta t < n[(i,q),t] < N(i,q)$, $M_u(i,q)$ is up and $M_d(i,q)$ is down at time t, and Buffer $B_{i,q}$ fills in $[t, t+\delta t]$, with probability $(1-p_u(i,q)\delta t - r_d(i,q)\delta t)$
- Buffer $B_{j_1,i}$ is empty, $M_d(j_1,i)$ is up, Buffer $B_{i,q}$ is full, $M_u(i,q)$ is up and $M_d(i,q)$ is up at time t, and $M_d(i,q)$ fails in $[t, t + \delta t]$, with probability $p_d(i,q)\delta t$.

It is not possible that Buffer $B_{i,q}$ is full and $M_d(i,q)$ down at time t because of the conditioning event $U_d(j_2,i,t)$. Nothing is implied with respect to Machine $M_u(j_1,i)$. It might be down at time t and it might also be up at time t if it is slower than $M_d(j_2,i)$ as this is necessary to have an empty priority one buffer $B_{j_1,i}$ at time t.

We can therefore express the third term on the right hand side of (B.160) in terms of the two possible states at time t and the respective transition probabilities:

$$
\text{prob}\left[\{n[(i,q),t+\delta t] = N(i,q)\} \text{ and } \{\alpha_d[(i,q),t+\delta t] = 0\}\Big|U_d(j_2,i,t)\right]
$$

$$
= (1 - p_u(i,q)\delta t - r_d(i,q)\delta t) \cdot
$$

$$
\text{prob}\left[\{N(i,q) - \mu_u(i,q)\delta t < n[(i,q),t] < N(i,q)\} \text{ and }\right.
$$
$$
\{\alpha_u[(i,q),t] = 1\} \text{ and } \alpha_d[(i,q),t] = 0\} \text{ and }
$$
$$
\{n[(j_1,i),t] = 0\} \text{ and }
$$
$$
\left\{\{\alpha_u[(j_1,i),t] = 0\} \text{ or } \{\alpha_u[(j_1,i),t] = 1\}\right\} \text{ and }
$$
$$
\left.\{\alpha_d[(j_1,i),t] = 1\}\Big|U_d(j_2,i,t)\right]
$$

$$
+ \ p_d(i,q)\delta t \cdot
$$

$$
\text{prob}\left[\{n[(i,q),t] = N(i,q)\} \text{ and }\right. \qquad\qquad (B.172)
$$
$$
\{\alpha_u[(i,q),t] = 1\} \text{ and } \alpha_d[(i,q),t] = 1\} \text{ and }
$$
$$
\{n[(j_1,i),t] = 0\} \text{ and }
$$
$$
\left\{\{\alpha_u[(j_1,i),t] = 0\} \text{ or } \{\alpha_u[(j_1,i),t] = 1\}\right\} \text{ and }
$$
$$
\left.\{\alpha_d[(j_1,i),t] = 1\}\Big|U_d(j_2,i,t)\right]
$$

If we assume as an approximation that Lines $L(j_1,i)$ and $L(i,q)$ are independent and use the definition of conditional probability, this can be written as

$$\text{prob}\ \Big[\{n[(i,q),t+\delta t]=N(i,q)\}\ \text{and}\ \{\alpha_d[(i,q),t+\delta t]=0\}\Big|U_d(j_2,i,t)\Big]$$

$$=\ \Big[(1-p_u(i,q)\delta t-r_d(i,q)\delta t)\cdot$$

$$\text{prob}\ \Big[\{N(i,q)-\mu_u(i,q)\delta t<n[(i,q),t]<N(i,q)\}\ \text{and}$$
$$\{\alpha_u[(i,q),t]=1\}\ \text{and}\ \alpha_d[(i,q),t]=0\}\Big]:$$
$$\text{prob}\Big[U_d(j_2,i,t)\Big]$$
$$+\ p_d(i,q)\delta t\cdot$$
$$\text{prob}\ \Big[\{n[(i,q),t]=N(i,q)\}\ \text{and}$$
$$\{\alpha_u[(i,q),t]=1\}\ \text{and}\ \alpha_d[(i,q),t]=1\}\Big]:$$
$$\text{prob}\Big[U_d(j_2,i,t)\Big]\Big]\cdot$$

$$\Big[\mathbf{p}[(j_1,i);011]+\mathbf{p}[(j_1,i);001]\Big] \tag{B.173}$$

where the first factor can be replaced using Equation (3.216) in (Gershwin, 1994, p. 122) to find

$$\text{prob}\ \Big[\{n[(i,q),t+\delta t]=N(i,q)\}\ \text{and}\ \{\alpha_d[(i,q),t+\delta t]=0\}\Big|U_d(j_2,i,t)\Big]$$

$$=\ \frac{r_d(i,q)\ \mathbf{p}[(i,q);N(i,q),10]\ \Big(\mathbf{p}[(j_1,i);011]+\mathbf{p}[(j_1,i);001]\Big)}{E_d(j_2,i)-\mathbf{p}[(j_2,i);011]\frac{\mu_u(j_2,i)}{\mu_d(j_2,i)}} \tag{B.174}$$

The interpretation of this equation is as follows: The part of the failure rate $p_d(j_2,i)$ of Machine $M_d(j_2,i)$ that is due to blocking of Machine M_i equals the rate of repairs of $M_d(i,q)$ when M_i is blocked, the priority one buffer $B_{j_1,i}$ is empty and $M_d(j_2,i)$ is up.

We make the usual substitutions to get the algorithm to converge:

$$\frac{\mu_d(i,i)}{\mu_i}\quad\Longrightarrow\quad\frac{\mu_d(i,q)}{\mu_u(i,q)} \tag{B.175}$$

$$E_d(j_2,i)-\mathbf{p}[(j_2,i);011]\frac{\mu_u(j_2,i)}{\mu_d(j_2,i)}\quad\Longrightarrow\quad E_d(j_2,i)$$

$$=\ \frac{PR(j_2,i)}{\mu_d(j_2,i)} \tag{B.176}$$

This leads to the following version of the interruption of flow equation for the downstream machine $M_d(j_2,i)$ in the priority two input buffer

$$p_d(j_2, i) = p_i + K_6 \, \mu_d(j_2, i) \qquad\qquad (B.177)$$

with

K_6

$$= \left[p_i \left(\frac{\mu_d(i,q)}{\mu_u(i,q)} - 1 \right) \mathbf{p}[(i,q); N(i,q); 11] \Big(\mathbf{p}[(j_1,i); 011] + \mathbf{p}[(j_1,i); 001] \Big) + \right.$$

$$\left[\begin{array}{ll} r_u(j_1, i) & \text{if } \mu_u(j_1, i) > \mu_d(j_1, i) \\ 0 & \text{otherwise} \end{array} \right] \mathbf{p}[(j_1,i); 001] +$$

$$\left. r_d(i,q) \, \mathbf{p}[(i,q); N(i,q); 10] \, \Big(\mathbf{p}[(j_1,i); 011] + \mathbf{p}[(j_1,i); 001] \Big) \right] \frac{1}{PR(j_2, i)}$$

Glossary of Notation

Notation	Definition
A	Initial investment at time 0 [$]
$\alpha_i(t)$	State of Machine M_i in the real system at time t
$\alpha_u[(i,q),t]$	State of (virtual upstream) Machine $M_u(i,q)$ in Line $L(i,q)$ at time t
$\alpha_d[(i,q),t]$	State of (virtual downstream) Machine $M_d(i,q)$ in Line $L(i,q)$ at time t
$B_{i,q}$	Buffer between Machines M_i and M_q in the real system
$\beta_i(t)$	Index of the machine to which a part is routed from Machine M_i at time t
$C_{i,q}$	Capacity of Buffer $B_{i,q}$ (maximum number of parts that can be stored between Machines M_i and M_q)
$\underline{C}$	Vector of all buffer capacities
c	Continuous cash flow between time 0 and time T [$/year]
$D(i)$	Set of buffers downstream of Machine M_i
$d_{i,q}$	Fraction of parts that are randomly routed from Machine M_i to Machine M_q
δt	Infinitesimally small length of a time interval
E	Fraction of time material passes through a machine or buffer
e_i	Isolated efficiency of Machine M_i
$e_u(i,q)$	Isolated efficiency of Machine $M_u(i,q)$
$e_d(i,q)$	Isolated efficiency of Machine $M_d(i,q)$
$K_1, ..., K_6$	Coefficients in decomposition equations
$L(i,q)$	Virtual two-machine line of the decomposition related to Buffer $B_{i,q}$ in the real system
L	Scrap value at time T [$]
M_i	Machine number i in the real system

$M_u(i,q)$	Virtual upstream machine of Line $L(i,q)$
$M_d(i,q)$	Virtual downstream machine of Line $L(i,q)$
$MTTF_i$	Mean time to failure of Machine M_i
$MTTR_i$	Mean time to repair of Machine M_i
μ_i	Processing rate of Machine M_i in the real system
$\mu_u[(i,q)]$	Processing rate of the virtual upstream machine $M_u(i,q)$
$\mu_d[(i,q)]$	Processing rate of the virtual downstream machine $M_d(i,q)$
$N(i,q)$	Size of the storage related to Line $L(i,q)$
NPV	Net present value of an investment [$]
$\bar{n}_{i,q}$	Expected work-in-process of Buffer $B_{i,q}$ [parts]
n	Storage level in a two-machine line
(n,α_1,α_2)	State of a two-machine line: the upstream machine is in state α_1, the downstream machine in state α_2, and the storage contains n parts
$\mathbf{p}[n,\alpha_1,\alpha_2]$	Steady-state probability of state (n,α_1,α_2)
p_i	Failure probability or rate of Machine M_i
$p_u(i,q)$	Failure probability or rate of (virtual upstream) Machine $M_u(i,q)$
$p_d(i,q)$	Failure probability or rate of (virtual downstream) Machine $M_d(i,q)$
PR	Expected production rate
r_i	Repair probability or rate of Machine M_i in the real system
$r_u(i,q)$	Repair probability or rate of (virtual upstream) Machine $M_u(i,q)$
$r_d(i,q)$	Repair probability or rate of (virtual downstream) Machine $M_d(i,q)$
T	Length of the time span over which a flow line is expected to operate [years]
$U(i)$	Set of buffers upstream of Machine M_i
u_i	Processing probability of Machine M_i in the real system
$u_u[(i,q)]$	Processing probability of the virtual upstream machine $M_u(i,q)$
$u_d[(i,q)]$	Processing probability of the virtual downstream machine $M_d(i,q)$
uc	Per unit cost of raw material [$/part]
ur	Per unit revenue of finished products [$/part]
Y	Number of production cycles per year

Bibliography

Altiok, T. (1996). *Performance Analysis of Manufacturing Systems*. Springer, New York et al.

Ammar, M. and Gershwin, S. B. (1989). Equivalence relations in queueing models of fork/join queueing networks with blocking. *Performance Evaluation*, 10:233–245.

Artamonov, G. (1977). Productivity of a two-instrument discrete processing line in the presence of failures. *Cybernetics*, 12:464–468.

Banks, J. (1996). Interpreting simulation software checklists. *OR/MS Today*, 22(3):74–78.

Bohuslav, K. and Stoyan, D. (1967). Eine Methode zur Dimensionierung von Bunkern in Bandbetrieben und ihre Anwendung bei der Untersuchung der gebrochenen Bandförderung (Band - Bunker - Band). *Bergbautechnik*, 17:368–373.

Bonvik, A. M. (1996). *Performance Analysis of Manufacturing Systems Under Hybrid Control Policies*. PhD thesis, Massachusetts Institute of Technology. Also available as Report LMP 96-003, MIT Laboratory for Manufacturing and Productivity.

Bronstein, I. and Semendjajew, K. (1983). *Taschenbuch der Mathematik*. Teubner, Leipzig, 21 edition.

Bürger, M. (1997). *Konfigurationsplanung flexibler Fließproduktionssysteme*. Galda+Wilch, Glienicke / Berlin.

Burman, M., Gershwin, S. B., and Suyematsu, C. (1998). Hewlett-Packard uses operations research to improve the design of a printer production line. *Interfaces*, 28(1):24–36.

Burman, M. H. (1995). *New results in flow line analysis*. PhD thesis, Massachusetts Institute of Technology. Also available as Report LMP-95-007, MIT Laboratory for Manufacturing and Productivity.

Buxey, G., Slack, N., and Wild, R. (1973). Production flow line system design - A review. *AIIE Transactions*, 5:37–48.

Buzacott, J. (1972). The effect of station breakdowns and random processing times on the capacity of flow lines. *AIIE Transactions*, 4:308–312.

Buzacott, J. and Kostelski, D. (1987). Matrix-geometric and recursive algorithm solution of a two-stage unreliable flow line. *IIE Transactions*, 19(4):429–438.

Buzacott, J. A. (1967). Automatic transfer lines with buffer stocks. *International Journal of Production Research*, 5(3):183–200.

Buzacott, J. A. and Hanifin, L. E. (1978). Models of automatic transfer lines with inventory banks - A review and comparison. *AIIE Transactions*, 10(2):197–207.

Buzacott, J. A. and Shanthikumar, J. G. (1993). *Stochastic Models of Manufacturing Systems*. Prentice Hall, NJ, USA.

Buzacott, L., Liu, X.-G., and Shanthikumar, J. (1995). Multistage flow line analysis with the stopped arrival queue model. *IIE Transactions*, 27(4):444–455.

Carrascosa, M. (1995). Variance of the output in a deterministic two-machine line. Master's thesis, Massachusetts Institute of Technology. Also available as Report LMP-95-010, MIT Laboratory for Manufacturing and Productivity.

Choong, Y. and Gershwin, S. (1987). A decomposition method for the approximate evaluation of capacitated transfer lines with unreliable machines and random processing times. *IIE Transactions*, 19:150–159.

Conway, R., Maxwell, W., McClain, J. O., and Thomas, J. L. (1988). The role of work-in-process inventory in serial production lines. *Operations Research*, 36(2):229–241.

Dallery, Y. (1994). On modeling failure and repair times in stochastic models of manufacturing systems using generalized exponential distributions. *Queuing Systems*, 15:199–209.

Dallery, Y., David, R., and Xie, X. (1989). Approximate analysis of transfer lines with unreliable machines and finite buffers. *IEEE Transactions on Automatic Control*, 34(9):943–953.

Dallery, Y., David, R., and Xie, X.-L. (1988). An efficient algorithm for analysis of transfer lines with unreliable machines and finite buffers. *IIE Transactions*, 20(3):280–283.

Dallery, Y. and Frein, Y. (1989). A decomposition method for approximate analysis of closed queueing networks with blocking. In Perros, H. and Altiok, T., editors, *Queueing Networks with Blocking*, pages 193–216. North Holland, Amsterdam.

Dallery, Y. and Gershwin, S. B. (1992). Manufacturing flow line systems: A review of models and analytical results. *Queuing Systems Theory and Applications*, 12(1-2):3–94. Special issue on queuing models of manufacturing systems.

Dallery, Y., Liu, Z., and Towsley, D. (1994). Equivalence, reversibility, symmetry and concavity properties in fork-join queuing networks with blocking. *Journal of the Association for Computing Machinery*, 41(5):903–942.

de Koster, M. (1989). *Capacity Oriented Analysis and Design of Production Systems*, volume 323 of *Lecture Notes in Economics and Mathematical Systems*. Springer, Berlin et al.

Di Mascolo, M., David, R., and Dallery, Y. (1991). Modeling and analysis of assembly systems with unreliable machines and finite buffers. *IIE Transactions*, 23(4):315–330.

Ferschl, F. (1991). Stochastic processes and optimization problems in assemblage systems. In Beckmann, M., Gopalan, M., and Subramanian, R., editors, *Stochastic Processes and their Applications*, volume 370 of *Lecture Notes in Economics and Mathematical Systems*, pages 211–221. Springer, Berlin et al. Proceedings, Bombay, India, December 27-30, 1990.

Frein, Y., Commault, C., and Dallery, Y. (1996). Modeling and analysis of closed-loop production lines with unreliable machines and finite buffers. *IIE Transactions*, 28:545–554.

Gaver, D. (1962). A waiting line with interrupted service, including priorities. *Journal of the Royal Statistical Society*, 24:73–90.

Gershwin, S. B. (1991). Assembly/disassembly systems: An efficient decomposition algorithm for tree-structured networks. *IIE Transactions*, 23(4):302–314.

Gershwin, S. (1987). An efficient decomposition algorithm for the approximate evaluation of tandem queues with finite storage space and blocking. *Operations Research*, 35:291–305.

Gershwin, S. and Schick, I. (1980). Continuous model of an unreliable two-stage material flow system with a finite interstage buffer. Technical Report LIDS-R-1039, Massachusetts Institute of Technology, Cambridge, Massachusetts.

Gershwin, S. and Schick, I. (1983). Modeling and analysis of three-stage transfer lines with unreliable machines and finite buffers. *Operations Research*, 31(2):354–380.

Gershwin, S. B. (1989). An efficient decomposition algorithm for unreliable tandem queueing systems with finite buffers. In Perros, H. G. and Altiok, T., editors, *Queueing Networks with Blocking*, pages 127–146. North Holland, Amsterdam.

Gershwin, S. B. (1993). Variance of output of a tandem production system. In Onvural, R. and Akyildiz, I., editors, *Queueing Network with Finite Capacity*, pages 291–304. North-Holland, Amsterdam.

Gershwin, S. B. (1994). *Manufacturing Systems Engineering*. PTR Prentice Hall, Englewood Cliffs, New Jersey.

Gershwin, S. B. and Berman, O. (1981). Analysis of transfer lines consisting of two unreliable machines with random processing times and finite storage buffers. *AIIE Transactions*, 13(1):2–11.

Gershwin, S. B. and Burman, M. H. (1995). The design and application of an assembly line decomposition analysis technique. Technical Report LMP 98-004, Massachusetts Institute of Technology, Cambridge, Massachusetts.

Gershwin, S. B. and Schor, J. E. (1997). Efficient algorithms for buffer space allocation. Unpublished manuscript. Massachusetts Institute of Technology, Cambridge, Massachusetts.

Gopalan, M. and Kannan, S. (1994). Expected duration analysis of a two-stage transfer-line production system subject to inspection and rework. *Journal of the Operational Research Society*, 45(7):797–805.

Harrison, J. M. (1973). Assembly-like queues. *Journal of Applied Probability*, 10:354–367.

Helber, S. (1995). Exact analysis of the two-machine transfer line with limited buffer capacity and geometrically distributed processing times, times to failure and times to repair. Technical report, Ludwig-Maximilians-Universität München, Ludwigstr. 28 RG/V., D-80539 München, Germany.

Helber, S. (1997a). Approximate analysis of unreliable transfer lines with rework or scrapping of parts. Technical report, Operations Research Center, Massachusetts Institute of Technology, Cambridge, Massachusetts. Working paper OR 320-97.

Helber, S. (1997b). Approximate analysis of unreliable transfer lines with splits in the flow of material. Technical report, Ludwig-Maximilians-Universität München, Ludwigstr. 28 RG/V., D-80539 München, Germany.

Helber, S. (1998). Decomposition of unreliable assembly/dissassembly networks with limited buffer capacity and random processing times. *European Journal of Operational Research*, 109(1):24–42.

Hillier, F. and Boling, R. (1967). Finite queues in series with exponential or Erlang service times - A numerical approach. *Operations Research*, 16:286–303.

Hillier, F. S., So, K. C., and Boling, R. W. (1993). Notes: Towards characterizing the optimal allocation of storage space in production line systems with variable processing times. *Management Science*, 39(1):126–133.

Jafari, M. A. and Shanthikumar, J. G. (1987). An approximate model of multistage automatic transfer lines with possible scrapping of workpieces. *IIE Transactions*, 19(3):252–265.

Jeong, K.-C. and Kim, Y.-D. (1998). Performance analysis of assembly/disassembly systems with unreliable machines and random processing times. *IIE Transactions*, 30:41–53.

Koenigsberg, E. (1959). Production lines and internal storage - A review. *Management Science*, 5:410–433.

Kuhn, H. (1998). *Fließproduktionssysteme: Leistungsbewertung, Konfigurations- und Instandhaltungsplanung.* Physica, Heidelberg.

Kuhn, H. and Tempelmeier, H. (1997). Analyse von Fließproduktionssystemen. *Zeitschrift für Betriebswirtschaft*, 67:561–586.

Küpper, H.-U. (1991). Multi-period production planning and managerial accounting. In Fandel, G. and Zäpfel, G., editors, *Modern Production Concepts. Theory and Applications*, pages 46–62, Berlin. Springer. Proceedings of an International Conference, Fernuniversität, Hagen, FRG, August 20-24, 1990.

Law, A. M. and Kelton, W. D. (1991). *Simulation Modeling and Analysis.* McGraw-Hill, New York et al., 2 edition.

Okamura, K. and Yamashina, H. (1977). Analysis of the effect of buffer storage capacity in transfer line systems. *AIEE Transactions*, 9:127–135.

Papadopoulus, H., Heavey, C., and Browne, J. (1993). *Queueing Theory in Manufacturing Systems Analysis and Design*. Chapman & Hall, London et al.

Petigk, J. (1987). Serielle Fertigungssysteme mit Zwischenspeichern. *messen - steuern - regeln*, 30(9):402–405.

Pourbabai, B. (1990). Optimal utilization of a finite capacity intergrated assembly system. *International Journal of Production Research*, 28(2):337–352.

Sastry, B. and Awate, P. (1988). Analysis of a two-station flow line with machine processing subject to inspection and rework. *Opsearch*, 25:89–97.

Schicht, K. (1996). Bewertung von Fließproduktionssystemen mit divergierendem Materialfluß durch analytische Dekompositionsansätze. Diplomarbeit, Fakultät für Betriebswirtschaft, Ludwig-Maximilians-Universität München.

Schmidbauer, H. (1995). Zur Zuverlässigkeit von stochastischen Fertigungssystemen. Habilitationsschrift, Fakultät für Philosophie, Wissenschaftstheorie und Statistik, Ludwig-Maximilians-Universität München.

Schmidbauer, H. and Rösch, A. (1994). A stochastic model of an unreliable kanban production system. *Operations Research Spectrum*, 16(2):155–160.

Scholl, A. (1995). *Balancing and Sequencing of Assembly Lines*. Physica, Heidelberg.

Schor, J. E. (1995). Efficient algorithms for buffer allocation. Master's thesis, Massachusetts Institute of Technology. Also available as MIT Laboratory for Manufacturing and Productivity report LMP-95-006.

Schweitzer, M. and Küpper, H.-U. (1998). *Systeme der Kosten- und Erlösrechnung*. Vahlen, München, 7 edition.

Sevast'yanov, B. A. (1962). Influence of storage bin capacity on the average standstill time of a production line. *Theory of Probability and Its Applications*, 7:429–438.

Shanthikumar, J. and Tien, C. (1983). An algorithmic solution to two-stage transfer lines with possible scrapping of units. *Management Science*, 29:1069–1086.

Swain, J. J. (1997). Simulation goes mainstream. *OR/MS Today*, 24(5):35–46.

Tempelmeier, H. (1997). *FlowEval*. Universität zu Köln, Seminar für Allgemeine Betriebswirtschaftslehre, Industriebetriebslehre und Produktionswirtschaft, Albertus Magnus-Platz, D-50923 Köln. Handbuch der Version 1.1.

Wijngaard, J. (1979). The effect of interstage buffer storage on the output of two unreliable production units in series, with different production rates. *AIIE Transactions*, 11(1):42–47.

Yeralan, S. and Muth, E. J. (1987). A general model of a production line with intermediate buffer and station breakdown. *IIE Transactions*, 19(2):130–139.

Yeralan, S. and Tan, B. (1997a). Analysis of multistation production systems with limited buffer capacity. Part 1: The subsystem model. *Mathematical and Computer Modelling*, 25(7):109–122.

Yeralan, S. and Tan, B. (1997b). A decomposition model for continuous materials flow production systems. *International Journal of Production Research*, 35(10):2759–2772.

Yeralan, S. and Tan, B. (1997c). A station model for continuous materials flow production systems. *International Journal of Production Research*, 35(9):2525–2542.

Yu, K.-Y. C. and Bricker, D. L. (1993). Analysis of a markov chain model of a multistage manufacturing system with inspection, rejection, and rework. *IIE Transactions*, 25(1):109–112.

Zimmern, B. (1956). Etudes de la propagation des arrêts aleatoires dans les chaines de production. *Review Statististical Applications*, 4:85–104.